AF597958

Methods in Molecular Biology

Series Editor
John M. Walker
School of Life and Medical Sciences
University of Hertfordshire
Hatfield, Hertfordshire, AL10 9AB, UK

For further volumes:
http://www.springer.com/series/7651

Wheat Rust Diseases

Methods and Protocols

Edited by

Sambasivam Periyannan

Research School of Biology, Australian National University, Canberra, ACT, Australia
Agriculture & Food, CSIRO, Canberra, ACT, Australia
Queensland Alliance for Agriculture and Food Innovation, University of Queensland, Brisbane, QLD, Australia
Plant Breeding Institute, University of Sydney, Sydney, NSW, Australia

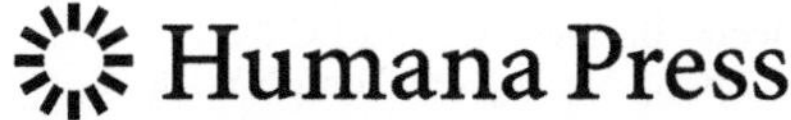

Editor
Sambasivam Periyannan
Research School of Biology
Australian National University
Canberra, ACT, Australia

Agriculture & Food
CSIRO
Canberra, ACT, Australia

Queensland Alliance for Agriculture and Food Innovation
University of Queensland
Brisbane, QLD, Australia

Plant Breeding Institute
University of Sydney
Sydney, NSW, Australia

ISSN 1064-3745 ISSN 1940-6029 (electronic)
Methods in Molecular Biology
ISBN 978-1-4939-7248-7 ISBN 978-1-4939-7249-4 (eBook)
DOI 10.1007/978-1-4939-7249-4

Library of Congress Control Number: 2017950234

© Springer Science+Business Media LLC 2017
This work is subject to copyright. All rights are reserved by the Publisher, whether the whole or part of the material is concerned, specifically the rights of translation, reprinting, reuse of illustrations, recitation, broadcasting, reproduction on microfilms or in any other physical way, and transmission or information storage and retrieval, electronic adaptation, computer software, or by similar or dissimilar methodology now known or hereafter developed.
The use of general descriptive names, registered names, trademarks, service marks, etc. in this publication does not imply, even in the absence of a specific statement, that such names are exempt from the relevant protective laws and regulations and therefore free for general use.
The publisher, the authors and the editors are safe to assume that the advice and information in this book are believed to be true and accurate at the date of publication. Neither the publisher nor the authors or the editors give a warranty, express or implied, with respect to the material contained herein or for any errors or omissions that may have been made. The publisher remains neutral with regard to jurisdictional claims in published maps and institutional affiliations.

Cover Caption: Photo courtesy of Carl Davies

Printed on acid-free paper

This Humana Press imprint is published by Springer Nature
The registered company is Springer Science+Business Media LLC
The registered company address is: 233 Spring Street, New York, NY 10013, U.S.A.

Preface

Rust disease, caused by strains of the fungus *Puccinia*, is a major threat to global wheat production and food security. For instance, the Ug99 stem rust that emerged in Eastern parts of Africa overcame the majority of genetic resistance present in commercial bread wheat. Along with the ability to evolve additional virulence, the pathogen spread rapidly to neighboring regions, and to date there are more than eight Ug99 lineage races spread across 13 countries. The emergence of highly virulent and temperature adaptable isolates of wheat stripe rust in parts of Europe, USA, Southeast Asia, Africa, and Australia has become an additional threat to worldwide wheat cultivation. To combat the threat posed by these rapidly evolving fungi, the global wheat rust research community came together through programs such as the Borlaug Global Rust Initiative and identified state-of-the-art techniques and tools for monitoring and preventing the spread of wheat rust diseases. Therefore, this Springer Protocol Series on "Wheat Rust Diseases" is a valuable collection of advanced tools that are currently used for characterization of rust, the host plant wheat, and the interactions between the two. Parts I and II of this volume consist of routinely used and advanced tools for characterizing rust pathogen where protocols for rust surveillance, genotyping, and molecular pathogenicity studies are discussed. Part III describes tools for genetic analysis of rust resistance while the subsequent Part IV covers new methods on rust resistance gene cloning which were based on next-generation sequencing and assembly tools. It also covers molecular assays for the functional analysis of cloned resistance genes. The last part (V) of the volume has a chapter on the isolation and screening of bacterial endophytes as biocontrol agents for rust disease management.

In summary, this volume covers a wide range of topics right from the rust pathogen to the genetics of the host plant wheat. Techniques covered in this volume are of value to both established and new generations of wheat rust researchers and to some extent to the whole of plant science and the microbial research community.

Canberra, ACT, Australia *Sambasivam Periyannan*

Preface

Contents

Preface . . . . *v*
Contributors . . . . *xi*

PART I WHEAT RUST SURVEILLANCE AND GENOTYPING

1 Wheat Rust Surveillance: Field Disease Scoring and Sample Collection for Phenotyping and Molecular Genotyping 3
Sajid Ali and David Hodson

2 Field Pathogenomics: An Advanced Tool for Wheat Rust Surveillance 13
Vanessa Bueno-Sancho, Daniel C.E. Bunting, Luis J. Yanes, Kentaro Yoshida, and Diane G.O. Saunders

3 Race Typing of *Puccinia striiformis* on Wheat 29
Mogens S. Hovmøller, Julian Rodriguez-Algaba, Tine Thach, and Chris K. Sørensen

4 Assessment of Aggressiveness of *Puccinia striiformis* on Wheat 41
Chris K. Sørensen, Tine Thach, and Mogens S. Hovmøller

5 Extraction of High Molecular Weight DNA from Fungal Rust Spores for Long Read Sequencing 49
Benjamin Schwessinger and John P. Rathjen

6 Microsatellite Genotyping of the Wheat Yellow Rust Pathogen *Puccinia striiformis* 59
Sajid Ali, Muhammad R. Khan, Angelique Gautier, Zahoor A. Swati, and Stephanie Walter

PART II MOLECULAR PATHOGENICITY

7 Computational Methods for Predicting Effectors in Rust Pathogens 73
Jana Sperschneider, Peter N. Dodds, Jennifer M. Taylor, and Sébastien Duplessis

8 Protein–Protein Interaction Assays with Effector–GFP Fusions in *Nicotiana benthamiana* 85
Benjamin Petre, Joe Win, Frank L.H. Menke, and Sophien Kamoun

9 Proteome Profiling by 2D–Liquid Chromatography Method for Wheat–Rust Interaction 99
Semra Hasançebi

10 Investigating Gene Function in Cereal Rust Fungi by Plant-Mediated Virus-Induced Gene Silencing 115
Vinay Panwar and Guus Bakkeren

11 Apoplastic Sugar Extraction and Quantification from Wheat Leaves Infected with Biotrophic Fungi 125
Veronica Roman-Reyna and John P. Rathjen

Part III Genetics of Rust Resistance

12 Genetic Analysis of Resistance to Wheat Rusts ... 137
Caixia Lan, Mandeep Singh Randhawa, Julio Huerta-Espino, and Ravi P. Singh

13 Advances in Identification and Mapping of Rust Resistance Genes in Wheat ... 151
Urmil Bansal and Harbans Bariana

14 Chromosome Engineering Techniques for Targeted Introgression of Rust Resistance from Wild Wheat Relatives ... 163
Peng Zhang, Ian S. Dundas, Steven S. Xu, Bernd Friebe, Robert A. McIntosh, and W. John Raupp

15 Applications of Genomic Selection in Breeding Wheat for Rust Resistance ... 173
Leonardo Ornella, Juan Manuel González-Camacho, Susanne Dreisigacker, and Jose Crossa

16 Rapid Phenotyping Adult Plant Resistance to Stem Rust in Wheat Grown under Controlled Conditions ... 183
Adnan Riaz and Lee T.Hickey

Part IV Rust Resistance Gene Isolation and Characterization

17 Generation of Loss-of-Function Mutants for Wheat Rust Disease Resistance Gene Cloning ... 199
Rohit Mago, Bradley Till, Sambasivam Periyannan, Guotai Yu, Brande B.H. Wulff, and Evans Lagudah

18 Isolation of Wheat Genomic DNA for Gene Mapping and Cloning ... 207
Guotai Yu, Asyraf Hatta, Sambasivam Periyannan, Evans Lagudah, and Brande B.H. Wulff

19 MutRenSeq: A Method for Rapid Cloning of Plant Disease Resistance Genes ... 215
Burkhard Steuernagel, Kamil Witek, Jonathan D.G. Jones, and Brande B.H. Wulff

20 Rapid Gene Isolation Using MutChromSeq ... 231
Burkhard Steuernagel, Jan Vrána, Miroslava Karafiátová, Brande B.H. Wulff, and Jaroslav Doležel

21 Rapid Identification of Rust Resistance Genes Through Cultivar-Specific De Novo Chromosome Assemblies ... 245
Anupriya K. Thind, Thomas Wicker, and Simon G. Krattinger

22 BSMV-Induced Gene Silencing Assay for Functional Analysis of Wheat Rust Resistance ... 257
Li Huang

23 Yeast as a Heterologous System to Functionally Characterize a Multiple Rust Resistance Gene that Encodes a Hexose Transporter.......... 265
Ricky J. Milne, Katherine E. Dibley, and Evans S. Lagudah

PART V BIOCONTROL AGENTS FOR RUST MANAGEMENT

24 Biocontrol Agents for Controlling Wheat Rust 277
Siliang Huang and Fahu Pang

Index *289*

Contributors

SAJID ALI • *Institute of Biotechnology & Genetic Engineering, University of Agriculture, Peshawar, Pakistan*

GUUS BAKKEREN • *Summerland Research and Development Centre, Agriculture and Agri-Food Canada, Summerland, BC, Canada*

URMIL BANSAL • *The University of Sydney Plant Breeding Institute, Cobbitty, NSW, Australia*

HARBANS BARIANA • *The University of Sydney Plant Breeding Institute, Cobbitty, NSW, Australia*

VANESSA BUENO-SANCHO • *Earlham Institute, Norwich Research Park, Norwich, UK; John Innes Centre, Norwich Research Park, Norwich, UK*

DANIEL C.E. BUNTING • *Earlham Institute, Norwich Research Park, Norwich, UK; John Innes Centre, Norwich Research Park, Norwich, UK*

JOSE CROSSA • *International Maize and Wheat Improvement Center (CIMMYT), Mexico, DF, Mexico*

KATHERINE E. DIBLEY • *CSIRO Agriculture and Food, Canberra, ACT, Australia*

PETER N. DODDS • *Black Mountain Laboratories, CSIRO Agriculture and Food, Canberra, ACT, Australia*

JAROSLAV DOLEŽEL • *Institute of Experimental Botany, Centre of the Regional Haná for Biotechnological and Agricultural Research, Olomouc, Czech Republic*

SUSANNE DREISIGACKER • *International Maize and Wheat Improvement Center (CIMMYT), Mexico, DF, Mexico*

IAN S. DUNDAS • *School of Agriculture, Food and Wine, University of Adelaide, Adelaide, Australia*

SÉBASTIEN DUPLESSIS • *INRA, Unité Mixte de Recherche INRA/Université de Lorraine 1136 Interactions Arbres-Microorganismes, INRA Centre Grand Est - Nancy, Champenoux, France*

BERND FRIEBE • *Department of Plant Pathology, Kansas State University, Manhattan, KS, USA*

ANGELIQUE GAUTIER • *UMR BIOGER, INRA, AgroParisTech, Université Paris-Saclay, Thiverval-Grignon, France*

JUAN MANUEL GONZÁLEZ-CAMACHO • *Colegio de Postgraduados, Texcoco, Mexico*

SEMRA HASANÇEBI • *Department of Genetics and Bioengineering, Trakya University, Faculty of Engineering, Edirne, Turkey*

ASYRAF HATTA • *John Innes Centre, Norwich Research Park, Norwich, UK; Department of Agriculture Technology, Universiti Putra Malaysia, Serdang, Malaysia*

LEE T. HICKEY • *Queensland Alliance for Agriculture and Food Innovation, The University of Queensland, Brisbane, QLD, Australia*

DAVID HODSON • *ILRI/CIMMYT, Addis Ababa, Ethiopia*

MOGENS S. HOVMØLLER • *Department of Agroecology, Aarhus University, Slagelse, Denmark*

LI HUANG • *Department of Plant Sciences and Plant Pathology, Montana State University, Bozeman, MT, USA*

SILIANG HUANG • *School of Agricultural Engineering, Nanyang Normal University, Nanyang, Henan, China*

JULIO HUERTA-ESPINO • *Campo Experimental Valle de México INIFAP, Chapingo, Mexico*
JONATHAN D.G. JONES • *The Sainsbury Laboratory, Norwich Research Park, Norwich, UK*
SOPHIEN KAMOUN • *The Sainsbury Laboratory, Norwich Research Park, Norwich, UK*
MIROSLAVA KARAFIÁTOVÁ • *Institute of Experimental Botany, Centre of the Regional Haná for Biotechnological and Agricultural Research, Olomouc, Czech Republic*
MUHAMMAD R. KHAN • *Institute of Biotechnology & Genetic Engineering, University of Agriculture, Peshawar, Pakistan*
SIMON G. KRATTINGER • *Department of Plant and Microbial Biology, University of Zurich, Zürich, Switzerland*
EVANS S. LAGUDAH • *CSIRO Agriculture and Food, Canberra, ACT, Australia*
CAIXIA LAN • *CIMMYT, Mexico, DF, Mexico*
ROHIT MAGO • *Agriculture and Food, CSIRO, Canberra, ACT, Australia*
ROBERT A. MCINTOSH • *Plant Breeding Institute, University of Sydney, Cobbitty, Australia*
FRANK L.H. MENKE • *The Sainsbury Laboratory, Norwich Research Park, Norwich, UK*
RICKY J. MILNE • *CSIRO Agriculture and Food, Canberra, ACT, Australia*
LEONARDO ORNELLA • *International Maize and Wheat Improvement Center (CIMMYT), Mexico, DF, Mexico*
FAHU PANG • *School of Agricultural Engineering, Nanyang Normal University, Nanyang, Henan, China*
VINAY PANWAR • *Morden Research and Development Centre, Agriculture and Agri-Food Canada, Morden, MB, Canada; National Research Council Canada, Plant Biotechnology Institute, Saskatoon, SK, Canada*
SAMBASIVAM PERIYANNAN • *Research School of Biology, Australian National University, Canberra, ACT, Australia; Agriculture & Food, CSIRO, Canberra, ACT, Australia; Queensland Alliance for Agriculture and Food Innovation, University of Queensland, Brisbane, QLD, Australia; Plant Breeding Institute, University of Sydney, Sydney, NSW, Australia*
BENJAMIN PETRE • *The Sainsbury Laboratory, Norwich Research Park, Norwich, UK*
MANDEEP SINGH RANDHAWA • *CIMMYT, Mexico, DF, Mexico*
JOHN P. RATHJEN • *Research School of Biology, Australian National University, Canberra, ACT, Australia*
W. JOHN RAUPP • *Department of Plant Pathology, Kansas State University, Manhattan, KS, USA*
ADNAN RIAZ • *Queensland Alliance for Agriculture and Food Innovation, The University of Queensland, Brisbane, Australia*
JULIAN RODRIGUEZ-ALGABA • *Department of Agroecology, Aarhus University, Slagelse, Denmark*
VERONICA ROMAN-REYNA • *Research School of Biology, Australian National University, Canberra, ACT, Australia*
DIANE G.O. SAUNDERS • *Earlham Institute, Norwich Research Park, Norwich, UK; John Innes Centre, Norwich Research Park, Norwich, UK*
BENJAMIN SCHWESSINGER • *Research School of Biology, Australian National University, Canberra, ACT, Australia*
RAVI P. SINGH • *CIMMYT, Mexico, DF, Mexico*
CHRIS K. SØRENSEN • *Department of Agroecology, Aarhus University, Slagelse, Denmark*
JANA SPERSCHNEIDER • *Centre for Environmental and Life Sciences, CSIRO Agriculture and Food, Floreat, WA, Australia*
BURKHARD STEUERNAGEL • *John Innes Centre, Norwich Research Park, Norwich, UK*

Zahoor A. Swati • *Institute of Biotechnology & Genetic Engineering, University of Agriculture, Peshawar, Pakistan*

Jennifer M. Taylor • *Black Mountain Laboratories, CSIRO Agriculture and Food, Canberra, ACT, Australia*

Tine Thach • *Department of Agroecology, Aarhus University, Slagelse, Denmark*

Anupriya K. Thind • *Department of Plant and Microbial Biology, University of Zurich, Zurich, Switzerland*

Bradley Till • *Plant Breeding and Genetics Laboratory, International Atomic Energy Agency, Vienna, Austria*

Jan Vrána • *Institute of Experimental Botany, Centre of the Regional Haná for Biotechnological and Agricultural Research, Olomouc, Czech Republic*

Stephanie Walter • *Department of Agroecology, Aarhus University, Slagelse, Denmark*

Thomas Wicker • *Department of Plant and Microbial Biology, University of Zurich, Zürich, Switzerland*

Joe Win • *The Sainsbury Laboratory, Norwich Research Park, Norwich, UK*

Kamil Witek • *The Sainsbury Laboratory, Norwich Research Park, Norwich, UK*

Brande B.H. Wulff • *John Innes Centre, Norwich Research Park, Norwich, UK*

Steven S. Xu • *USDA-ARS Cereal Crops Research Unit, Northern Crop Science Laboratory, Fargo, ND, USA*

Luis J. Yanes • *Earlham Institute, Norwich Research Park, Norwich, UK*

Kentaro Yoshida • *Graduate School of Agricultural Science, Kobe University, Kobe, Japan*

Guotai Yu • *John Innes Centre, Norwich Research Park, Norwich, UK*

Peng Zhang • *Plant Breeding Institute, University of Sydney, Cobbitty, Australia*

Part I

Wheat Rust Surveillance and Genotyping

Chapter 1

Wheat Rust Surveillance: Field Disease Scoring and Sample Collection for Phenotyping and Molecular Genotyping

Sajid Ali and David Hodson

Abstract

Long-distance migration capacity, emergence of invasive lineages, and variability in adaptation to a wide range of climatic conditions make wheat rusts the most important threat to wheat production worldwide. Efficient and coordinated efforts are required for surveillance of the pathogen population at different geographical levels to enable tracking of rust pathogen populations at local, regional, continental, and ultimately worldwide scale. Here we describe a standard procedure for rust surveillance to enable comparison across various research groups for a final compilation. The procedure described would enable tracking of disease severity, field level expression of host resistance, and collection of samples for further virulence phenotyping and molecular genotyping.

Key words Rust pathogens, Disease epidemics, Tracking populations

1 Introduction

Wheat rust pathogens cause severe epidemics on wheat worldwide [1–4]. The long-distance dispersal capacity of rusts has resulted in continental and global spread of invasive lineages, representing an important threat to wheat production [5–7]. Efficient and coordinated efforts must be made to track the pathogen population across various geographical levels. Surveillance is usually carried out on a local scale in various parts of the world. To enable an overall compilation, surveillance data must be shared and must be undertaken in a coordinated manner following predefined procedures. This enables tracking of rust pathogens at country, regional, continental, and ultimately worldwide scale [8]. Appropriate scoring of disease severity and host response also provides other useful information such as disease epidemics status and field level expression of host resistance [9]. Here we describe a procedure for wheat rust surveillance and sample collection, which is based on the methods used successfully in the Borlaug Global Rust Initiative (BGRI) rust monitoring system and during rust surveillance project at the

Sambasivam Periyannan (ed.), *Wheat Rust Diseases: Methods and Protocols*, Methods in Molecular Biology, vol. 1659, DOI 10.1007/978-1-4939-7249-4_1, © Springer Science+Business Media LLC 2017

University of Agriculture, Peshawar, Pakistan. The surveillance procedure is defined with the objectives to track disease severity, field level expression of host resistance, and sampling for further virulence phenotyping and molecular genotyping (detailed in separate chapters). The sampling procedure of live sample collection is described for multiplication and phenotyping, while the procedure of dead sample collection is described for molecular work only.

2 Materials

1. Scissors.
2. Notebook, pen, and pencil.
3. Permanent marker.
4. Paper bags, glycine bags.
5. Ethanol.
6. 2 mL Eppendorf tubes/Cryotubes.
7. Parafilm or Scotch tape.
8. GPS.
9. Standardized survey form (paper or electronic).

3 Methods

3.1 Selection of Location/Fields and Rust Scoring

Considering the aerial dispersal of rust pathogens, it is recommended to cover the maximum representative area for a given wheat growing region. The selection of site will depend on the objective of surveillance such as phenotyping or genotyping and of course considering the time and cost limitations. However, to cover the maximum region and for efficient sample collection, we recommend the following points:

1. Select locations to be surveyed in a given wheat production region and identify representative sites within each location (Fig. 1).
2. The selection of sites must be representative of the area, though considering logistic and other limitations. Consider geography, types of cultivars. and other attributes in selection of survey site (*see* **Note 1**).
3. At each site, make a survey at 5–15 fields, dependent on the heterogeneity in host varieties and prior information on pathogen diversity. Alternatively, the semirandom survey could be made while following a specific route and surveying each 10–20 km.

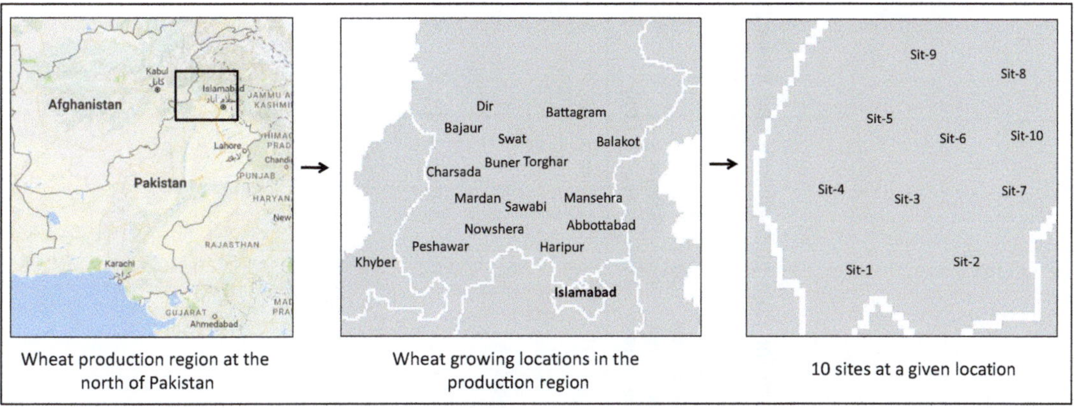

Fig. 1 Schematic representation of a geographical region of wheat production potentially divided into locations and surveillance sites to maximize representation. The survey sites could also be based on the semirandom sites while following a specific route and surveying each 10–20 km

4. Depending on the size of field plot, make a survey at 3–5 spots within the field.
5. Record maximum information for each field plot (like GPS points, host varieties, geographical barriers, etc. as in Table 1). Alternatively, a standardized survey form (paper or electronic) could be used as provided by the Rust Tracker website (http://rusttracker.cimmyt.org/?page_id=279).
6. Record if the plots are naturally infected or inoculated, or near to an inoculated trial. For population biology studies, naturally infected plots must be surveyed.
7. Also record additional information such as alternate hosts and other cereals, if present, nearby the sites (*see* **Note 2**).
8. Record rust severity scores using modified cob scale [9]. Similarly record the host resistance response following the scale described [10] and detailed in Table 2.

3.2 Sampling for Rust Multiplication and Phenotyping

Live rust samples are collected for multiplication and for phenotyping various traits such as virulence and aggressiveness.

1. For stripe or leaf rust, select a freshly sporulating young leaf with multiple lesions. Avoid old leaves or the lesions which are dry or wet due to rain. For stem rust, select heavily infected stems.
2. For stripe or leaf rust, collect the infected samples from leaves, while for stem rust use stem segments of 4–5 cm length and remove the core stem tissue (to facilitate drying of samples).

Table 1
Information needed to be recorded during cereal rusts field surveillance for field scoring and sample collection for race phenotyping and molecular genotyping

Isolate code	Crop year	Country	Location	Site at a given location[a]	Field at a given location	Latitude	Longitude	Elevation (m)	Host species	Host crop	Host variety	Growth stage	Yellow rust score	Leaf rust score	Stem rust score	Surveyor name	Date	Other information
PK16_1	2016	Pakistan	Peshawar	UAP	1	34.01862	71.46704	359	*T. aestivum*	Spring wheat	Sehar	Milk	0	10 MR	0	R. Khan	03/04/2016	After heavy rains
PK16_48	2016	Pakistan	Abbottabbad	HARI	5	34.37454	73.19683	1256	*T. aestivum*	Winter wheat	Breeding line	Milk	20 MS	0	0	S. Ali	15/04/2016	Berberis zone
ET16_23	2016	Ethiopia	Bale	Tibo	1	7.02817	40.30898	2278	*T. aestivum*	Spring wheat	Digalu	MIlk	0	0	40 S	D. Hodson	19/1/2016	None
ET16_24	2016	Ethiopia	Ginnir	Jame	1	7.16655	40.72022	1891	*T. aestivum*	Spring wheat	Kubsa	Dough	50 S	0	0	D. Hodson	20/1/2016	None

[a]Subsite designation within location could be useful if the GPS data is not available during sampling effort

Table 2
Host reaction categories and its symptoms to be recorded during field scoring of rust infection

Host reaction	Host reaction	Symptoms	Increment for CI values[a]
I	Immune	No visible infection	0.00
R	Resistant	Necrotic areas with or without small pustules	0.10
MR	Moderately resistant	Small pustules surrounded by necrotic areas	0.25
M	Moderately resistant-Moderately susceptible	Combination of both MR and MS	0.50
MS	Moderately susceptible	Medium-sized pustules, no necrosis, some chlorosis	0.75
S	Susceptible	Large pustules, no necrosis or chlorosis.	1.00

[a]Coefficient of infection = increment × severity (e.g., 70 MS = 70 × 0.75 = 52.5)

3. Place the leaf or stem samples into the glycine bag. If glycine bags are not available, use a paper bag as an alternative. Never use plastic bags, and also keep the envelopes in dry condition.
4. Properly label all sample envelopes—it is recommended to include date, location, variety (if known), rust disease type, disease score, and a unique sample code. The same code must be recorded on the survey form (or in Table 1) where all the information related to the samples are recorded (*see* **Note 3**).
5. Several leaves or stems from the same field site could be placed into the same envelope.
6. Disinfect your fingers and the scissors with ethanol between samples and at the end.
7. Keep the samples at room temperature and air-dry for at least 24 h and then pack all the samples in a large paper envelope representing a location. While drying, you can place a light weight on the bags to avoid further rolling of the leaves.
8. The dry samples could be dispatched, under a permit if necessary, to the laboratories which can work with the live samples. Ideally, samples should be sent within 7–10 days of collection (*see* **Note 4**).
9. The number of live samples collected may vary according to the objectives and capacity to revive them, but 2–5 leaf samples or 5–10 stem samples per field plot could be collected to give a representative number of isolates.

3.3 Sampling for Molecular Genotyping

Molecular genotyping could be done using DNA from live rust samples generated through spore multiplication or from dead samples [11]. To avoid the requirement of quarantine measures, the use of dead samples is usually recommended for molecular genotyping (particularly for microsatellite analysis).

1. For stripe rust, infected leaves with a single lesion collected from the rust infected hot spot of the field are used. For stem rust, use infected stems with a clear single pustule. Similarly, for leaf rust leaf, samples with a single isolated pustule are used (*see* **Note 5**).
2. For stripe or leaf rust, excise the lesion/pustule from the whole leaf by discarding the healthy part around (Fig. 2). Keep only the infected part so as to get more DNA from rust than the wheat plant. For stem rust, cut either side of the pustule and then remove the core stem tissue (Fig. 2).
3. Place the excised lesion or pustule into a Cryo tube or Eppendorf tube.
4. Add ethanol to kill the rust pathogen.
5. Seal the tube with Parafilm or Scotch tape to avoid opening while handling.
6. Label the tubes properly with a code (prelabeled barcoded tubes can also be used), with an identical code used on the survey form or in Table 1 (*see* **Note 3**).
7. Disinfect your hands and scissors with ethanol between handling of two samples.
8. It is recommended to sample more than ten samples per field plot, which could be stored, and the exact number to be genotyped could be decided later considering the exact scientific question and available resources [12].

4 Notes

1. Geographical barriers like mountains and valleys can influence pathogen dispersal and population structure, so adjust the survey accordingly and record due information. Similarly, host barriers like distribution of durum and bread wheat in different sites in a given wheat growing area may also influence the pathogen and could preferably be considered during the surveillance effort.
2. There is an increasing interest in understanding the role of non-wheat hosts in rusts epidemiology, particularly the role of alternate hosts [12]. Thus, information of these species must be recorded, and sampling of rusts on these species could be useful, if found. Details on their sampling could be found in

Fig. 2 Characteristic multiple and single pustule lesions of wheat rust diseases (wheat stem rust, leaf rust, and yellow rust)

relevant publications and the BGRI training video "Going to the Source of New Virulence: Isolating Cereal Rust Pathogens from Barberry Species" (*see* http://www.globalrust.org/page/videos).

3. It is important to code the bags and record maximum information in a precise manner. Along with the code it is also recommended to record date, location name, host variety/type (if known), disease and disease score on the bags to enable tracking.
4. Live samples should be treated in the local laboratories having facilities to multiply and phenotype the pathogen. These could also be shared with foreign laboratories, if required permits are obtained prior to shipping, and the international labs should have quarantine facilities and permission to receive such exotic samples, e.g., Global Rust Reference Centre, Denmark; Cereal Disease Laboratory-USDA, USA; and AAFC laboratory at Morden, Canada. It is essential to contact international labs prior to sampling to obtain an import permit and to follow shipping protocols.
5. For the molecular work, use a single lesion or pustule as multiple lesions arise from different spores, which could be potentially different genotypes. Genotyping of multiple lesions will give mixed genotypes. Along with the above mentioned labs for live samples, Dr. Ali lab at The University of Agriculture, Peshawar, Pakistan, also receive and process samples for molecular genotyping, particularly from Central, West and South Asia.

References

1. Beddow JM, Pardey PG, Chai Y, Hurley TM, Kriticos DJ, Braun J-C, Park RF, Cuddy WS, Yonow T (2015) Research investment implications of shifts in the global geography of wheat stripe rust. Nat Plants 1:15132–15135
2. Hovmøller MS, Walter S, Bayles R, Hubbard A, Flath K, Sommerfeldt N, Leconte M, Rodriguez-Algaba J, Hansen JG, Lassen P, Justesen AF, Ali S, de Vallavieille-Pope C (2016) Replacement of the European wheat yellow rust population by new races from the centre of diversity in the near-Himalayan region. Plant Pathol 65:402–411
3. Hovmøller MS, Yahyaoui AH, Milus EA, Justesen AF (2008) Rapid global spread of two aggressive strains of a wheat rust fungus. Mol Ecol 17(17):3818–3826
4. Ali S, Gladieux P, Leconte M, Gautier A, Justesen AF, Hovmoller MS, Enjalbert J, De Vallavieille-Pope C (2014) Origin, migration routes and worldwide population genetic structure of the wheat yellow rust pathogen *Puccinia striiformis* f. sp. *tritici*. PLoS Pathog 10:e1003903
5. Singh RP, Hodson DP, Jin Y, Huerta-Espino J, Kinyua MG, Wanyera R, Njau P, Ward WR (2006) Current status, likely migration and strategies to mitigate the threat to wheat production from race Ug99 (TTKS) of stem rust pathogen. CAB Rev 1:054
6. Walter S, Ali S, Kemen E, Nazari K, Bahri B, Enjalbert J, Brown JK, Sicheritz-Pontén T, Jones J, de Vallvieille-Pope C, Hovmøller MS, Justesen AF (2016) Molecular markers for tracking the origin and worldwide distribution of invasive strains of *Puccinia striiformis*. Ecol Evol 6:2790–2804
7. Ali S, Rodriguez-Algaba J, Thach T, Sørensen CK, Hansen JG, Lassen P, Nazari K, Hodson DP, Justesen AF and Hovmøller MS (2017)

Yellow Rust Epidemics Worldwide Were Caused by Pathogen Races from Divergent Genetic Lineages. Front. Plant Sci. 8:1057. doi: 10.3389/fpls.2017.01057

8. Hovmøller MS, Walter S, Bayles R, Hubbard A, Flath K, Sommerfeldt N, Leconte M, Rodriguez-Algaba J, Hansen JG, Lassen P, Justesen AF, Ali S, de Vallavieille-Pope C (2015) The need for coordinated European efforts to fight invasive crop pathogens. Paper presented at the Conference: XVIII. International Plant Protection Congress, Berlin, Germany
9. Peterson RF, Campbell AB, Hannah AE (1948) A diagrammatic scale for rust intensity on leaves and stems of cereals. Can J Res 26:496–500
10. Singh RP (1993) Resistance to leaf rust in 26 Mexican wheat cultivars. Crop Sci 33: 633–637
11. Ali S, Gautier A, Leconte M, Enjalbert J, de Vallavieille-Pope C (2011) A rapid genotyping method for an obligate fungal pathogen, *Puccinia striiformis* f. sp. *tritici*, based on DNA extraction from infected leaf and Multiplex PCR genotyping. BMC Res Notes 4:240
12. Ali S, Gladieux P, Rahman H, Saqib MS, Fiaz M, Ahmed H, Leconte M, Gautier A, Justesen AF, Hovmøller MS, Enjalbert J, de Vallavieille-Pope C (2014) Inferring the contribution of sexual reproduction, migration and off-season survival to the temporal maintenance of microbial populations: a case study on the wheat fungal pathogen *Puccinia striiformis* f. sp. *tritici*. Mol Ecol 23:603–617

Chapter 2

Field Pathogenomics: An Advanced Tool for Wheat Rust Surveillance

Vanessa Bueno-Sancho, Daniel C.E. Bunting, Luis J. Yanes, Kentaro Yoshida, and Diane G.O. Saunders

Abstract

Traditionally, diagnostic tools for plant pathogens were limited to the analysis of purified pathogen isolates subjected to phenotypic characterization and/or PCR-based genotypic analysis. However, these approaches detect only already known pathogenic agents, may not always recognize novel races, and can introduce bias in the results. Recent advances in next-generation sequencing technologies have provided new opportunities to integrate high-resolution genotype data into pathogen surveillance programs. Here, we describe some of the key bioinformatics analysis used in the newly developed "Field Pathogenomics" pathogen surveillance technique. This technique is based on RNA-seq data generated directly form pathogen-infected plant leaf samples collected in the field, providing a unique opportunity to characterize the pathogen population and its host directly in their natural environment. We describe two main analyses: (1) a phylogenetic analysis of the pathogen isolates that have been collected to understand how they are related to each other, and (2) a population structure analysis to provide insight into the genetic substructure within the pathogen population. This provides a high-resolution representation of pathogen population dynamics directly in the field, providing new insights into pathogen biology, population structure, and pathogenesis.

Key words Filamentous plant pathogens, Population genetics, Pathogen surveillance, Phylogenomics, Emerging pathogens

1 Introduction

Emerging and re-emerging pathogens pose a continuous threat to food security and human health. Recent disease outbreaks in plants have been associated with expansions of pathogen geographic distribution and increased virulence of known pathogens, such as the European outbreak of ash dieback and wheat stem rust across Africa and the Middle East. Independent of the host organism, the scale

Electronic supplementary material: The online version of this chapter (doi:10.1007/978-1-4939-7249-4_2) contains supplementary material, which is available to authorized users.

Sambasivam Periyannan (ed.), *Wheat Rust Diseases: Methods and Protocols*, Methods in Molecular Biology, vol. 1659, DOI 10.1007/978-1-4939-7249-4_2, © Springer Science+Business Media LLC 2017

and frequency of emerging diseases have increased with the globalization and industrialization of food production systems [1]. Such invasive species can cause significant economic and ecological problems within the new habitat as the population grows. Therefore, new surveillance tools to identify and track the spread of these pathogens are urgently required. Traditionally, for plant pathogens diagnostic tools were limited to the analysis of purified pathogen isolates subjected to phenotypic characterization and/or PCR-based genotypic analysis. However, these approaches detect only already known pathogenic agents, may not always recognize novel races and can introduce bias in the results [2]. Recent advances in next-generation sequencing technologies have provided new opportunities to integrate high-resolution genotype data into pathogen surveillance programs [3, 4].

In this chapter, we describe some of the bioinformatic analysis incorporated into the newly developed "Field Pathogenomics" technique that has been recently used to reveal a complete shift in the population of the wheat yellow rust pathogen, *Puccinia striiformis f. sp. tritici* (PST), in the UK [4]. The speed of this method was recently demonstrated during the response to the wheat blast outbreak in Asia early in 2016 where the causal agent was determined to be a wheat-infecting lineage of *Magnaporthe oryzae*, and the origin determined to be South America, within just 6 weeks of sample collection in the field [5]. This approach has the potential to substantially accelerate pathogen diagnostics, while providing detailed genotypic substructure of pathogen populations at an unprecedented resolution. Furthermore, the approach uses RNA-seq data that is generated directly from pathogen-infected leaves collected in the field, providing a unique opportunity to characterize the pathogen population and its host directly in their natural environment.

In order to study the pathogen population, we perform two main analyses: (1) a phylogenetic analysis of the isolates that have been collected to understand how they are related to each other (Subheading 3.2), and (2) a population structure analysis to provide insight into the genetic groups that constitute the pathogen population (Subheading 3.3). These analyses are based on the genetic variation between isolates, which is encapsulated by the single nucleotide polymorphism (SNP) variant sites. Therefore, the SNP sites need to be determined before running these analyses (Subheading 3.1).

The basic workflow of this method is divided into three main sections (Fig. 1):

1. SNP calling
2. Phylogenetic analysis
3. Population structure

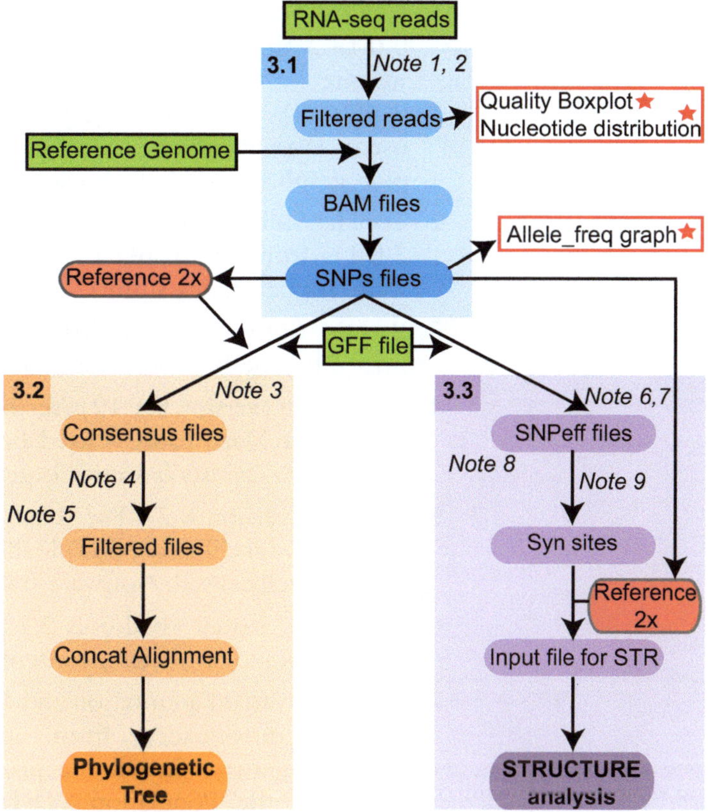

Fig. 1 Overview of the 'Field Pathogenomics' approach. *Numbers* refer to subheadings of Subheading 3. There are three main pipelines, highlighted in three different colors; SNP calling in *blue*, phylogenetic analysis in *orange* and population structure in *purple*. The *green boxes* are input files required for the pipeline and the *red boxes* are outputs. The *stars* indicate checkpoints explained in Subheading 3.1. Notes are also included in the different steps

The first step in this method is the SNP calling (Subheading 3.1). The input for this pipeline is the raw RNA-seq reads that are filtered to remove low-quality reads and subsequently aligned to the reference genome to extract variant sites. These sites are then used in the phylogenetic analysis to generate a phylogenetic tree using the RAxML software (Subheading 3.2). Finally, synonymous SNP sites from each sample are extracted and compared across all samples to define the genetic groups within the population using the software STRUCTURE (Subheading 3.3).

2 Materials

All scripts included herein are to be executed using the Linux command line. These commands are shown below with a '*$*' prefix in *italics*. Before starting, all required programs and packages need to be installed.

2.1 Software and Tools

1. FASTX_Toolkit: This method uses the FASTX_Toolkit version 0.0.13.2 that can be downloaded from the FASTX-Toolkit website (http://hannonlab.cshl.edu/fastx_toolkit/download.html). The following packages from this software are used: fastx_trimmer, fastx_quality_stats, fastq_quality_boxplot_graph.sh, and fastx_nucleotide_distribution_graph.sh.
2. TopHat Software: For the TopHat software, download version 2.0.11. from the TopHat website (https://ccb.jhu.edu/software/tophat/index.shtml).
3. Bowtie Software: Download version Bowtie-0.1.19 from http://bowtie-bio.sourceforge.net/index.shtml.
4. SAM Tools (version 0.1.19): Version 0.1.19 of SAM (Sequence Alignment Map) tools is used here and can be downloaded from http://samtools.sourceforge.net/.
5. RAxML (version 8.2): For RAxML (Randomized Axelerated Maximum Likelihood), download version 8.2.9 (http://sco.h-its.org/exelixis/web/software/raxml/index.html).
6. Bedtools: Install bedtools-2.17.0 from http://bedtools.readthedocs.io/en/latest/content/installation.html.
7. SnpEff: A variant annotation and effect prediction toolbox that can be downloaded from the website: http://snpeff.sourceforge.net/. Install the program within the SCRIPTS folder, in the Population_STRUCTURE directory. Follow the installation instructions from the SnpEff manual (http://snpeff.sourceforge.net/SnpEff_manual.html#run).
8. Make sure that the database includes the organism you will use. If not, you can build a database manually as described in the instruction manual.
9. STRUCTURE Software: Version 2.3.4 can be downloaded from the Structure Software website (http://pritchardlab.stanford.edu/structure.html)

2.2 Programming Languages and Libraries

All scripts used for the "Field Pathogenomics" approach are included in the SCRIPTS directory (Supplemental File 1). Copy this folder into your working directory. Within this folder, you will find three subdirectories containing the scripts needed for each section (SNP_calling, Phylogenetic_Analysis, and Population_STRUCTURE) and five bash scripts required for this demonstration. These bash scripts will call other scripts located in the corresponding directory.

This method uses scripts that are written in several programming languages and utilize different packages. It is necessary to download the appropriate version and the given packages before running the scripts.

The following programming languages are required:

1. R version 3.2.2: Used in the command line or using the R-studio software (https://www.rstudio.com/products/rstudio/download3/). The ggplot2 package is also required for Step 2 of Subheading 3.1.2 (https://cran.r-project.org/web/packages/ggplot2/index.html).
2. Perl version 5.22.1: Downloaded from https://www.perl.org/get.html, and Bioperl-1.6.922 from (http://bioperl.org/INSTALL.html). Make sure that all the following packages are included: Solexa::Parser; Solexa::Fastq; File::Basename; Bio:SeqIO; Bio::AlignIO; Bio::SimpleAlign; Bio::LocatableSeq; and Bio::Align::Utilities.
3. Python version 2.7.9: Installed from https://www.python.org/downloads/.
4. Java version 7.11: Install Java Runtime Environment (JRE) from https://www.neowin.net/news/java-runtime-environment-7-update-11.

2.3 RNA-seq Data

For demonstration purposes of this method, a sample dataset is provided. This dataset consists of RNA-seq data that has been extracted from PST-infected wheat and triticale samples [4]. A total of 39 samples will be used (Supplemental Table 1). The raw RNA-seq reads are available on the NCBI (National Centre for Biotechnology Information) website (www.ncbi.nlm.nih.gov/sra).

1. Install SRA Toolkit tools from the following website: http://trace.ncbi.nlm.nih.gov/Traces/sra/sra.cgi?view=toolkit_doc&f=std.
2. Access and download the data on the website using the accession numbers given in Supplemental Table 1.
3. In your working directory create a directory with the name of the library. LIB4468 (Sample 13/23) is used here as an example:

```
$ mkdir LIB4468
```

4. Unzip the paired-end reads and copy them to the directory:

```
$cp *LIB4468*R1.fastq LIB4468/
$cp *LIB4468*R2.fastq LIB4468/
```

5. Carry out this step for each of the libraries in Supplemental Table 1 until a total of 39 directories in your working directory containing both paired-end reads per library has been created.

2.4 Reference Genome

The reference genome used here to align the raw RNA-seq reads in Subheading 3.1 **step 2** is the PST130 genome [6] (Supplemental File 2).

To create the consensus files for the phylogenetic analysis in Subheading 3.2, **step 2**, the reference genome in gff3 format is also required (Supplemental File 3).

1. Create a directory to store the PST130 genome in your working directory:

```
$ mkdir -p Genomes/PST130/
```

2. Copy the PST130 genome to the recently created directory:

```
$ cp PST130.fasta Genomes/PST130/
$ cp PST130.gff3 Genomes/PST130/
```

3 Methods

This approach has three main pipelines: (Subheading 3.1) SNP calling, (Subheading 3.2) phylogenetic analysis, and (Subheading 3.3) population structure analysis.

3.1 SNP Calling

This pipeline describes the procedure to extract the single nucleotide polymorphisms (SNPs) from the RNA-seq raw reads. It is important to make sure that the reads included are of good quality; therefore, the first step in the pipeline is to filter the reads, verifying their length and quality. Then proceed to trim any adaptor sequences as required (*see* **Note 1**). Since the paired-end reads used here were sequenced using Illumina, we need to add the parameter -Q33 to specify that the reads use the Sanger FASTQ encoding with quality score offset of 33. The quality of the reads as well as the nucleotide distribution will be plotted to check the samples reach a sufficient quality threshold. The filtered reads are then aligned against the reference genome and will be sorted and indexed to proceed with the SNP calling. For SNP calling, heterokaryotic sites with allelic frequencies ranging from 0.2 to 0.8, and 20× depth of coverage will be extracted.

As mentioned in Subheading 2.3, before running this pipeline, the raw reads need to be located inside a directory in your working area. Then proceed with the following method:

1. Index the reference genome. To align the raw RNA-seq reads Bowtie requires a reference genome that has been previously indexed. This needs to be carried out just once per genome:

```
$ bowtie2-build Genomes/PST130/PST130.fasta Genomes/
PST130/PST130
```

2. To run the SNP_pipeline.sh script, the length of the reads needs to be specified, alongside the name of the library and

the genome that will be used for the alignment (*see* **Note 2**). Do this for each one of the 39 libraries:

```
$ bash SNP_pipeline.sh -l 101 -n LIB4468 -g Genomes/PST130/
PST130
```

3.1.1 Quality Checkpoint

1. Check the quality of the samples by inspecting the plots located in LIB4486/Plots (Fig. 2). Inside this directory, there will be two subdirectories: Quality_plots and Nt_distribution.
2. Go to the Quality_plots directory. Look at the graphs to ensure the quality of the reads is at least above 20, ideally above 30 (Fig. 2a).
3. Go to the Nt_distribution directory and verify that the chart follows a random distribution, that is, there is no preference for a specific nucleotide in a given position as shown in Fig. 2b.
4. Check the quality of the alignments. Open the alignment summary created inside the top_hat directory to see if the percentage of aligned data is sufficiently high, and therefore the amount the data is sufficient to be included in the downstream analyses (Supplemental Table 2).

```
$ less LIB*/top_hat/align_summary.txt
```

3.1.2 Heterokaryotic Sites Checkpoint

When the SNP_pipeline.sh script has successfully finished, a file with heterokaryotic sites and their allele frequencies can be found inside each library directory. This file is the input file for the Allele_freq_ggplot.R script that will plot in pdf format an allele frequency graph to make sure that the sample consists of a single PST genotype (Fig. 2c).

1. Copy all the allele frequency files into a directory:

```
$ mkdir Allele_freq
$ cp LIB*/Allele_freq/*ggplot2input_allele_frequency.txt
Allele_freq/
```

2. Go to the Allele_freq directory and run the following R script to check the results:

```
$ cd Allele_freq/
$ R ../SCRIPTS/SNP_calling/Allele_freq_ggplot.r
```

3. Open the pdf files that are generated to check that each one of the samples contains a single genotype. For PST that is a dikaryon, with two haploid nuclei per cell, the mean read counts at heterokaryotic positions should have a single mode at 0.5.

3.2 Phylogenetic Analysis

To establish the genetic relationship between the collected isolates, we perform a phylogenetic analysis. Here, we will only include

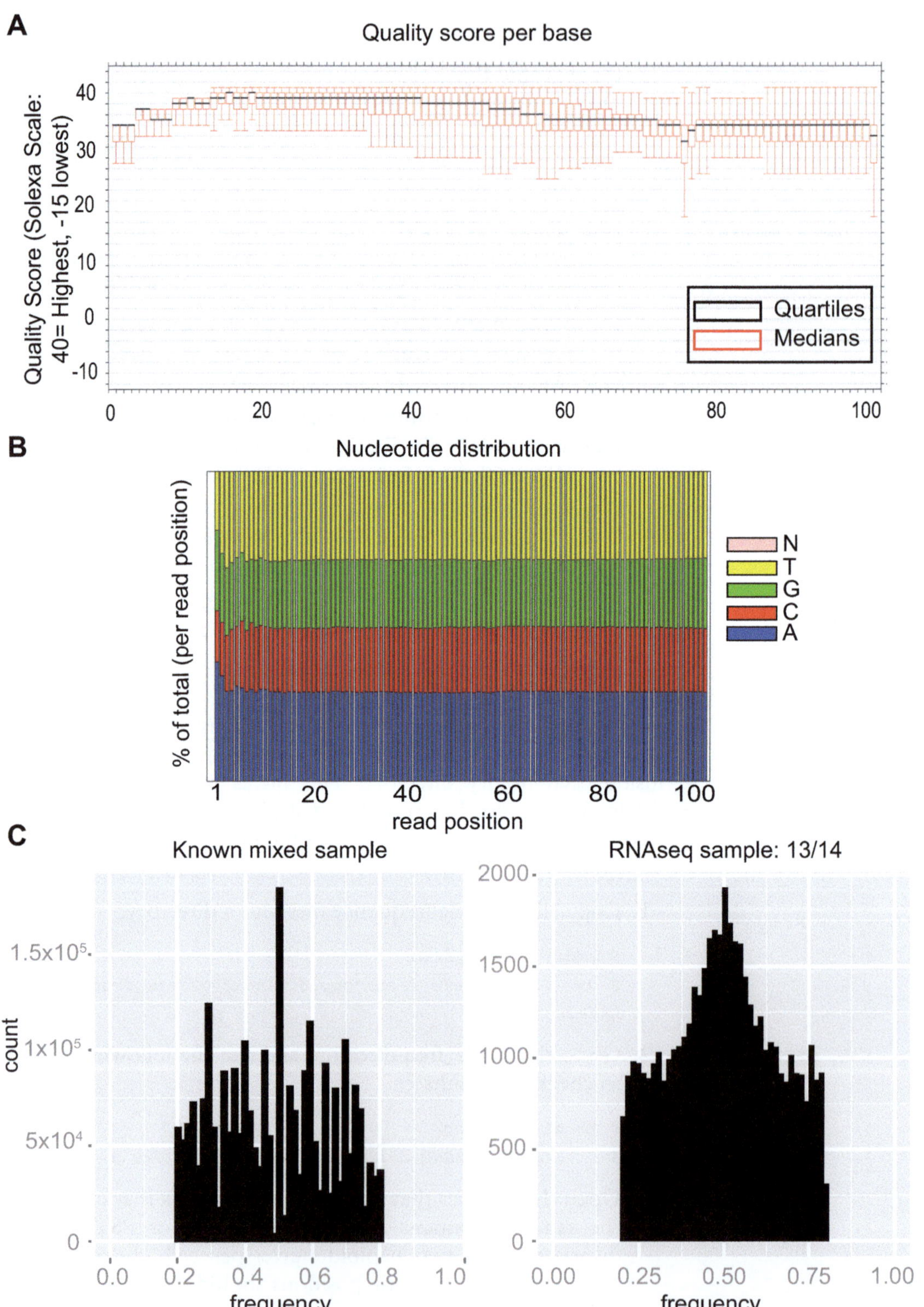

Fig. 2 Quality checkpoint graphs. (**a**) Quality per base for 101 base pairs of sequence. The quality score needs to be higher than 20 throughout the read. (**b**) The nucleotide distribution follows a random distribution, that is, there is no preference for a specific nucleotide in a given position. (**c**) Comparison of biallelic read counts for a known PST sample of mixed genotypes and sample consisting of a single genotype. When read counts at biallelic sites have a single mode at 0.5 (*right*), this is indicative of the sample consisting of a single PST genotype. Mixed genotypes provide an uneven distribution. Taken from Hubbard et al. [4]

libraries of high quality (Subheading 3.1.1) and that consist of a single PST genotype (Subheading 3.1.2). To run the phylogenetic analysis, the first step is to obtain consensus files for the libraries containing the nucleotide residues per gene. If a site is identical to the reference base then it must have a minimum of 2× depth of coverage to be included. However, if a site differs from the reference then to be included it must satisfy a minimum of 20× depth of coverage. These consensus files will then be used to generate a phylip alignment that is the input for the RAxML software to construct the phylogenetic tree.

1. Create a list with all the names of the libraries to be included in the analysis. All of them should be in your current working directory as stated in Subheading 2.3. This file can be created manually or using the following command:

```
$ ls | grep "LIB" > libraries.txt
```

2. Get the consensus files. Run the get_consensus.sh script. The reference genome needs to be specified in gff3 format, and the file containing the names of the libraries included in the analysis provided (libraries.txt) (*see* **Note 3**). A new directory called "Reference2x" will be created with a file per library that contains each position that matches the reference genome and has at least 2× depth of coverage. Then, the Reference2x, the SNP_freq_20x files (that contain all the positions that differ from the reference with at least 20× coverage) and the gff3 file will be used to create the consensus files.

```
$ bash SCRIPTS/get_consensus.sh -g Genomes/PST130/PST130.
gff3 -l libraries.txt -o consensus
```

3. **Step 2** will create a directory with the output consensus files in it. Now, create a new directory where the Tree_pipeline.sh script will be executed and generate the phylogenetic tree (*see* **Note 4**):

```
$ mkdir Tree
```

4. Run the Tree_pipeline.sh script. There are a few parameters that can be given to the program (*see* **Note 5**). First, the consensus files of all the libraries are sorted by contig name and the ones with sufficient coverage pulled out, depending on the parameters specified when running the pipeline (-s, and –l). The specified codon is used for making the alignment (codon3 by default), and then the phylip alignments are extracted and merged into a single file that is called PST_concatinatelan.phy:

```
$ bash SCRIPTS/Tree_pipeline.sh -o Tree -i consensus-n PST
-c codon3 -s 0 -l 0 -d SCRIPTS
```

5. Once the merging step is complete, run the RAxML package to generate the tree. The RAxML package allows the construction of the phylogenetic tree using several threads to parallelize the process. Remember to also generate a tree with bootstrapping values that provides confidence estimates for the individual branches of the tree [7]:

```
$ cd Tree
$ bash raxmlHPC-PTHREADS-SSE3 -T 10 -s PST_concatinatelan.
phy -m GTRGAMMA -n PST_3rd_codon_tree.phy -p 100
$ bash raxmlHPC-PTHREADS-SSE3 -T 10 -s PST_concatinetalan.
phy -m GTRGAMMA -n PST_3rd_codon_bootstraps.phy -p 100 -b
1234 -N 10
```

6. Use the two files generated in **Step 2** to obtain the branch labels file with the bootstrap values:

```
$ bash raxmlHPC-PTHREADS-SSE3 -T 10 -m GTRGAMMA -p 100 -f b
-t RAxML_result.PST_all_3rd_codon_tree.phy -z RAxML_boot-
strap.PST_all_3rd_codon_tree_boots.phy -n PST_all_3rd_co-
don_Tree_bipart.phy
```

The RAxML software provides the tree in phy format, but this can be changed to nwk to be visualized with any standard visualization tool such as MEGA [8]. A total of 3,883,679 sites were used in the example phylogenetic analysis, and the final tree obtained is shown in Fig. 3.

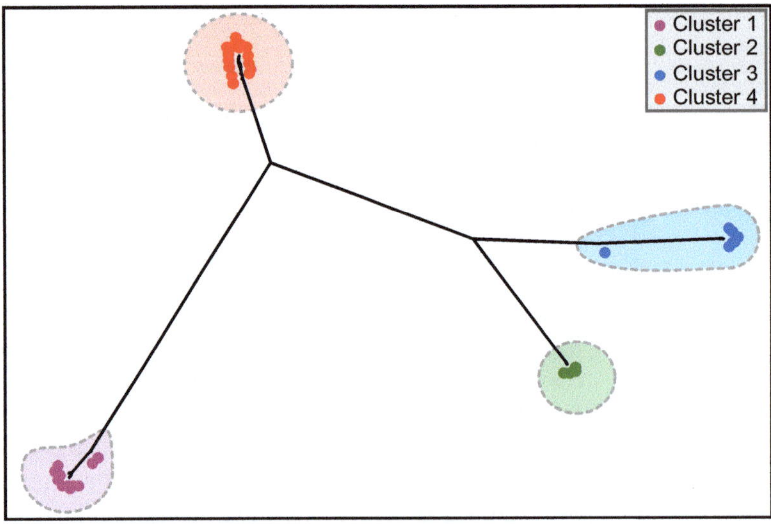

Fig. 3 Phylogenetic analysis of the 39 PST isolates analyzed herein indicates the presence of four different genetic groups within the population. Phylogenetic analysis was carried out on a total of 39 PST UK isolates collected in 2013 using the third codon position of 18,023 PST-130 gene models (3,883,678 sites) using a maximum-likelihood model. Clusters indicate the potential grouping of isolates based on the phylogenetic analysis

3.3 Population Structure

As in Subheading 3.2, only the libraries with sufficient quality and consisting of a single genotype should be included in the analysis. In this section, the synonymous SNP sites are extracted (*see* **Note 6**) for each library and used to study the structure of the pathogen population.

1. Create a file with the name of the libraries, as done in Subheading 3.2, **step 1** that are going to be included in the STRUCTURE analysis:

```
$ ls | grep "LIB" > libraries_structure.txt
```

2. Obtain SnpEff input files (*see* **Note 7**). To do so, the SNP frequency 20x files from Subheading 3.1 **step 2** need to be converted to BED format, and then to SnpEff format using the gff3 file (*see* **Note 8**). Once all the SnpEff files have been generated, they can be found inside their library directory within the SNPs directory. To run the script, the gff3 file (-g) must be specified, and the file that contains the name of the libraries included in the analysis (-l):

```
$ bash SCRIPTS/create_snpeffs.sh -g Genomes/PST130/PST130.
gff3 -l libraries_structure.txt
```

3. Create a new directory in which to execute the Structure pipeline:

```
$ mkdir Structure
```

4. Create a directory to store the SnpEff files created in **Step 2**, and copy them into the new directory:

```
$ mkdir SNPEFF
$ cp LIB*/SNPs/LIB*_snpEff.txt SNPEFF/
```

5. Run the STRUCTURE pipeline (*see* **Note 9**). Remember to give the list of libraries (-l) and the directory where the SnpEff input files are located (-i).
   (a) Extract the synonymous SNP sites from the snpEff files.
   (b) Match the extracted sites to the original file (the SNP ratios files will be used for this) and generate a 'positions' file with all the positions that will be used for the analysis (this will depend on the samples used).
   (c) Extract all data where there is either a syn/non-syn SNP at 20× coverage or covered by at least 2× in the reference. This step uses the "Reference 2x file" created in Subheading 3.2, **step 2**.

(d) Merge all the individual files into a single file, the multi-allelic sites are then removed and the matrix is transposed to obtain the STRUCTURE input file.

```
$ bash Structure_pipeline.sh -l libraries_for_structure.
txt -i SNPEFF
```

6. Now run STRUCTURE using the output file obtained in **Step 5** as input (Structure_syn_all_data_biallelic_flip.str) and the python StrAuto program, version 3.1 using the guidance in the associated manual [9].

The results of the STRUCTURE software (Supplemental Table 3) show a stabilization at $k = 4$ (Fig. 4a), which indicates the existence of four subdivisions within the population (Fig. 4b). These results agree with the phylogenetic tree generated above.

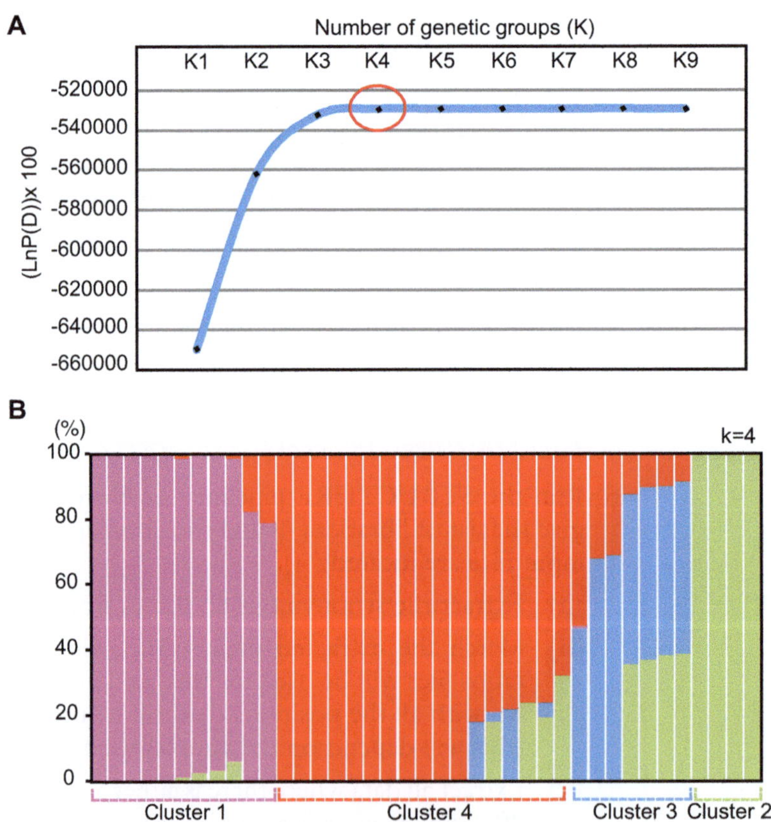

Fig. 4 Population Structure Analysis of 39 PST isolates from the UK in 2013. (**a**) Distribution of the average log probability (LnP(D)) of all k values for the PST population. Population analysis of the 39 PST isolates indicated stabilization at $k = 4$, that is, there are four subpopulations within the population. (**b**) Four clusters were found within the population ($k = 4$). A total of 30,226 biallelic synonymous SNP sites were used to define subdivisions within the population, using the average log probability (LnP(D)) of each k value in the model-based clustering program STRUCTURE

4 Notes

1. The SNP_pipeline.sh script will remove adaptors automatically. By default, the first 14 nucleotides will be removed (that is given in the fastx_trimmer command by the –f parameter). Change this number, or delete this step according to your data.
2. The SNP_pipeline.sh script accepts three arguments:
   (a) –l, Length of the reads, in the example this is 101 bp, which is the default value. This value is used for filtering, and it has to be replaced by the length of the input reads.
   (b) –n, Name of the library, in our example LIB4468. Change this when running the script for each of the libraries. The name also needs to be the same name of the directory where the reads are located.
   (c) –g, Genome, previously indexed (Subheading 3.1 **step 1**) against which the reads will be aligned, that is: Genomes/PST130/PST130. This can be changed if the pipeline is being used for another pathogen.
3. All SNP ratio files that were created in Subheading 3.1 **step 2** should be located in the SNPs folder, inside each library directory and will be created there by default. All the library directories should be located in your working directory.
4. It is possible to change the name of the directory where the tree will be made; to do this, specify the directory when running the Tree_pipeline.sh script.
5. The Tree_pipeline.sh script has seven arguments.
   (a) –o, Output directory ('Tree' by default). *See* **Note 3**.
   (b) –i, Input directory, where the consensus files for each library are located. This argument needs to be given for the script to run. This will be the directory specified with the argument –o when running the get_consensus.sh script in Subheading 3.2, **step 2**.
   (c) –n, Name of the species you are working with (PST by default). This will only be used to name the output file.
   (d) –c, Codon you want to use for the tree. The third codon is given by default, but the first; second; third; both first and second; or all codons can be used (using the options: codon1, codon2, codon3, codon12, or codon123, respectively).
   (e) –s. Minimum percentage of Sequence required to be known per contig. For each contig in the consensus files, specify the minimum percentage of non-missing data for the contig to be considered as valid (0 by default). Give a

value from 0 to 100. A value of 0 means there is no threshold (a contig is accepted even if all sites are missing), whereas when 100 is specified it will only consider a contig if there is no missing data in the sequence (100% of the sequence is known).

(f) –l, Minimum percentage of accepted Libraries. For each contig in the reference genome, specify the minimum percentage of libraries that need to have that contig covered to be included in the analysis (0 by default). Give a value from 0 to 100. A value of 0 means there is no threshold (a contig is included as long as at least 1 library has it), whereas with 100 it will only consider a contig that exists in all samples (100% of samples have that contig).

(g) –d, Scripts directory, this is SCRIPTS by default. Give the path to the scripts directory in case the directory SCRIPTS is not located in your working directory.

6. The synonymous sites are extracted by default when obtaining the SnpEff files, using the '*SNPs_extract_snpEff_syn_coding.pl*' script. To work with the non-synonymous sites or with all sites, the script that is called in the first step of the Structure_pipeline.sh shell script needs to be modified.

   (a) To use the non-synonymous sites, use the *SNPs_extract_snpEff_NonSyn_coding.pl* script.

   (b) To include all sites, call the *SNPs_extract_snpEff_ALL_coding.pl* script.

   All scripts are located in the same directory within the SCRIPTS folder (Population_STRUCTURE/snpeff_and_structure/).

7. To obtain the files, SnpEff is needed. SnpEff is a variant annotation and effect prediction toolbox (see documentation here: http://snpeff.sourceforge.net/) that includes databases for over 2500 genomes. *See* Subheading 2.1, **item** 7 to install it. Make sure that the snpEff program is located in SCRIPTS/Population_STRUCTURE.

8. The create_snpeff.sh shell script calls 'PST-130' by default, as it is the species used in this demonstration. To work with another organism, change this by passing the argument –n <species> when running the script. To check if a particular organism is included in the database use the following command (run it in the directory where the snpEff.jar is located, that is, SCRIPTS/Population_STRUCTURE/snpEff):

```
$ java -jar snpEff.jar database | grep -i <species>
```

If it is not included, you will need to build the database manually. To do so, you will need the reference genome (PST130.fa in this case) and the annotation (you can use the GFF3 file provided here or a GTF file). Add the new genome to the configuration file, editing the snpEffect.config file and create the database using the -gff3 flag:

```
$ java -jar snpEff.jar build -gff3 -v PST-130
```

You will then have to pass the name of the new genome you just added to the create_snpeff.sh script using the –n flag.

9. The Structure_pipeline.sh script has three arguments:
   (a) –l, File containing the names of the libraries that will be included in the analysis.
   (b) –i, Input directory. This is the directory where the SnpEff files that are created in Subheading 3.2, **step 2** are located. This is 'SNPEFF' by default.
   (c) –d, Working directory. The directory where all the files are going to be created. This is 'Structure' by default.

Acknowledgments

This work was funded by an Industrial Partnership Award (BB/M025519/1) from the Biotechnology Biological Sciences Research Council (BBSRC), the BBSRC Institute Strategic Programme (BB/J004553/1 and BB/P012574/1) and the John Innes Foundation.

References

1. Firth C, Lipkin WI (2013) The genomics of emerging pathogens. Annu Rev Genomics Hum Genet 14:281–300
2. Naccache SN, Federman S, Veeraraghavan N, Zaharia M, Lee D, Samayoa E, Bouquet J, Greninger AL, Luk KC, Enge B, Wadford DA, Messenger SL, Genrich GL, Pellegrino K, Grard G, Leroy E, Schneider BS, Fair JN, Martinez MA, Isa P, Crump JA, DeRisi JL, Sittler T, Hackett J Jr, Miller S, Chiu CY (2014) A cloud-compatible bioinformatics pipeline for ultrarapid pathogen identification from next-generation sequencing of clinical samples. Genome Res 24(7):1180–1192
3. Derevnina L, Michelmore RW (2015) Wheat rusts never sleep but neither do sequencers: will pathogenomics transform the way plant diseases are managed? Genome Biol 16:44
4. Hubbard A, Lewis CM, Yoshida K, Ramirez-Gonzalez RH, de Vallavieille-Pope C, Thomas J, Kamoun S, Bayles R, Uauy C, Saunders DGO (2015) Field pathogenomics reveals the emergence of a diverse wheat yellow rust population. Genome Biol 16:23
5. Islam MT, Croll D, Gladieux P, Soanes DM, Persoons A, Bhattacharjee P, Hossain SM, Gupta DR, Rahman MM, Mahboob MG, Cook N, Salam MU, Surovy MZ, Sancho VB, Maciel JLN, NhaniJúnior A, Castroagudín VL, Reges JTA, Ceresini PC, Ravel S, Kellner R, Fournier E, Tharreau D, Lebrun M-H, McDonald BA, Stitt T, Swan D, Talbot NJ, Saunders DGO, Win J, Kamoun S (2016) Emergence of wheat blast in Bangladesh was caused by a South American lineage of *Magnaporthe oryzae*. BMC Biol 14:84

6. Cantu D, Segovia V, MacLean D, Bayles R, Chen X, Kamoun S, Dubcovsky J, Saunders DG, Uauy C (2013) Genome analyses of the wheat yellow (stripe) rust pathogen *Puccinia striiformis* f. sp. *tritici* reveal polymorphic and haustorial expressed secreted proteins as candidate effectors. BMC Genomics 14:270
7. Stamatakis A (2014) RAxML version 8: a tool for phylogenetic analysis and post-analysis of large phylogenies. Bioinformatics 30 (9):1312–1313
8. Kumar S, Tamura K, Nei M (1994) MEGA: Molecular Evolutionary Genetics Analysis software for microcomputers. Comput Appl Biosci 10(2):189–191
9. Chhatre VE (2012) StrAuto ver0.3.1: A Python utility to automate Structure analysis. http://www.crypticlineage.net/pages/software.html

Chapter 3

Race Typing of *Puccinia striiformis* on Wheat

Mogens S. Hovmøller, Julian Rodriguez-Algaba, Tine Thach, and Chris K. Sørensen

Abstract

A procedure for virulence phenotyping of isolates of yellow (stripe) rust using spray inoculation of wheat seedlings by spores suspended in an engineered fluid, Novec™ 7100, is presented. Differential sets consisting of near-isogenic Avocet lines, selected lines from the "World" and "European" sets, and additional varieties showing race-specificity facilitate a robust assessment of race, irrespectively of geographical and evolutionary origin of isolates. A simple procedure for purification of samples consisting of multiple races is also presented.

Key words Race, Infection type, Yellow (stripe) rust, Spray inoculation, Genetic interpretation, Purification

1 Introduction

In cereal rust pathology, "race" is often defined by the pattern of compatible and incompatible interactions between host and pathogen. The pathogen phenotype is described as "virulent" in case of a compatible interaction, conferred by "high" infection type scores on host differential lines, and "avirulent" in case of incompatible interactions conferred by "low" infection types ([1], Fig. 1). The genetic interpretation of such race (pathotype) data depend on the extent that R-genes have been identified in the host differential lines, and additional resistance specificities have been resolved by exposure of the lines to a wide array of pathogen isolates of diverse geographical and evolutionary origin.

Historically, the study of race dynamics in *Puccinia striiformis* populations in Europe [2–4], North America [5, 6], China [7], and Australia [8] has often focused on virulence dynamics of national or regional relevance, with limited overlap of host differential lines among laboratories. Significant influences of experimental conditions such as light quality, intensity and duration, temperature regimes, local procedures, interpretation of data, and the presence

Sambasivam Periyannan (ed.), *Wheat Rust Diseases: Methods and Protocols*, Methods in Molecular Biology, vol. 1659, DOI 10.1007/978-1-4939-7249-4_3, © Springer Science+Business Media LLC 2017

Fig. 1 Disease scoring scale (0–9) of infection types (IT) for *yellow* (*stripe*) rust according to McNeal et al. Distinct appearance of chlorosis and necrosis for individual IT may vary depending on R-genes involved; IT 0–6 generally considered incompatible (avirulent) and IT 7–9 compatible (virulent). IT from left to right, *0*: no visible disease symptoms (immune), *1*: minor chlorotic and necrotic flecks, *2*: chlorotic and necrotic flecks without sporulation, *3–4*: chlorotic and necrotic areas with limited sporulation, *5–6*: chlorotic and necrotic areas with moderate sporulation, *7*: abundant sporulation with moderate chlorosis, *8–9*: abundant and dense sporulation without notable chlorosis and necrosis

of undetected resistance specificities toward particular isolates, have further restricted the comparison of race typing results from different parts of the world. The initiative by Ron W. Stubbs and colleagues in Wageningen, the Netherlands, is one significant exception since they offered yellow rust race typing for many countries worldwide between the 1960s and 1980s [9]. The service was stopped around 1990, and in 2010 the collection of more than 5000 spore samples preserved in liquid nitrogen was transferred to the Global Rust Reference Center, Aarhus University, Denmark [10]. The escalating yellow rust epidemics worldwide in recent years [11], and the spread of epidemics to new areas, where the disease have previously been absent or nonsignificant, have increased the need for understanding spread, establishment, and evolution of yellow rust races at a global scale [12, 13].

The methodologies presented in this chapter are based on the experiences of virulence phenotyping at the Global Rust Reference Center of more than 1000 isolates from recent years, representing 46 countries and five continents, and linking to results in the past by the recovery of more than 200 spore samples from the Stubbs collection [14]. Procedures for purification of spore samples of mixed races and the rationale for a robust phenotyping by taking into account the results from multiple differential lines with shared R-genes are also presented.

2 Materials

1. Six to eight seeds of each wheat differential line per set (Table 1; *see* **Note 1**).
2. Plastic pots (7 × 7 × 8 cm).

Table 1
Current standard and extended set of wheat differential lines used for race typing of *P. striiformis* isolates at the Global rust Reference Center (GRRC), www.wheatrust.org

Differential set	Differential line	Yellow rust resistance genes (*Yr*)[a]	GRRC standard set	GRRC extended set
World (W)	Chinese 166	*1*	X	X
	Vilmorin 23	*3, +*	X	X
	Heines Kolben	*6, +*	X	X
	Lee	*7, +*	X	X
	Moro	*10*	X	X
	Strubes Dickkopf	*Sd, 25, +*		X
	Suwon 92/Omar	*Su*		X
European (E)	Hybrid 46	*4, +*	X	X
	Heines Peko	*2, 6, 25, +*		X
	Heines VII	*2, 25, +*		X
	Compair	*8, +*		X
	Carstens V	*32, 25, +*	X	X
	Spaldings Prolific	*Sp, 25, +*		X
Avocet near-isogenic lines	Avocet S	*AvS*	X	X
	Avocet/*Yr1*	*1, 18*[b]*, AvS*		X
	Avocet/Yr5	*5, 18*[b]*, AvS*		X
	Avocet/Yr6	*6, AvS*	X	X
	Avocet/Yr7	*7, AvS*		X
	Avocet/Yr8	*8*	X	X
	Avocet/Yr9	*9, AvS*	X	X
	Avocet/Yr10	*10, 18*[b]*, AvS*		X
	Avocet/Yr15	*15, 18*[b]*, AvS*		X
	Avocet/Yr17	*17, AvS*	X	X
	Avocet/Yr24	*24, AvS*		X

(continued)

Table 1
(continued)

Differential set	Differential line	Yellow rust resistance genes (*Yr*)[a]	GRRC standard set	GRRC extended set
	Avocet/Yr27	*27, AvS*		X
	Avocet/Yr32	*32, AvS*		X
	Avocet/YrSp	*Sp, 18*[b]*, AvS*	X	X
Additional	Ambition	*Amb*[c]	X	X
	Anja	*25, +*		X
	Brigadier	*17, 9, +*		X
	Cortez	*15*	X	X
	Kalyansona	*2, +*	X	X
	Opata	*27, 18*[b]*, +*	X	X
	Sleipner	*9, +*		X
	TP 981	*25, +*	X	X
	VPM1	*17, +*	X	X
References	Cartago	*Unknown*[d]	X	X
	Morocco	*Unknown*[e]		X
Number of entries			20	36

[a]According to [8, 15, 16] and the present study.
[b]*Yr18* detected by PCR test at GRRC according to [17]
[c]Resistance specificity of variety Ambition.
[d]Generally susceptible except for particular isolates from Himalayan region.
[e]Resistance specificity toward *Yr2* avirulent isolates.

3. Standard peat-based substrate with slow release nutrients optimized for cereal growth.
4. Plastic trays with transparent lids: Standard set trays of the size 50 × 40 × 8 cm (L:W:H) were used to fit up to 36 pots, one pot for each differential line.
5. Airbrush spray gun, vacuum pump and glass flask.
6. Novec™ 7100, a hydroflourether engineered fluid.
7. Urediniospore sample from a single isolate (10–20 mg dried spores).
8. Hand mist sprayer with distilled water.

3 Methods

3.1 Inoculation and Incubation

1. Seeds of each wheat differential line are sown in individual pots containing Pindstrup peat-based substrate. One pot of each line is placed in a tray and grown in spore-proof greenhouse cabins (*see* **Note 2**) with 16 h of natural light supplemented with sodium light at 100 mE/m^2/s when daylight is <10,000 lux. Temperature is set to 17–12 °C (day–night) and relative humidity to 70–80%.
2. The seedlings are kept in the greenhouse until inoculation at the time when the second leaf is halfway unfolded (12–14 days).
3. Connect the airbrush spray gun to the vacuum pump. Ten to twenty milligrams of urediniospores of a single isolate (*see* **Note 3**) is suspended in 5 mL Novec™ 7100 in a glass flask and connected to the airbrush spray gun (Fig. 2a).
4. Inoculate seedlings in fume hood using the airbrush spray gun from 10 to 15 cm distance on all sides and from the top by turning the tray (Fig. 2b).
5. Mist the transparent lid and the seedlings with water before covering the tray to ensure dew formation during incubation at 10 °C in the dark (Fig. 2c, d). Transfer the tray to a spore proof greenhouse cabin after 20–24 h of incubation, after which the lid is removed (Fig. 2e).
6. The plants are scored for infection types after 15–18 days (*see* **Note 4**; Fig. 2f).

3.2 Scoring and Interpretation of Infection Types

1. Visual assessment of infection type is carried out according to McNeal et al. [18]. Scores are from 0 to 9 which represent no or less infection to severe infection as detailed in Fig. 1.
2. Infection type is scored individually on the first and second leaf of each plant within a pot. Number of leaves with a particular infection type is noted.
3. Leaves showing no signs of infection (including chlorosis and necrosis) are categorized as escape.
4. The general disease level is assessed based on susceptible lines. The disease level is categorized as low, medium or high where low represent cases with many escapes per pot, in which case the results may not be reliable.
5. If one or few plants within a pot show distinct different responses than the rest this is noted as seed contamination (*see* **Note 5**).

Fig. 2 Procedure for inoculation of differential sets for virulence phenotyping. (**a**) Pipette, glass flask, spore sample, airbrush spray gun, vacuum pump, and Novec™ 7100 engineered fluid, (**b**) 12–14-day-old seedlings inoculated using an airbrush spray gun in a fume hood, (**c**) seedlings misted with water before incubation to ensure dew formation, (**d**) seedlings covered and ready for incubation, 20–24 h at 10 °C in darkness, (**e**) inoculated plants transferred to spore-proof greenhouse cabins with automatic watering, (**f**) virulence phenotyping 15–18 days after inoculation when infection types are well developed

Fig. 3 Signs of multiple races revealed by contrasting infections types (IT) of compatible and incompatible interactions on the second leave of the near-isogenic lines Avocet *S* and Avocet/*Yr9*. (**a**) Avocet *S* displaying IT 1 (*lower* to *middle part* of the leaf) and IT 6–7 (*upper part* of the leaf), (**b**) Avocet/*Yr9* displaying IT 1–2 (*lower* to *middle part* of the leaf) and IT 7 (*upper part* of the leaf)

6. Contrasting infection types conferred by clearly compatible and incompatible interactions within a leaf may indicate the presence of more than one race (Fig. 3; *see* **Note 6**). Contamination involving intermediate infection types (IT 4–6) of individual races may be difficult to resolve.
7. The virulence phenotype of individual isolates is generally inferred based on infection types across multiple wheat differential lines carrying shared host R-genes (Table 2). The complementarity of Avocet near-isogenic lines and additional differential lines is illustrated by three *Yr6*-virulent isolates of different origin (Fig. 4) and three *Yr17*-virulent isolates and one avirulent isolate (Fig. 5).
8. Avirulence to resistance genes present in multiple differential lines, e.g., the resistance specificity in Avocet S (present in all Avocet lines considered except Avocet/*Yr8*) and *Yr25* present in seven differential lines, generally restrict the genetic resolution of results (Tables 1 and 2).

Table 2

Rationale for the assessment of virulence phenotype of isolates of *P. striiformis* of diverse origin based on infection type scores on wheat differential lines

Virulence inferred	Differential lines	Resistance genes	Refinement comment
v1	Chinese 166 Avocet/*Yr1*	*Yr1* *Yr1, Yr18, YrAvS*	*Avr*AvS: consider Chinese 166
v2	Heines VII Kalyansona Heines Peko	*Yr2, Yr25*, + *Yr2*, + *Yr2, Yr6, Yr25*, +	*Avr25* and *Avr6*: consider Kalyansona
v3	Vilmorin 23	*Yr3*, +	High and intermediate IT (5–6) imply *v3*
v4	Hybrid 46	*Yr4*, +	High and intermediate IT (4–6) imply *v4,* often associated with high IT on Suwon/Omar
v5	Avocet/*Yr5* *Triticum spelta album*	*Yr5, Yr18, YrAvS* *Yr5*	*Avr*AvS: consider *Triticum spelata album*; *v5* rarely observed
v6	Avocet/*Yr6* Heines Kolben Heines Peko	*Yr6, AvS* *Yr6*, + *Yr2, Yr6, Yr25*, +	*Avr*AvS: consider Heines Kolben and Heines Peko. Low IT (1–3) on Heines Kolben and Heines Peko: consider Avocet/Yr6
v7	Lee Avocet/*Yr7*	*Yr7*, + *Yr7, AvS*	*Avr*AvS: consider Lee
v8	Compair Avocet/*Yr8*	*Yr8*, + *Yr8*	Intermediate IT (4–6) on Compair: consider Avocet/*Yr8*,
v9	Sleipner	*Yr9*, +	Low IT on Sleipner: consider Avocet/*Yr9*
	Avocet/*Yr9*	*Yr9, AvS*	*Avr*AvS: consider Sleipner
v10	Moro Avocet/*Yr10*	*Yr10* *Yr10, Yr18, YrAvS*	*Avr*AvS: consider Moro
v15	Cortez Avocet/*Yr15*	*Yr15* *Yr15, Yr18, AvS*	*Avr*AvS: consider Cortez; *v15* rarely observed
v17	VPM1 Avocet/*Yr17*	*Yr17*, + *Yr17, AvS*	*Avr*AvS: consider VPM1 Low IT (1–3) on VPM1: consider Avocet/*Yr17*
v24	Avocet/*Yr24*	*Yr24, AvS*	*Avr*AvS: *v24/Avr24* not accessible
v25	TP 981 Anja	*Yr25*, + *Yr25*, +	Intermediate IT (5–6) on TP981: consider Anja Intermediate IT (3–5): consider TP981
v27	Opata Avocet/*Yr27*	*Yr27, Yr18*, + *Yr27, YrAvS*	*Avr*AvS: consider Opata Intermediate IT (5–6) on Opata: consider Avocet/Yr27
v32	Carstens V Avocet/*Yr32*	*Yr32, Yr25*, + *Yr32, AvS*	*Avr*AvS: consider Carstens V Low IT (0-2) on Carstens V: consider Avocet/*Yr32*

(continued)

Table 2 (continued)

Virulence inferred	Differential lines	Resistance genes	Refinement comment
vSp	Spaldings Prolific Avocet/*YrSp*	*YrSp, Yr25, +* *YrSp, YrAvS*	*Avr*AvS: consider Spaldings Prolific Low and intermediate IT on Spaldings Prolific: consider Avocet/*YrSp*
vAvs	Avocet S	*YrAvS*	*Avr*AvS conferred by IT 0–1
vAmb	Ambition	*YrAmb*	Resistance component(s) in Ambition conferred by IT 1–5
Reference	Cartago	None	Susceptible check, avirulence only observed in particular isolates from Pakistan. IT 6–7 observed for isolates of Warrior race

Virulence is generally inferred by the highest infection type within groups of two or more differential lines sharing a considered resistance gene

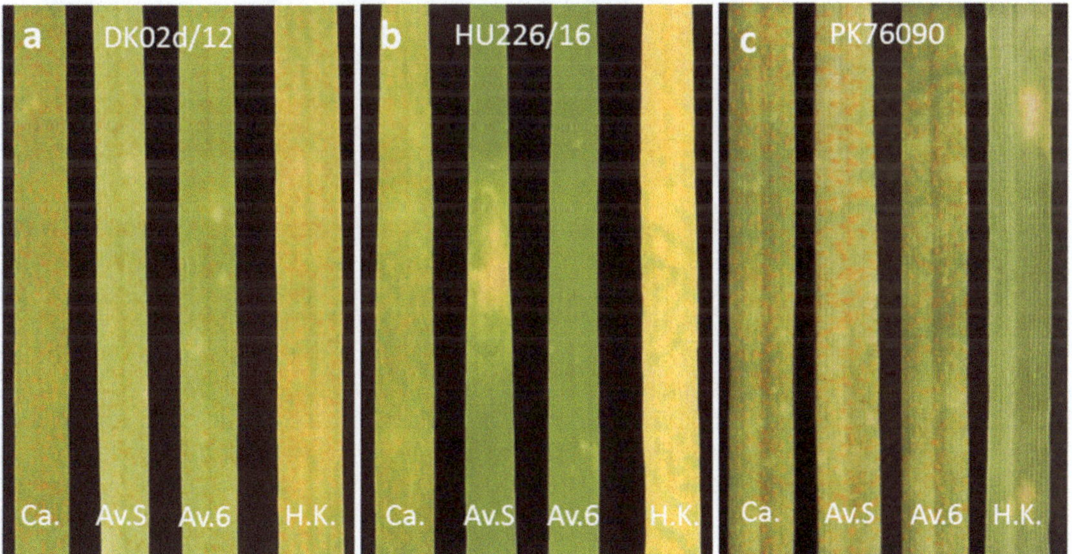

Fig. 4 Infection types (IT) on a 0–9 scale for the isolates DK02d/12, HU226/16, and PK7690 inoculated on wheat lines carrying *Yr6*, i.e., Avocet/*Yr6* (Av.6) and Heines Kolben (H.K.). Cartago (Ca.) and Avocet S (Av.*S*) are included as susceptible controls. (**a**) DK02d/12 displaying a compatible interaction (IT 7–8), (**b**) HU226/16 displaying IT 8 on Ca., IT 2 on Av.S, IT 1 on Av.6, and IT 9 on H.K., (**c**) PK76090 displaying IT 7-8 on Ca., Av.S, and Av.6 and IT 2 on H.K

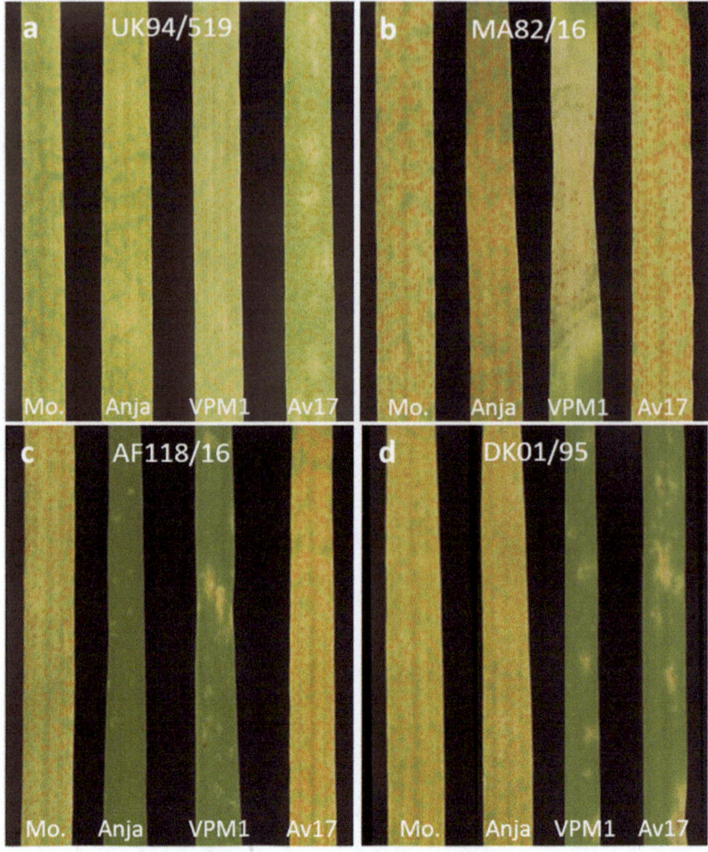

Fig. 5 Infection types (IT) on a 0–9 scale for the isolates AF118/16, DK01/95, UK94/519, and MA82/16 on *Yr17* resistant wheat varieties, i.e., VPM1, Avocet/*Yr17* (Av.17). Morocco (Mo.) and Anja are included as susceptible controls. (**a**) UK94/519 displaying a compatible interaction (IT 7–8); note chlorotic flecks on Avocet/*Yr17* which is consistent for *Yr17*-virulent isolates from the NW-European *P. striiformis* population, (**b**) MA82/16 displaying IT 8-9 on Mo., Anja and Av.17, and IT 4 on VPM1, (**c**) AF118/16 displaying IT 7-8 on Mo. and Av.17 and IT 1-2 on Anja, and VPM1, (**d**) DK01/95 displaying IT 8-9 on Mo. and Anja, and IT 1-2 on VPM1 and Av.17

4 Notes

1. Commercial local varieties or other lines of special interest may be included in the set. New races are often identified because they overcome resistance in widely grown varieties.
2. It is important that the plants are grown in disease-free environment prior to inoculation to avoid unintentional contamination.

3. Safe handling of spore samples to minimize the risk of unintentional spread is vital.
4. Optimal time of scoring may vary depending on the season and isolate.
5. Seed contamination is based on plant morphology and infection type. Off-type plants are not considered in the analyses.
6. Samples containing more than one race can be purified as follows: single lesions of high infection types are collected from at least two different differential lines, rinsed with water, transferred to petri dish with moist filter paper, and incubated for 2–3 days at 12–14 °C under light to allow germination of detached spores of the off-type and production of new spores from the lesion. The single lesions are subsequently used as basis for new spore multiplications using susceptible lines. Purity of the new isolates can be confirmed on new differential sets.

References

1. McIntosh RA, Welling CR, Park RF (1995) Wheat rusts: an atlas of resistance genes. CSIRO, Melbourne, VIC. 200 pp
2. Hovmøller MS (2001) Disease severity and pathotype dynamics of *Puccinia striiformis* f. sp. *tritici* in Denmark. Plant Pathol 50:181–189
3. de Vallavieille-Pope C, Ali S, Leconte M, Enjalbert J, Delos M, Rouzet J (2012) Virulence dynamics and regional structuring of *Puccinia striiformis* f. sp. *tritici* in France between 1984 and 2009. Plant Dis 96:131–140
4. Johnson R (1992) Reflection of a plant pathologist on breeding for disease resistance, with emphasis on yellow rust and eyespot of wheat. Plant Pathol 41:239–254
5. Chen X, Penman L (2005) Stripe rust epidemic and races of *Puccinia striiformis* in the United States in 2004. Phytopathology 95(6): S19–S19
6. Line RF, Qayoum A (1992) Virulence, aggressiveness, evolution and distribution of races of *Puccinia striiformis* (the cause of stripe rust of wheat) in North America, 1968-1987. US Dep Agric Tech Bull 1788:44
7. Wan A, Zhao Z, Chen X, He Z, Jin S, Jia Q, Yao G, Yang J, Wang B, Li G, BI Y, Yuan Z (2004) Wheat stripe rust epidemic and virulence of *Puccinia striiformis* f. sp. *tritici* in China in 2002. Plant Dis 88:896–904
8. Wellings CR (2007) *Puccinia striiformis* in Australia: a review of the incursion, evolution, and adaptation of stripe rust in the period 1979-2006. Aust J Agr Res 58:567–575
9. Stubbs R (1988) Pathogenicity analysis of yellow (stripe) rust of wheat and its significance in a global context. In: Simmonds NW, Rajaram S (eds) Breeding strategies for resistance to the rusts of wheat. CIMMYT, Mexico, DF, pp 23–38
10. Hovmøller MS, Yahyaoui A, Singh RP (2009) A global reference centre for wheat yellow rust: pathogen variability, evolution and dispersal pathways at regional and global levels. BGRI Technical Workshop, Cd. Obregón, Sonora, Mexico, March 17–20, 2009
11. Hovmøller MS, Walter S, Justesen AF (2010) Escalating Threat of Wheat Rusts. Science 329 (5990):369–369
12. Hovmøller MS, Sørensen CK, Walter S, Justesen AF (2011) Diversity of *Puccinia striiformis* on cereals and grasses. Annu Rev Phytopathol 49:197–217
13. Walter S, Ali S, Kemen E, Nazari K, Bahri BA, Enjalbert J, Hansen JG, Brown JKM, Sicheritz-Pontén T, Jones J, de Vallavieille-Pope C, Hovmøller MS, Justesen AF (2016) Molecular markers for tracking the origin and worldwide distribution of invasive strains of *Puccinia striiformis*. Ecol Evol 6(9):2790–2804
14. Thach T, Ali S, Justesen AF, Rodriguez-Algaba J, Hovmøller MS (2015) Recovery and virulence phenotyping of the historic 'Stubbs collection' of the yellow rust fungus *Puccinia striiformis* from wheat. Ann Appl Biol 167 (3):314–326
15. Hovmøller MS, Justesen AF (2007) Appearance of atypical *Puccinia striiformis* f. sp. *tritici*

phenotypes in north-western Europe. Aust J Agr Res 58:518–524

16. Nazari K, El Amil R (2013) First report of resistance of wheat line Avocet 'S' to stripe rust caused by *Puccinia striiformis* f. sp. *tritici* (Pst) in Syria. Plant Dis 97(7):996–996
17. Lagudah ES, Krattinger SG, Herrera-Foessel S, Singh RP, Huerta-Espino J, Spielmeyer W, Brown-Guedira G, Selter LL, Keller B (2009) Gene-specific markers for the wheat gene *Lr34/Yr18/Pm38* which confers resistance to multiple fungal pathogens. Theor Appl Genet 119:889–898
18. McNeal FM, Konzak CF, Smith EP, Tate WS, Russell TS (1971) A system for recording and processing cereal research data. US Agric Res Serv 42:34–121

Chapter 4

Assessment of Aggressiveness of *Puccinia striiformis* on Wheat

Chris K. Sørensen, Tine Thach, and Mogens S. Hovmøller

Abstract

A simple point-inoculation method using Novec™ 7100, a volatile engineered fluid, is presented for the assessment of aggressiveness of *Puccinia striiformis* isolates on seedlings of wheat. The method allows for quantification of the applied inoculum with a minimal risk of cross-contamination of rust from leaves grown side by side. The method is also applicable for the assessment of qualitative differences inferred by compatible and incompatible host–pathogen interactions, and it can be adjusted to other cereal rust and powdery mildew fungi on other host species, and other plant growth stages as appropriate.

Key words Aggressiveness, Epidemiological parameters, Plant pathogen, Fungus, *Puccinia*, Point inoculation, Latent period, Lesion growth

1 Introduction

Accurate assessment of pathogen aggressiveness often represents a significant challenge in host-pathogen systems. This is particularly the case for biotrophic pathogens, where the assessment is restricted to observations on live host plants [1, 2]. Aggressiveness can be quantified by parameters like infection efficiency, latent period, pathogen growth, and spore production, i.e., parameters defining the ability of a virulent isolate to cause disease on a susceptible host [3].

Pathogen growth and host response may be influenced by factors like nutritional status of the host, light and shading regimes, temperature and humidity conditions [3–6]. It is therefore important to use a standardized experimental setup to reduce variability caused by environment.

Here, we present a simple experimental setup and inoculation method, which is based on materials available in most laboratories. Seedling leaves are fixed on acrylic pedestals and point inoculated by applying a drop of spore suspension with a pipette. This setup allows assessment of several parameters associated with

Sambasivam Periyannan (ed.), *Wheat Rust Diseases: Methods and Protocols*, Methods in Molecular Biology, vol. 1659,
DOI 10.1007/978-1-4939-7249-4_4, © Springer Science+Business Media LLC 2017

aggressiveness in an integrated manner. For instance, the latent period and lesion growth can be continuously monitored on the same individual leaves. The setup is flexible and can be modified to other experimental designs. It is also expected that the method can be easily adapted to other pathosystems and be adjusted for assessment of other parameters than those described in this protocol.

A number of experiments were performed for evaluating the most reliable and efficient inoculation method and the assessment of both quantitative and qualitative parameters [7]. The presented method allowed quantification of inoculum and resulted in high infection efficiency with a minimal risk of cross-contamination of leaves grown side by side. The system has been used for assessment of aggressiveness in segregating progeny isolates of *Puccinia striiformis* [8] and for assessment of infection types caused by various pathogen races on selected host genotypes. In this respect the method may complement traditional methods, using settling towers and spray guns for inoculation [9, 10].

2 Materials

1. Plastic trays with clear lids: Trays of different sizes can be used depending on the purpose and experimental setup. Here trays of the size 55 × 45 × 7 cm (Length–Width–Height) are used.
2. Standard peat based substrate with slow release nutrients optimized for cereal growth.
3. Acrylic U-shaped pedestals of appropriate size for the experimental setup: Here we used pedestal of the size 22.5 × 15 × 15 cm (L:W:H).
4. Transparent double-sided tape.
5. Water- and air-permeable transparent tape: Tape with a width of 2.5 cm is used although this type of tape comes in different sizes.
6. 20 μl pipette and compatible tips.
7. Spores are suspended in 3M™ Novec™ 7100 engineered fluid.
8. 10× hand lens.

3 Methods

3.1 Fixation and Inoculation of Wheat Seedlings with Urediniospores of Puccinia striiformis

1. Acrylic pedestals are placed in plastic trays and soil is added. Seeds are sown along the pedestals (*see* **Note 1**). The soil is watered, covered with the lid, and transferred to the greenhouse where the plants are grown for approx. 18 days (17 °C day–12 °C night and 50–100 μE/m^2/s of artificial light applied when daylight <10,000 lux).

Fig. 1 The second leaf of wheat seedlings fixed on acrylic pedestals and inoculated with a pipette containing spores of *Puccinia striiformis* suspended in Novec™ 7100

2. The second leaf of the seedlings is fixed on the acrylic pedestals when it is fully emerged (*see* **Note 2**). The double-sided tape is placed near the edges of the pedestal (*see* **Note 3**), and leaves are then gently pressed onto the tape and overlaid with the permeable tape (*see* **Note 4**) (Fig. 1).
3. Spores samples are mixed with Novec™ 7100 at a concentration of 2 mg of spores per milliliter of Novec™ 7100 (*see* **Note 5**). It is important to keep the suspension in a closed container, e.g., a glass flask, after preparation as Novec™ 7100 evaporates very fast (*see* **Note 6**).
4. Two microliter of the spore suspension is applied to each leaf with a pipette (Fig. 1).
5. After inoculation of the leaves, the soil is watered and the tray is covered with a lid to ensure passive dew formation during incubation (*see* **Note 7**).
6. Plants are incubated in a dark cold room at 10–12 °C for approx. 24 h.

3.2 Experimental Design and Method for Assessment of Infection Type

1. Leaves of plants grown side by side are fixed on the acrylic pedestals and inoculated with the considered pathogen isolates.
2. The progression of disease symptoms on the fixed leaves can be easily monitored based on for example the timing of appearance of disease symptoms and sporulation.

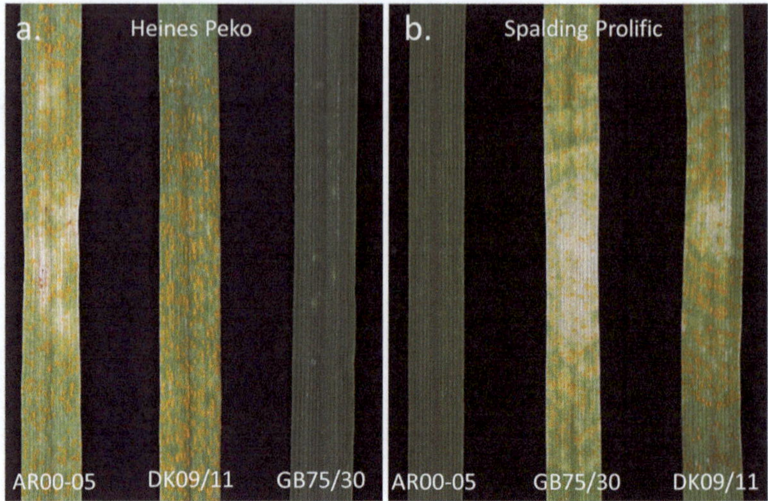

Fig. 2 Seedling of two different cultivars, (**a**) Heines Peko and (**b**) Spalding Prolific, assessed for response to three different isolates of *Puccinia striiformis*: AR00-05, GB75/30, and DK09/11

3. The infection type on the host plants can be assessed from 16 to 19 days after inoculation (dai) using the 0–9 scale [11] (Fig. 2).
4. Photos can be taken for additional image analysis and documentation (Fig. 2), e.g., assessment of the size of the affected leaf area and the number and size of the pathogen pustules.

3.3 Experimental Design and Method for Assessment of Aggressiveness

1. Assessment of aggressiveness parameters, e.g., latent period and lesion growth, can be carried out for isolates side by side.
2. For *Puccinia striiformis*, the latent period can be recorded as the time from inoculation until the first pustules with visible spores break through the leaf epidermis (Fig. 3a). Under optimal conditions the first pustules of a fast isolate will emerge approx. 9 dai. Daily monitoring is recommended to start 7 dai and continue at 12–24 h intervals. Observations can be done efficiently with a 10× hand lens.
3. Lesion growth can be assessed as the length of the full lesion at a specified date after inoculation or as the lesion expansion between two time points (Fig. 3b) (*see* **Note 8**).
4. Measuring lesion growth between two consecutive dates is done by marking the edges of the lesion with a felt pen (*see* **Note 9**) at specific time points, e.g., 4 and 8 days after the latent period. The difference between the markings represents the lesion growth in millimeter/4 days (Fig. 3b).

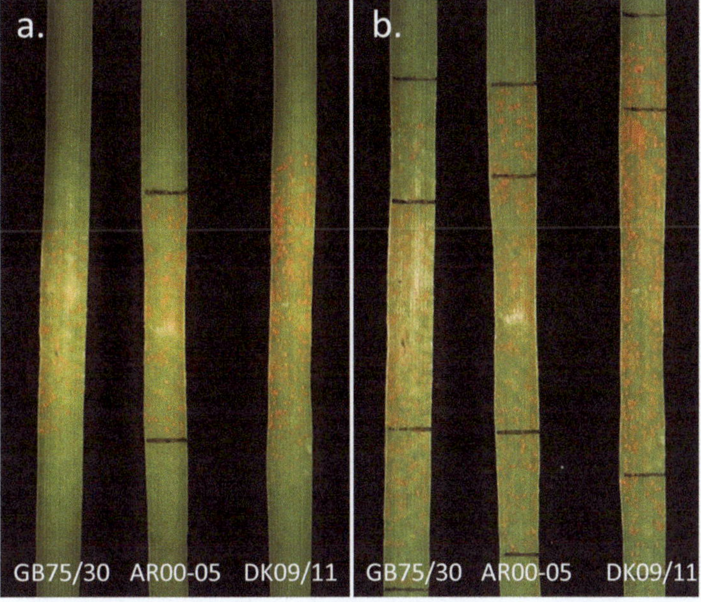

Fig. 3 Comparison of disease progression of three *Puccinia striiformis* isolates of different races (GB75/30, AR00-05, and DK09/11) on seedling of Cropland Genetic 514W. (**a**) Spore producing lesion 15 dai., Latent period was recorded as the time from inoculation until the emergence of the first spore-bearing pustules. AR00-05 started sporulation 264 hours after inoculation (hai), DK09/11 started 276 hai., whereas GB75/30 started 288 hai. (**b**) Lesion growth was estimated by marking the edges of the lesions 4 and 8 days after the latent period. The length of the full lesion is measured as the distance between the outer markings

4 Notes

1. Important to use high-quality seeds and to sow more plants than needed to be able to produce and pick the best plants for the experiment.
2. Different types of stands can be used for leaf fixation, but it is important to align the height of the plants with the height of the stand to avoid plant/leaf damage.
3. Double-sided tape can be left out. In some designs it may be easier to fix the leaves without this tape.
4. It is important to use permeable tape for the top fixation of the leaves to allow leaf transpiration. Use of conventional tape will result in leaf yellowing under the tape which will most likely affect the health status of the whole leaf.
5. This spore concentration gives a high infection rate and results in very clear disease symptoms for *Puccinia striiformis* on

wheat. The concentration may need to be optimized if the method is used for other pathosystems.

6. Based on our experience the urediniospores can stay for several hours in Novec™ 7100 without losing viability.
7. Avoid direct misting of leaves as this may lead to movement of spores between and within leaves. Instead, passive dew formation can be achieved by wetting the soil before incubation in the tray with a lid on.
8. More variation can be expected in data when assessing the full length of the lesion due to variation in the size of the inoculated area.
9. The front of the lesions can sometimes be skewed, i.e., more progressed in one side of the leaf than in the other. In this case, different criteria can be used to define the lesion, but it is important that the same criteria are used for consecutive markings. Suggested criteria are (1) marking the lesion front where you observe the first spore-bearing pustule regardless of the shape of the front, or (2) marking the lesion halfway between the most and least advanced part of the lesion front.

Acknowledgments

This method was developed under the RUSTFIGHT (Grant number: 11-116241) funded by Innovation Fund Denmark, The Ministry of Higher Education and Science, Denmark, and GUDP, Ministry of Environment and Food of Denmark (grant no. 34120902567). The authors thank Steen Meier, Ellen Jørgensen, and Janne Hansen for technical support, and Annemarie Fejer Justesen, Mehran Patpour, and Julian Rodriguez-Algaba for valuable discussions.

References

1. Milus EA, Seyran E, McNew R (2006) Aggressiveness of *Puccinia striiformis* f. sp. *tritici* isolates in the South-Central United States. Plant Dis 90:847–852
2. Milus EA, Kristensen K, Hovmøller MS (2009) Evidence for increased aggressiveness in a recent widespread strain of *Puccinia striiformis* f. sp. *tritici* causing stripe rust of wheat. Phytopathology 99:89–94
3. Pariaud B, Ravigné V, Halkett F, Goyeau H, Carlier J, Lannou C (2009) Aggressiveness and its role in the adaptation of plant pathogens. Plant Pathol 58:409–424
4. de Vallavieille-Pope C, Huber L, Leconte M, Bethenod O (2002) Preinoculation effects of light quantity on infection efficiency of *Puccinia striiformis* and *P. triticina* on wheat seedlings. Phytopathology 92:1308–1314
5. de Vallavieille-Pope C, Huber L, Leconte M, Goyeau H (1995) Comparative effects of temperature and interrupted wet periods on germination, penetration, and infection of *Puccinia recondita* f. sp. *tritici* and *P. striiformis* on wheat seedlings. Phytopathology 85:409–415
6. Singh DP (2015) Plant nutrition in the management of plant diseases with particular reference to wheat. In: Awasthi LP (ed) Recent advances in the diagnosis and management of plant diseases. Springer, New Delhi, pp 273–284

7. Sørensen CK, Thach T, Hovmøller MS (2016) Evaluation of spray and point inoculation methods for the phenotyping of *Puccinia striiformis* on Wheat. Plant Dis 100(6):1064–1070
8. Sørensen CK, Rodriguez-Algaba J, Justesen AF. Hovmøller MS Segregation for aggressiveness in sexual ofspring of the yellow rust pathogen *Puccinia striiformis*. In: BGRI Technical Workshop, Sydney, Australia, 17–20 September 2015.
9. Eyal Z, Clifford BC, Caldwell RM (1968) A settling tower for quantitative inoculation of leaf blades of mature small grain plants with urediospores. Phytopathology 58(4):712–714
10. Andres MW, Wilcoxson RD (1984) A device for uniform deposition of liquid-suspended urediospores on seedlings and adult cereal plants. Phytopathology 74:550–552
11. McNeal FM, Konzak CF, Smith EP, Tate WS, Russell TS (1971) A system for recording and processing cereal research data. US Agric Res Serv 42:34–121

Chapter 5

Extraction of High Molecular Weight DNA from Fungal Rust Spores for Long Read Sequencing

Benjamin Schwessinger and John P. Rathjen

Abstract

Wheat rust fungi are complex organisms with a complete life cycle that involves two different host plants and five different spore types. During the asexual infection cycle on wheat, rusts produce massive amounts of dikaryotic urediniospores. These spores are dikaryotic (two nuclei) with each nucleus containing one haploid genome. This dikaryotic state is likely to contribute to their evolutionary success, making them some of the major wheat pathogens globally. Despite this, most published wheat rust genomes are highly fragmented and contain very little haplotype-specific sequence information. Current long-read sequencing technologies hold great promise to provide more contiguous and haplotype-phased genome assemblies. Long reads are able to span repetitive regions and phase structural differences between the haplomes. This increased genome resolution enables the identification of complex loci and the study of genome evolution beyond simple nucleotide polymorphisms. Long-read technologies require pure high molecular weight DNA as an input for sequencing. Here, we describe a DNA extraction protocol for rust spores that yields pure double-stranded DNA molecules with molecular weight of >50 kilo-base pairs (kbp). The isolated DNA is of sufficient purity for PacBio long-read sequencing, but may require additional purification for other sequencing technologies such as Nanopore and 10× Genomics.

Key words Rust fungi, *Puccinia striiformis* f. sp. *tritici*, *Puccinia graminis* f. sp. *tritici*, Long-read sequencing, PacBio, Nanopore, 10× Genomics, HMW DNA

1 Introduction

Puccinia striiformis f. sp. *tritici* (*Pst*), *P. triticina* (*Pt*), and *P. graminis* f. sp. *tritici* (*Pgt*) are the three major rust fungi that infect wheat [1–3]. All three are heteroecious and undergo five different spore stages to complete their entire life cycles. Urediniospores are asexually produced spores that are able to reinfect wheat repeatedly throughout the growing season, causing major crop losses globally. The continuing threat to global wheat production from rusts is caused by the stochastic appearance of novel hypervirulent rust isolates and the rapid local adaptation of these isolates to locally employed resistance wheat cultivars [1–3]. It is likely that the dikaryotic genetic organization of urediniospores is fundamental

Sambasivam Periyannan (ed.), *Wheat Rust Diseases: Methods and Protocols*, Methods in Molecular Biology, vol. 1659, DOI 10.1007/978-1-4939-7249-4_5, © Springer Science+Business Media LLC 2017

to this evolutionary success. Our insights into the genetic architecture of wheat rusts are limited as all genomes except for *Pgt* have been produced using short read sequencing data only. This leads to highly fragmented genome assemblies with over 10,000 contigs per genome that lack any significant structural or haplotype specific information [4–9]. These fragmented genomes limit evolutionary studies as they reduce the analysis of genetic diversity to small polymorphisms of only up to several base pairs in length. In addition, they lack higher order structure and miss most of the repetitive regions. Current long-read single molecule sequencing technologies such as PacBio [10, 11] and Nanopore [12], or synthetic long reads produced by 10× Genomics [13], produce much longer reads with average size >10 kbp and reads of 200–300 kbp reported. Such long reads together with novel assembly algorithms have provided more contiguous genomes with rich haplotype-specific information [14–16]. All long-read sequencing technologies require high quality high molecular weight (HMW) DNA to harness their full potential. Extraction of pure high quality DNA is especially difficult for obligate biotrophic organisms such as wheat rusts, because starting biomass and material is restricted to spores or germinated spores. Here, we report an improved DNA extraction method [17–19] that yields double-stranded (ds) DNA fragments with an average size above 50 kbp using dried wheat rust spores as starting material. This DNA is of sufficient purity to generate 20 kbp insert size libraries for PacBio sequencing. For other technologies such as Nanopore sequencing and 10× Genomics, an additional purification step such as size selection and gel purification using BluePippin is required to further purify the dsDNA from residual contaminants.

2 Materials

2.1 Main Buffer Components

2.1.1 Buffer A

1. 0.35 M sorbitol.
2. 0.1 M Tris–HCl, pH 9.
3. 5 mM EDTA, pH 8.
4. Autoclave to sterilize.

2.1.2 Buffer B

1. 0.2 M Tris–HCl, pH 9.
2. 50 mM EDTA, pH 8.
3. 2 M NaCl.
4. 2% [w/v] CTAB.
5. Autoclave to sterilize.

2.1.3 Buffer C

5% Sarkosyl (*N*-lauroylsarcosine sodium salt).

2.2 Additional Solutions

1. 5 M potassium acetate (KAc) pH 7.5.
2. 1% [w/v] polyvinylpyrrolidone (40,000 MW) (PVP40).
3. 3 M sodium acetate (NaAc) pH 5.2, filter-sterilized.
4. 100% isopropanol.
5. 70% [v/v] fresh ethanol.
6. Buffered phenol–chloroform–isoamyl alcohol (P/C/I) (25:24:1)
7. Autoclaved acid-washed sand.
8. AMPure beads (Beckman).

2.3 Enzymes

1. RNAse T1 (1000 U/mL).
2. Proteinase K (800 U/mL).

2.4 Additional Materials

1. Washed and autoclaved mortar and pestle.
2. Vacuum or freeze-dried rust spores.
3. Qubit Broad Range (BR) DNA quantification dye.
4. NanoDrop DNA quantification.
5. 0.8% agarose gels TBE.
6. λ DNA–HindIII marker.

2.5 Optional Materials

Pulsed-field gel electrophoresis apparatus.

3 Methods

3.1 Crude DNA Extraction from Dried Rust Spores (See Note 1)

1. Prepare lysis buffer by mixing buffer components A + B + C (2.5:2.5:1 + 0.1% PVP final) and heat to 64 °C for 30 min. Let cool on bench to room temperature. Use a 50 mL Falcon tube for each extraction containing 17.5 mL lysis buffer. Use up to 500 mg starting spore material for 17.5 mL lysis buffer.
2. Add 10 μL (10 kU) RNAse T1 to lysis buffer.
3. Add sand to prechilled mortar and cool with liquid nitrogen. Add spores to sand once all liquid nitrogen has dissipated. Use 0.5 g of sand per 100 mg of starting material. Grind for 2 min in four 15-s bursts, adding liquid nitrogen after each 15-s-grinding burst. (*see* **Note 2**)
4. Transfer powder to a 50 mL Falcon tube containing lysis buffer and RNAse, and mix well by shaking.
5. Incubate at room temperature for 30 min. Mix by inversion every 5 min.
6. Add 200 μL Proteinase K, incubate at room temperature for 30 min. Mix by inversion every 5 min.

7. Cool on ice for 5 min.
8. Add 3.5 mL (0.2 vol) of KAc 5 M, mix by inversion, incubate on ice for a maximum of 5 min.
9. Centrifuge at 5000 × *g* for 12 min at 4 °C.
10. Transfer supernatant to fresh Falcon tube containing 17.5 mL (1 vol) P/C/I and mix by inversion for 2 min.
11. Centrifuge at 6000 × *g* for 10 min at 4 °C.
12. Transfer supernatant to a fresh Falcon tube containing 17.5 mL (1 vol) (P/C/I) and mix by inversion for 2 min.
13. Centrifuge at 6000 × *g* for 10 min at 4 °C.
14. Transfer supernatant (~17 mL) to fresh Falcon tube and add 5 μL RNAse T1. Incubate for 20–30 min at RT.
15. Add 1.8 mL (~0.1 vol) NaAc and mix by inversion.
16. Add 18 mL (~1 vol) isopropanol at room temperature and mix by inversion.
17. Incubate at room temperature for 5–10 min.
18. Centrifuge at 10,000 × *g* for 30 min at 4 °C.
19. Carefully remove supernatant using a pipette until about 1–2 mL remain. The DNA and other precipitates will form a translucent to white film pellet at the bottom of the tube.
20. Use a 1 mL pipette tip to transfer the pellet and the remaining supernatant into a fresh 1.7 mL Eppendorf tube. If the pellet loses integrity during transfer, add 1.5 mL fresh 70% EtOH to the Falcon tube and centrifuge it at 4000 × *g* for 5 min. Discard 1 mL of the supernatant and transfer the remaining volume containing the DNA pellet to the same 1.7 mL Eppendorf tube.
21. Centrifuge the Eppendorf tube in a table top centrifuge at 13,000 × *g* for 5 min.
22. Remove the supernatant with a pipette and wash with 1.5 mL fresh 70% ethanol. Invert several times to dislodge pellet.
23. Repeat **step 22**.
24. Remove supernatant with pipette and wash the pellet with 1.5 mL fresh 70% ethanol, inverting several times to dislodge the pellet.
25. Remove the supernatant with a pipette.
26. Repeat **step 22**.
27. Remove remaining ethanol with a pipette.
28. Air-dry the pellet for 7 min on the bench.
29. Add 200 μL of 10 mM Tris-HCl pH 9 and leave at room temperature for 3 h to dissolve the DNA.

Table 1
Purity control measurements on isolated dsDNA using NanoDrop and Qubit

			NanoDrop			Qubit	*c* ratio
	Input (mg)	Volume (μL)	*c* (ng/μL)	260/280	260/230	*c* (ng/μL)	Qubit/ NanoDrop
DNA extraction 3.1	380	400	1023.26	2.09	2.32	58.4	0.06
DNA extraction 3.3	380	80	251	1.87	1.37	157	0.63

30. Flick tube slightly to mix, and add 200 μL of TE buffer. *Do not vortex* as it shears the DNA.
31. Leave at room temperature overnight.
32. Incubate for 1 h at 28 °C on a rotary shaker at 1400 rpm.
33. Keep 10–20 μL of the DNA solution for quality control analysis.

3.2 DNA Quantification and Purity Analysis (See Note 3)

1. Measure the quantity of dsDNA on a Qubit BR using 2 μL of the sample (Table 1).
2. Measure UV absorbance profile on the NanoDrop using 2 μL of sample. Note estimated concentration, absorbance profile, and 260/230 and 280/230 ratios (Table 1).

3.3 Second Step DNA Enrichment and Purification Using AMPure Beads

1. Add 0.45 vol of room temperature AMPure beads to the DNA solution. Mix gently and completely by flicking the tube with your finger. Incubate on a rotary wheel for 10 min.
2. Place in a magnetic rack and wait until the solution clears. This may take a long time (>10 min) depending on your sample. Do not remove the tube from the magnetic rack.
3. Pipette off all liquid while leaving the beads undisturbed.
4. Add 1.5 mL of fresh 70% ethanol while not disturbing the beads, and incubate for 1 min.
5. Pipette off all liquid leaving the beads undisturbed.
6. Add 1.5 mL of fresh 70% ethanol without disturbing the beads, and incubate for 1 min.
7. Pipette off all liquid while not disturbing the beads.
8. Remove the tube from the magnetic rack. Centrifuge briefly for 2 s to collect any residual liquid.
9. Place tube on the magnetic rack and remove all remaining liquid.

10. Remove the tube from the rack and let beads dry with the lid open for a maximum of 45–60 s.
11. Add sufficient 10 mM Tris pH 8 for a final dsDNA concentration of 150 ng/μL (Qubit), assuming a recovery rate of 50% in respect to input dsDNA. Mix gently and completely by flicking the tube with your finger. Incubate on a rotary wheel for 10 min.
12. Place the tube on the magnetic rack and transfer the liquid to a fresh DNA-free low-bind tube, without disturbing the pellet.

3.4 DNA Quantification, and Quality and Purity Analysis (See Note 3)

1. Measure the quantity of dsDNA quantity on a Qubit BR using 2 μL of the sample.
2. Measure the UV absorbance profile on the NanoDrop using 2 μL of the sample. Note the estimated concentration, absorbance profile, and 260/230 and 280/230 ratios (Table 1).
3. Load 50 ng (Qubit) dsDNA from the samples retained after 3.1 and 3.3 on a 0.8% agarose gel, including 1 kb and λ DNA–HindIII marker (Fig. 1) (*see* **Notes 3–5**).
4. Optionally, run 200 ng of the samples retained after 3.1 and 3.4 overnight on pulsed-field gel electrophoresis (Fig. 2) (*see* **Notes 3–5**).

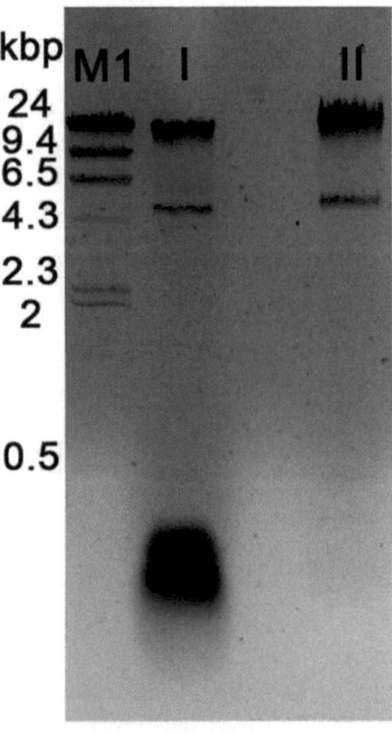

Fig. 1 DNA quality control on a 0.8% agarose gel. Fifty nanograms of dsDNA was resolved on a 0.8% agarose gel in 0.5% TBE buffer. No smearing could be observed for either sample. The large low molecular weight band in sample I is partially digested RNA carried over from the DNA purification described in Subheading 3.1. M1, λ DNA–HindIII marker; I, the DNA sample described in Subheading 3.1; II, DNA sample described in Subheading 3.3

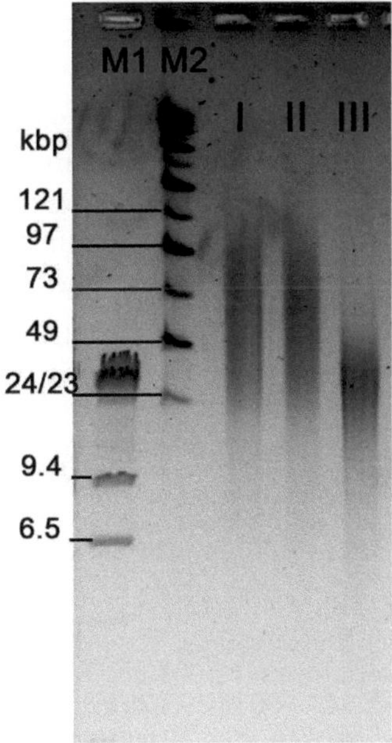

Fig. 2 DNA quality control on a 1% agarose gel using pulsed-field gel electrophoresis. Two hundred nanograms of dsDNA resolved on a 1% agarose gel using pulsed-field gel electrophoresis. Most of the DNA sample of I and II runs above 49 kbp. Sample III is sample I cleaned over a DNA silica spin column and is shown as a comparison highlighting the shearing induced by spin columns (*see* **Note 5**). M1, λ DNA–HindIII marker; M2, Mid Range II PFG marker NEB; I, DNA sample described in Subheading 3.1; II, DNA sample described in Subheading 3.3; III, DNA sample I cleaned over a DNA silica spin column

4 Notes

1. Never vortex DNA samples as this will shear the DNA. Only use flicking or inversion to resuspend and mix solutions.
2. For how to best add spores to the liquid nitrogen-cooled sand, see this video online.
3. The final aim for all sequencing platforms is to obtain the following purity values of high molecular DNA: *c* ratio Qubit/NanoDrop >0.5, 260/230 ~2, 280/230 1.8–2.2.
4. The DNA integrity of the samples shown in Figs. 1 and 2 was above 50 kbp when measured by Bioanalyzer or BluePippin as alternatives to in-house pulsed-field gel electrophoresis.
5. The DNA quality and purity described in this protocol is sufficient to generate PacBio 20 kbp insert libraries even though the 280/230 ratios are below 1.8. The average and median read length for these libraries was >11 kb.

Acknowledgments

We thank Drs Megan McDonald, Melania Figueroa, Claire Anderson, Andrii Gryganskyi, and David Hayward for useful input while developing this protocol. We would like to thank several users for their positive feedback on successful application of this protocol to various fungal and oomycete species. This was enabled by early access to previous versions of this protocol on the platform "protocols.io," which provides the opportunity to share and update protocols online. B.S. is supported by an Australian Research Council Discovery Early Career Research Award (DE150101897).

References

1. Singh RP, Hodson DP, Jin Y, Lagudah ES, Ayliffe MA, Bhavani S, Rouse MN, Pretorius ZA, Szabo LJ, Huerta-Espino J, Basnet BR, Lan C, Hovmøller MS (2015) Emergence and spread of new races of wheat stem rust fungus: continued threat to food security and prospects of genetic control. Phytopathol 105:872–884
2. Schwessinger B (2017) Fundamental wheat stripe rust research in the 21st century. New Phytol 213:1625–1631
3. Kolmer J (2013) Leaf rust of wheat: pathogen biology, variation and host resistance. Forests. 4:70–84
4. Cantu D, Segovia V, MacLean D, Bayles R, Chen X, Kamoun S, Dubcovsky J, Saunders DG, Uauy C (2013) Genome analyses of the wheat yellow (stripe) rust pathogen *Puccinia striiformis* f. sp. *tritici* reveal polymorphic and haustorial expressed secreted proteins as candidate effectors. BMC Genomics 14:270
5. Kiran K, Rawal HC, Dubey H, Jaswal R, Devanna BN, Gupta DK, Bhardwaj SC, Prasad P, Pal D, Chhuneja P, Balasubramanian P, Kumar J, Swami M, Solanke AU, Gaikwad K, Singh NK, Sharma TR (2016) Draft genome of the wheat rust pathogen (*Puccinia triticina*) unravels genome-wide structural variations during evolution. Genome Biol Evol 8:2702–2721
6. Cuomo CA, Bakkeren G, Khalil HB, Panwar V, Joly D, Linning R, Sakthikumar S, Song X, Adiconis X, Fan L, Goldberg JM, Levin JZ, Young S, Zeng Q, Anikster Y, Bruce M, Wang M, Yin C, McCallum B, Szabo LJ, Hulbert S, Chen X, Fellers JP (2017) Comparative analysis highlights variable genome content of wheat rusts and divergence of the mating loci. Genes Genomes Genet 7:361–376
7. Upadhyaya NM, Garnica DP, Karaoglu H, Sperschneider J, Nemri A, Xu B, Mago R, Cuomo CA, Rathjen JP, Park RF, Ellis JG, Dodds PN (2015) Comparative genomics of Australian isolates of the wheat stem rust pathogen *Puccinia graminis* f. sp. *tritici* reveals extensive polymorphism in candidate effector genes. Front Plant Sci 5:759
8. Duplessis S, Cuomo CA, Lin YC, Aerts A, Tisserant E, Veneault-Fourrey C, Joly DL, Hacquard S, Amselem J, Cantarel BL, Chiu R, Coutinho PM, Feau N, Field M, Frey P, Gelhaye E, Goldberg J, Grabherr MG, Kodira CD, Kohler A, Kues U, Lindquist EA, Lucas SM, Mago R, Mauceli E, Morin E, Murat C, Pangilinan JL, Park R, Pearson M, Quesneville H, Rouhier N, Sakthikumar S, Salamov AA, Schmutz J, Selles B, Shapiro H, Tanguay P, Tuskan GA, Henrissat B, Van de Peer Y, Rouze P, Ellis JG, Dodds PN, Schein JE, Zhong S, Hamelin RC, Grigoriev IV, Szabo LJ, Martin F (2011) Obligate biotrophy features unraveled by the genomic analysis of rust fungi. Proc Natl Acad Sci USA 108 (22):9166–9171
9. Zheng W, Huang L, Huang J, Wang X, Chen X, Zhao J, Guo J, Zhuang H, Qiu C, Liu J, Liu H, Huang X, Pei G, Zhan G, Tang C, Cheng Y, Liu M, Zhang J, Zhao Z, Zhang S, Han Q, Han D, Zhang H, Zhao J, Gao X, Wang J, Ni P, Dong W, Yang L, Yang H, Xu J, Zhang G, Kang Z (2013) High genome heterozygosity and endemic genetic recombination in the wheat stripe rust fungus. Nat Commun 4:2673
10. Eid J, Fehr A, Gray J, Luong K, Lyle J, Otto G, Peluso P, Rank D, Baybayan P, Bettman B, Bibillo A, Bjornson K, Chaudhuri B, Christians F, Cicero R, Clark S, Dalal R, deWinter A, Dixon J, Foquet M, Gaertner A, Hardenbol P, Heiner C, Hester K, Holden D, Kearns G,

Kong X, Kuse R, Lacroix Y, Lin S, Lundquist P, Ma C, Marks P, Maxham M, Murphy D, Park I, Pham T, Phillips M, Roy J, Sebra R, Shen G, Sorenson J, Tomaney A, Travers K, Trulson M, Vieceli J, Wegener J, Wu D, Yang A, Zaccarin D, Zhao P, Zhong F, Korlach J, Turner S (2009) Real-time DNA sequencing from single polymerase molecules. Science 323:133–138

11. Wang M, Beck CR, English AC, Meng Q, Buhay C, Han Y, Doddapaneni HV, Yu F, Boerwinkle E, Lupski JR, Muzny DM, Gibbs RA (2015) PacBio-LITS: a large-insert targeted sequencing method for characterization of human disease-associated chromosomal structural variations. BMC Genomics 16:214
12. Jain M, Olsen HE, Paten B, Akeson M (2016) The Oxford Nanopore MinION: delivery of nanopore sequencing to the genomics community. Genome Biol 17:239
13. Zheng GXY, Lau BT, Schnall-Levin M, Jarosz M, Bell JM, Hindson CM, Kyriazopoulou-Panagiotopoulou S, Masquelier DA, Merrill L, Terry JM, Mudivarti PA, Wyatt PW, Bharadwaj R, Makarewicz AJ, Li Y, Belgrader P, Price AD, Lowe AJ, Marks P, Vurens GM, Hardenbol P, Montesclaros L, Luo M, Greenfield L, Wong A, Birch DE, Short SW, Bjornson KP, Patel P, Hopmans ES, Wood C, Kaur S, Lockwood GK, Stafford D, Delaney JP, Wu I, Ordonez HS, Grimes SM, Greer S, Lee JY, Belhocine K, Giorda KM, Heaton WH, McDermott GP, Bent ZW, Meschi F, Kondov NO, Wilson R, Bernate JA, Gauby S, Kindwall A, Bermejo C, Fehr AN, Chan A, Saxonov S, Ness KD, Hindson BJ, Ji HP (2016) 0 Haplotyping germline and cancer genomes with high-throughput linked-read sequencing. Nat Biotechnol 34:303–311
14. Spies N, Weng Z, Bishara A, McDaniel J, Catoe D, Zook JM, Salit M, West RB, Batzoglou S, Sidow A (2016) Genome-wide reconstruction of complex structural variants using read clouds. bioRxiv:74518. doi:10.1101/074518
15. Mostovoy Y, Levy-Sakin M, Lam J, Lam ET, Hastie AR, Marks P, Lee J, Chu C, Lin C, Džakula Z, Cao H, Schlebusch SA, Giorda K, Schnall-Levin M, Wall JD, Kwok PY (2016) A hybrid approach for de novo human genome sequence assembly and phasing. Nat Methods 13:587–590
16. Chin CS, Peluso P, Sedlazeck FJ, Nattestad M, Concepcion GT, Clum A, Dunn C, O'Malley R, Figueroa-Balderas R, Morales-Cruz A, Cramer GR, Delledonne M, Luo C, Ecker JR, Cantu D, Rank DR, Schatz MC (2016) Phased diploid genome assembly with single-molecule real-time sequencing. Nat Methods 13:1050–1054
17. Fulton TM, Chunwongse J, Tanksley SD (1995) Microprep protocol for extraction of DNA from tomato and other herbaceous plants. Plant Mol Biol Report 13:207–209
18. High quality DNA from fungi for long read sequencing e.g. PacBio (2016) https://www.protocols.io/view/High-quality-DNA-from-fungi-for-long-read-sequencing-ewtbfen. Accessed on February 2, 2017
19. Extraction of high quality DNA for genome sequencing, 1000 fungal genomes project (2012). http://1000.fungalgenomes.org/home/protocols/high-quality-genomic-dna-extraction/. Accessed on February 2, 2017

Chapter 6

Microsatellite Genotyping of the Wheat Yellow Rust Pathogen *Puccinia striiformis*

Sajid Ali, Muhammad R. Khan, Angelique Gautier, Zahoor A. Swati, and Stephanie Walter

Abstract

To combat the ever-increasing threat of wheat yellow rust worldwide, understanding of the pathogen (*Puccinia striiformis*) population biology is indispensable. Molecular markers, particularly microsatellites, have been reported to be important tools for deciphering pathogen population structure, invasion sources, and migration history. The utility of these DNA-based markers and sequencing has been increased by the direct DNA extraction from infected leaves with subsequent multiplex-based SSR genotyping. In this chapter we describe the protocol for direct DNA extraction and its genotyping with microsatellite markers in multiplex reactions. We describe the procedure for allele scoring, and various troubles faced during microsatellite scoring and potential solutions for them.

Key words Simple sequence repeats, Wheat stripe rust, Fragment analyses

1 Introduction

Wheat yellow rust represents a serious threat to the worldwide production of wheat, a crop cultivated worldwide across diverse agroecological zones [1–3]. This worldwide threat could be tackled through understanding the pathogen population structure both at regional and worldwide level using various molecular markers. Microsatellite markers have recently been widely utilized to determine the regional and worldwide population structure of pathogens [4–6], sources of invasions [4, 7–9], ancestral migration history [4], temporal maintenance [10], recombination signature [11, 12], and centers of diversity [4, 6]. The utility of these markers is further increased due to their applicability to isolates based on direct DNA extraction from infected lesions, which avoids the prerequisite of several cycles of spore multiplication [13]. In this chapter we describe the detailed protocol for microsatellite genotyping for wheat yellow rust using a multiplex technique advocated

Sambasivam Periyannan (ed.), *Wheat Rust Diseases: Methods and Protocols*, Methods in Molecular Biology, vol. 1659, DOI 10.1007/978-1-4939-7249-4_6, © Springer Science+Business Media LLC 2017

and used in our previous studies [13, 14]. The present chapter thus describes the procedure for DNA extraction from lesions and spores followed by microsatellite genotyping.

2 Materials

1. Deionized or distilled water.
2. Geno/Grinder or any other instruments appropriate for grinding of fungal mycelium/spores (*see* **Note 1**).
3. Liquid nitrogen.
4. Prepare the CTAB buffer as follows: For 200 mL of CTAB buffer, add 5.0 g D-sorbitol, 2.0 g sarkosyl (*N*-lauroylsarcosine), 1.6 g CTAB (cetyltrimethylammonium bromide), 9.4 g NaCl, 1.6 g EDTA, 2.0 g PVPP (polyvinylpolypyrrolidone) insoluble, and 0.1 M Tris-HCl, pH 8 (we dissolve all ingredients in 0.1 M Tris-HCl, pH 8) and heat when stirring. The solution will have a milky/grayish color. The buffer should be stirred (because of the PVPP) before use and must be prewarmed at 65 °C before use every time.
5. Prepare the chloroform–iso-amyl alcohol at 24:1 ratio.
6. Isopropanol, absolute ethanol, and 70% ethanol.
7. 1.5 and 2 mL Eppendorf tubes.
8. RNase A (optional).
9. Microsatellite markers as listed in Table 1 and prepare their working solutions as mentioned in Table 2.
10. PCR amplification plates, appropriate for later use in a sequencer.
11. Thermocycler.
12. PCR kit.
13. Access to a sequencing platform to perform fragment separation.
14. GeneMarker or any other software to read the microsatellite peaks output generated by a sequencer.

3 Methods

3.1 DNA Extraction

Yellow rust microsatellite genotyping could be made on DNA extracted from single spore infected lesions or spores originating from single pustule [13]. DNA extraction could be made through various protocols with numerous commercial kits and automated robots available. We described here the preextraction step which will be helpful for all the protocols and then the modified CTAB method, which could be used in many labs without expensive kits.

Table 1
Details of 20 microsatellite markers to be amplified in two multiplex PCR reactions for genotyping of *P. striiformis*

	SSR locus	Oligo forward (5′–3′) fluorescent labeled	Oligo reverse (5′–3′) non-labeled	Florescence label	Allele size range[a]	Motif size (bp)	Reference
Multiplex-1	RJO4	GTGGGTTGGGCTGGAGTC	GCTAATCCATTCCACGCACC	6FAM	199–208	3	[15]
	RJO24	TTGCTGAGTAGTTTGCGGTGAG	CTCAAGCCCATCCTCCAACC	6FAM	275–308	3	[15]
	RJN12	TGTTGACAAACACGACGACC	ACTTATTGCAGCTTGAGTAAAACG	VIC	187–202	3	[5]
	RJN8	ACTGGGCAGACTGGTCAAC	TCGTTTCCCTCCAGATGGC	VIC	300–315	3	[5]
	RJN13	TTAGCTCAGCCGGTTCCTC	CAGGTGTAGCCCCATCTCC	NED	144–153	2	[5]
	RJO3	GCAGCACTGGCAGGTGG	GATGAATCAGGATGGCTCC	NED	198–208	2	[15]
	RJN3	TGGTGGTGCTCCTCTAGTC	AGGGGTCTTGTAAGATGCTC	NED	330–344	2	[5]
	RJN11	CCGACACCTCCTCTGATCG	ACTCCGCTGCTCATCTTCG	PET	172–196	2	[5]
	RJO27	CGTCCCGACTAATCTGGTCC	ATGAGTTAGTTTAGATCAGGTCGAC	PET	217–239	2	[15]
	RJN6	CAATCTGGCGGACAGCAAC	CACCTAGGATACCACCGCC	PET	309–321	3	[5]
Multiplex-2	RJO21	TTCCTGGATTGAATTCGTCG	CAGTTCTCACTCGGACCCAG	6FAM	164–188	3	[15]
	RJN10	ACGTGCCAGCTCAACTCTC	AAGGGGCGGATGGATTACG	6FAM	221–230	3	[5]
	RJO18	CTGCCCATGCTCTTCGTC	GATGAAGTGGGTGCTGCTG	6FAM	331–363	3	[15]
	WU6	CAGCTCTGTTGATTTCTTCC	GGTTTGACATTGATTCACCT	VIC	206–212	2	[13]
	RJO20	AGAAGATCGACGCACCCG	CCTCCGATTGGCTTAGGC	VIC	281–293	3	[15]
	RJN2	TTGTGGCGGAAGGGAACG	GCATGAAACGATCAAAGAAGATAGC	NED	164–196	2	[5]
	RJN4	CATTCATGACCCTCGCCTC	TGGACGAGATGTGTGCAAG	NED	253–259	2	[5]
	RJN9	TTAGGCGCTCAACAAGCAG	ACAACAACCTTTCAACCCG	NED	330–334	2	[5]
	RJN5	AACGGTCAACAGCACTCAC	AGTTGGTCGCGTTTTGCTC	PET	222–232	2	[5]
	WU12	GGAAACTGTAGCACCTTCAC	TTGATTTGTGGATTGAGTTG	PET	323–334	3	[13]

[a]Allele range as observed in Pakistani *P. striiformis* population. Other population could have alleles below or above the range for few multiples of motifs

Table 2
Concentration of individual primers to set up the two SSR multiplex primer mixes for *P. striiformis* genotyping

SSR Multiplex-1 primer mix (final volume 750 μL)			SSR Multiplex-2 primer mix (final volume 750 μL)		
SSR locus	Primer[a]	μL of primer stock (100 pmol/μl)	SSR locus	Primer[a]	μL of primer stock (100 pmol/μl)
RJO4F	RJO4F RJO4R	14.25 14.25	RJO21F	RJO21F RJO21R	14.25 14.25
RJO24F	RJO24F RJO24R	14.25 14.25	RJN10F	RJN10F RJN10R	14.25 14.25
RJN12F	RJN12F RJN12R	28.50 28.50	RJO18F	RJO18F RJO18R	42.75 42.75
RJN8F	RJN8F RJN8R	14.25 14.25	WU6F	WU6F WU6R	28.50 28.50
RJN13F	RJN13F RJN13R	14.25 14.25	RJO20F	RJO20F RJO20R	14.25 14.25
RJN3F	RJN3F RJN3R	14.25 14.25	RJN2F	RJN2F RJN2R	28.50 28.50
RJO3F	RJO3F RJO3R	42.75 42.75	RJN4F	RJN4F RJN4R	28.50 28.50
RJN11F	RJN11F RJN11R	14.25 14.25	RJN9F	RJN9F RJN9R	28.50 28.50
RJO27F	RJO27F RJO27R	42.75 42.75	RJN5F	RJN5F RJN5R	14.25 14.25
RJN6F	RJN6F RJN6R	14.25 14.25	WU12F	WU12F WU12R	14.25 14.25
H_2O to make final 750 μL		322.50	H_2O to make final 750 μL		294.00

[a]In case of weak signals for some loci, we recommend to increase the concentration of primers (e.g., 2× or 3×). Otherwise, individual single PCR reaction for the locus in question could be made for a subset of isolate

3.1.1 Grinding the Samples (Lesion/Spores)

The first step is to crush the spores or lesion (including spores and mycelium) for subsequent DNA extraction.

1. Prepare the single stripe lesion from infected plants and place it in a 2 mL Eppendorf tube (Fig. 1). A lesion of 3–10 cm will be sufficient, while a minimum of 2–5 mg of spores will be required for spore DNA extraction.
2. Add four mini steel balls (~1.5 mm in diameter) per Eppendorf tube containing (freeze-dried, if possible) infected leaves or spores.
3. Grind twice for 45 s at 1500 strokes/min speed using grinding machines like Geno/Grinder (*see* **Note 1**).

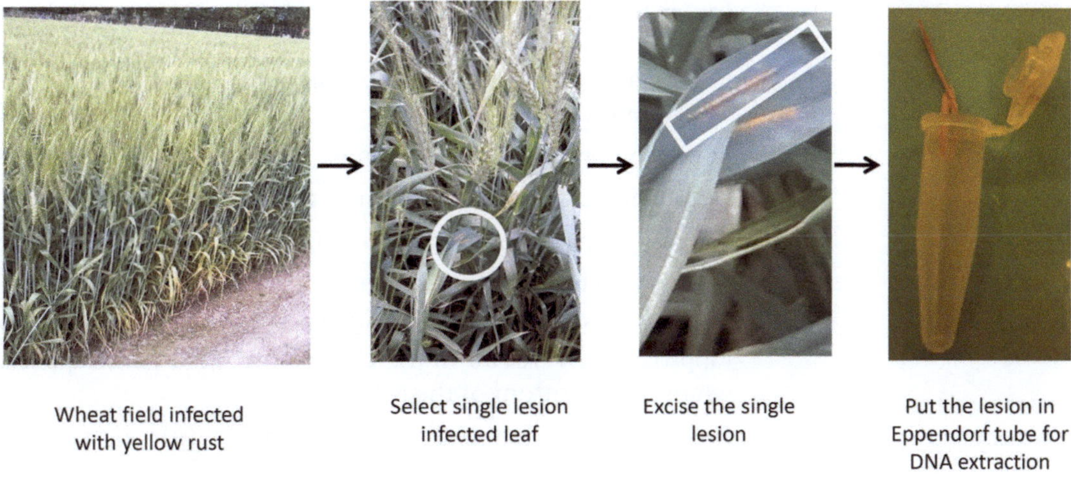

Fig. 1 Lesion isolation for DNA extraction for subsequent genotyping from infected wheat plants from field samples

4. In case of first calibration for spores, it is recommended to check the maceration of spores under a microscope.
5. The above-described steps would be the same for any DNA extraction method. After this step, DNA extraction could be done according to various protocols, like CTAB-based method or commercial available kits for DNA extraction from plant tissues.

3.1.2 DNA Extraction with CTAB Method

Steps for a modified CTAB-based DNA extraction [16] are detailed below, while for commercial kits the manufacturer's procedure must be followed:

1. Grind single lesions or spores in Eppendorf tube as described above.
2. Add 400 mL preheated CTAB buffer to an Eppendorf tube containing spore/lesion extract.
3. Incubate the CTAB–spore/lesion extract mixture for about 60 min at 65 °C.
4. After incubation, spin the tube at 2500 × *g* for 10 min to spin down cell debris. Transfer the supernatant to clean Eppendorf tubes.
5. To each tube add 0.6 μL of chloroform–iso-amyl alcohol (24:1) and mix the solution by inversion. After mixing, spin the tubes at 1000 rcf for 1 min.
6. Transfer only the upper aqueous phase to a clean Eppendorf tube.
7. To each tube add 0.6 volume of prechilled (−20 °C) isopropanol (*see* **Note 2**).

8. Invert the tubes gently 20 times to let the DNA precipitate. Generally, the DNA can be seen to precipitate out of solution. For a better outcome, the DNA can be precipitated on ice for 10–15 min.
9. Following precipitation, centrifuge at 2500 × *g* for 15 min.
10. Pellet will be formed at bottom of the Eppendorf tube.
11. Carefully remove supernatant and retain the pellet.
12. Wash the pellet with 70% ethanol.
13. Remove all the supernatant and allow the DNA pellet to dry (approximately 15–30 min). Usually the dried pellet appears white at the bottom of the tube.
14. Resuspend pellet in 100 μL 1× TE buffer (pH 8), store at −20 °C.
15. Make a working solution of 50 ng/μL in autoclaved dist. H_2O (*see* **Note 3**).

3.2 PCR Amplification for Microsatellite Screening

The PCR amplification protocol described here is based on multiplex-based microsatellite genotyping for subsequent sequencer based fragment analyses [13, 14]. The set of 19 SSRs are set up in two multiplex reactions for PCR amplification [14] from the original three multiplex setup [13]. We present the protocol using the Type-it Microsatellite PCR Kit (Qiagen, Germany), based on our own experience, but any other appropriate kit could be used.

Prepare the PCR mixes and run the PCR as detailed here:

1. Prepare the SSR multiplex primer mix as described in Table 2.
2. Set up the PCR reaction to a final volume of 15 μL per reaction for each multiplex as detailed in Table 3.

Table 3
Details of the PCR setup for microsatellite genotyping of *P. striiformis* to be amplified in two multiplex PCR reactions

PCR ingredient	Single reaction (μL)	Full plate (96 reactions)
H_2O	1.95	235.20
Q-solution (Qiagen)	1.00	96.00
SSR Multiplex Primer Mix[a]	1.05	100.80
Type-it premix (Qiagen)	5.00	480.00
DNA (50 ng/μL)[b]	1.00	1 × 96
Total	10	

[a]SSR Multiplex Primer Mix for each multiplex as prepared according to Table 2
[b]DNA quantity must be calibrated (adjust H_2O quantity accordingly)

3. Run the PCR reaction as follows:
   - 95 °C for 5 min—for 1 cycle
   - 95 °C for 30 s, 63 °C (with −0.5 °C/cycle) for 60 s and 72 °C for 30 s—for 17 cycles
   - 95 °C for 30 s, 55 °C for 60 s and 72 °C for 30 s—for 25 cycles
   - 72 °C for 10 min—for 1 cycle
   - 4 °C for storage
4. It is recommended to check the amplification on a 1% agarose gel before sending it for fragment analysis on an ABI capillary sequencer system such as ABI 3730XL.
5. Prepare the plate or tubes for sequencer (*see* **Note 4**).

3.3 Fragment Analyses for Microsatellite

The sequencer output format may vary according to various sequencer technologies. Various software programs are available for scoring allele sizes in the sequencer outputs. Our comments on fragments reading are based on GeneMarker software, which is freely available as a demo version. However, following the developer guidelines any software could be used. It is recommended to not use software for automatic reading of alleles, instead score each allele visually (Fig. 1).

1. Open GeneMarker or other software and upload the sequencer output files, and select appropriate color for the length marker (orange).
2. Check for the length marker if the peaks are as expected without any shift.
3. Then select the color according to the applied fluorescence for the desired microsatellite marker and look at the peak in the expected allele size range (given in Table 1).
4. It is recommended to have first an overall view of all the peaks of a given multiplex and then look at a single locus at a time. Some of the loci could be in the same range, but during establishment of the multiplex reaction, they have been labeled with different fluorescent dyes and thus will appear with different colors. This is the example of RJN-12 and RJO-4 in Fig. 2.
5. As microsatellites are co-dominant markers and *P. striiformis* is a dikaryotic organism, the single peak will reflect a homozygous individual and a double peak will reflect a heterozygous genotype. Figure 3 shows a homozygous and a heterozygous genotype for two Pakistani isolates for the locus RJO-4.
6. Some of the isolates may have more than two peaks. If it is several loci in both multiplexes (Fig. 4a, b), it could be a mix of genotypes, which means that the isolate needs to be purified.

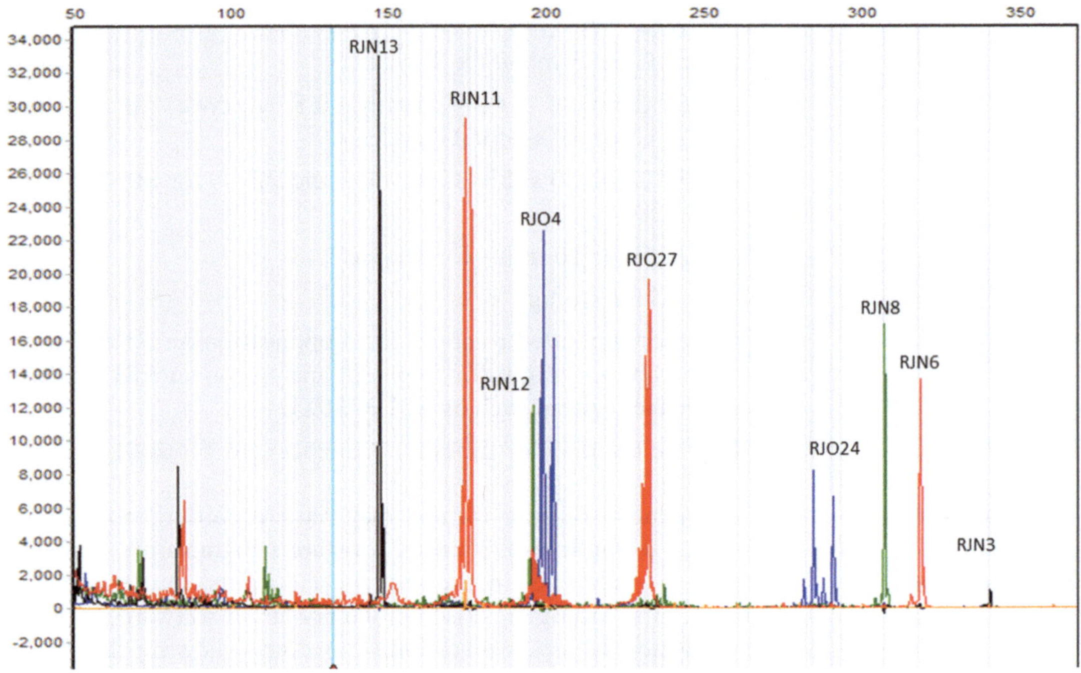

Fig. 2 Sequencer chromatogram as visualized in GeneMarker® for Multiplex-1 amplified for a Pakistani isolate, representing 9 SSR loci labeled with different fluorescent dyes, i.e., RJO-4 and RJO-24 are labeled *blue*; RJN-12 and RJN-8 are labeled *green*; RJN-13 and RJN-3 are labeled *black*; and RJN-11, RJO-27 and RJN-6 are labeled *red*

If it is three or less loci, then it will be better to rerun those markers separately in individual PCRs.

7. In cases where the PCR and sequencer separation have not worked correctly, no clear peaks will be observed (Fig. 4c). In some other cases a very low signal will be detected most probably due to low PCR product concentration (Fig. 4d). Rerunning the PCR will be helpful in such cases.
8. In some cases there will be an overall shift in allele sizes (e.g., 20 bases) for all loci. Rerunning the sequencer through adding size ladder could solve the problem in such cases (Fig. 4e, f). In some cases oversaturation due to more DNA content could also lead to this problem; a dilution could be useful in such cases.
9. Some microsatellite primers reveal some additional very minor peaks, which should not be included in scoring.
10. Based on our experience, RJO-3, RJN-2, and RJO-27 generally show peaks of low intensity when included in the multiplex and could therefore be tested in single PCR as well.

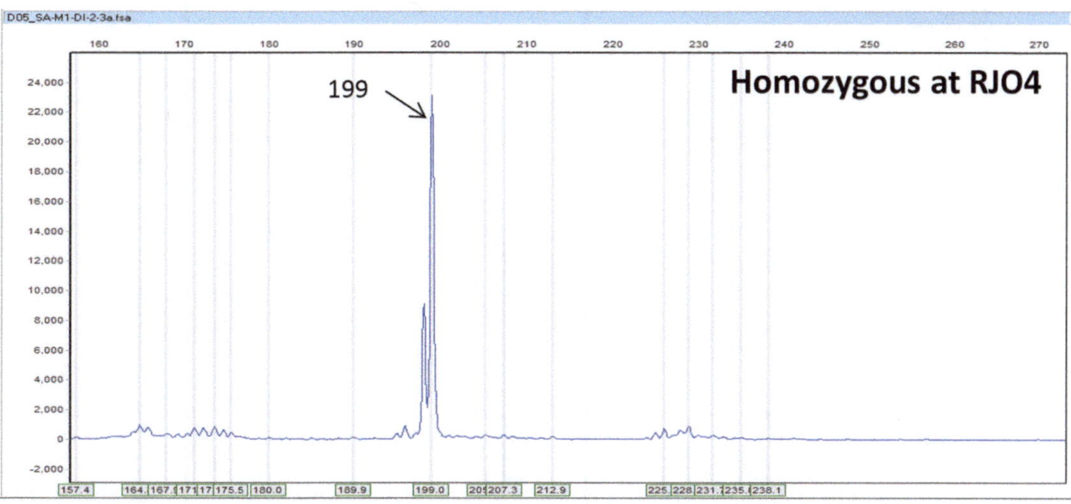

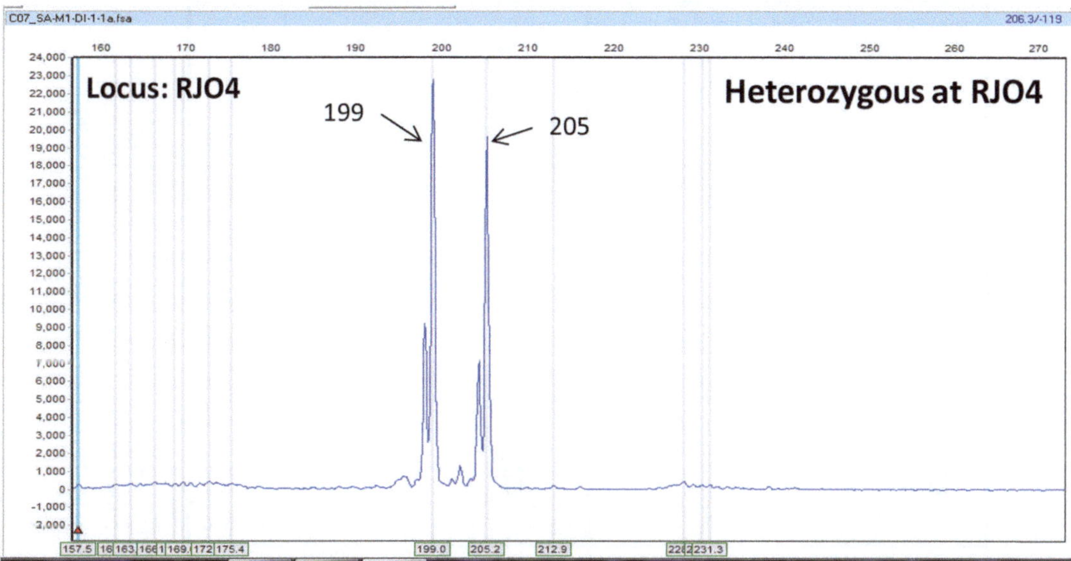

Fig. 3 Sequencer chromatogram for RJO-4 (visualized blue in GeneMarker®) for isolates having a homozygous (*top*) or a heterozygous genotype (*bottom*) at this locus

11. RJN-3 and RJ0-18 generally show peaks of lower intensity for which the primer concentration could be further increased (Table 2).
12. RJN-3 may have null alleles in South Asian populations; however, this must be confirmed in individual PCR reaction.
13. Note the score in an Excel file or any other data file to be later organized as infile for various population genetics software programs to carry out various population genetics analyses [10].

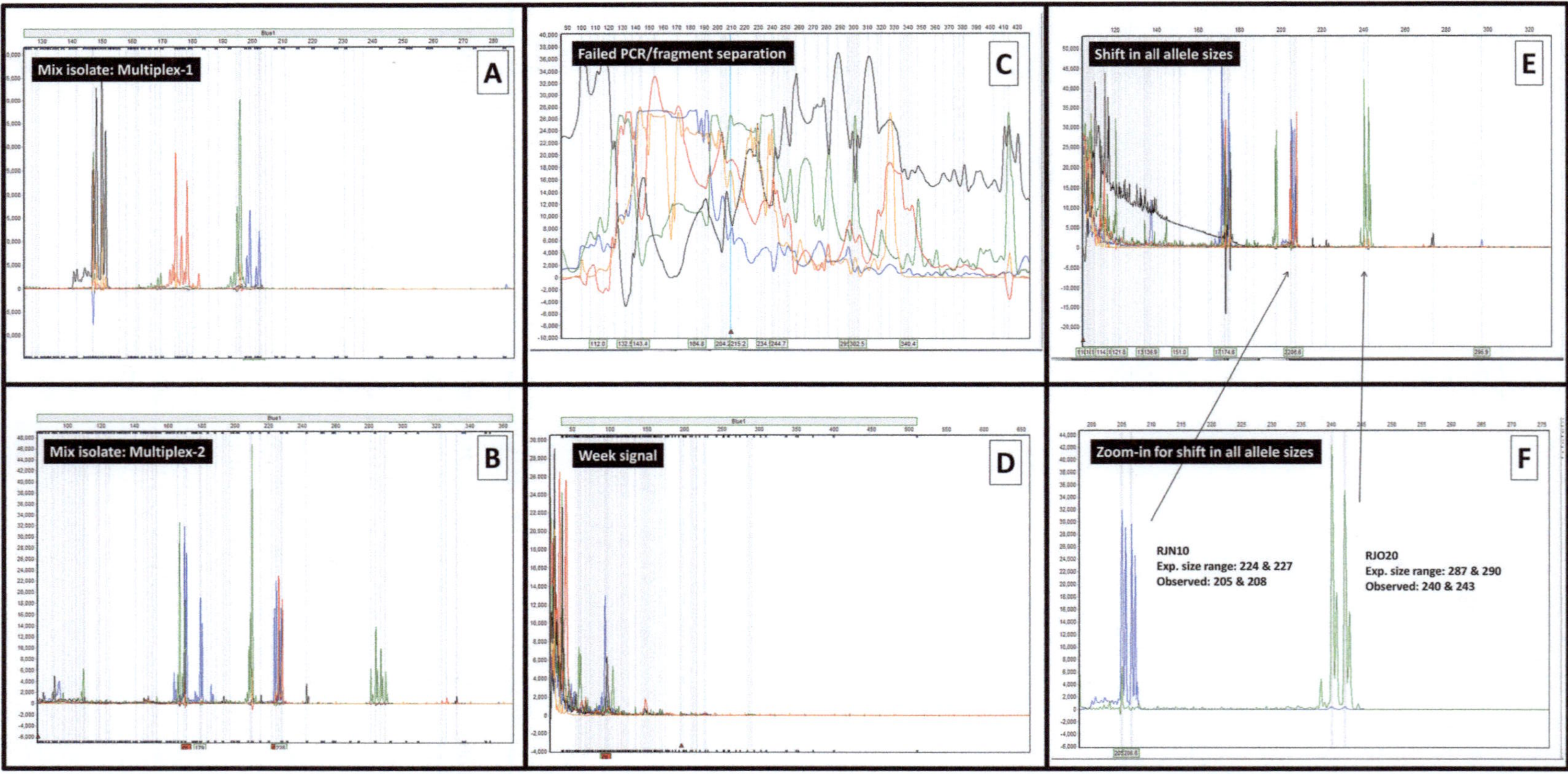

Fig. 4 Various problems in fragment analyses as visualized in a sequencer chromatogram in GeneMarker®. Same isolate which is potentially a mixed isolate for both Multiplex-1 (**a**) and Multiplex-2 (**b**); a failed PCR reaction and/or fragment separation through sequencer (**c**); low signal detection (**d**); and shifts in allele size from their expected range for all loci in a multiplex (**e**) and the zoom in showing shift for RJN-10 and RJ0-20

4 Notes

1. If no instrument is available, grinding in liquid nitrogen with a mortar and pestle should work.
2. We do not use an RNase step, but if RNA is a problem (e.g., as shown through the gel) an RNase step could be followed between **steps 6** and 7. Add 10 μL RNase A (10 mg/mL) and mix, incubate for 1 h at 37 °C, Add 0.6 μl of chloroform–iso-amyl alcohol (24:1), mix samples by inverting gently 20 times, spin at 2500 × *g* at 18 °C for 5 min. Transfer supernatant to a new 1.5 mL tube.
3. Quality of the extracted DNA could be checked with NanoDrop using 2× 1 μL DNA against the DNA solvent (e.g., 1× TE) as well as on 1.5% agarose gel.
4. There are different requirements according to the sequencer service provider. Some will take it directly after PCR amplification; some require ladder markers to be added to the plate. In case of saturation of allele peaks due to high DNA concentration (>500 ng/μL), dilution of DNA and/or PCR reaction is recommended.

Acknowledgments

We would like to thank J. Enjalbert and C. de-Vallavieille-Pope (INRA, France), and A.F. Justesen and M.S. Hovmøller (AU, Denmark), as the protocol presented here has been benefited from the research work conducted under their guidance. We also acknowledge T. Thach and A.F. Justesen (AU, Denmark) for suggestions on the chapter text. The mention of trade names for instruments used in the protocol does not imply their performance over other related products.

References

1. Brown JKM, Hovmøller MS (2002) Aerial dispersal of fungi on the global and continental scales and its consequences for plant disease. Science 297:537–541
2. Dean R, Van Kan J, Pretorius ZA, Hammond-Kosack KE, Pietro A, Spanu PD, Rudd JJ, Dickman M, Kahmann R, Ellis J, Foster GD (2012) The top 10 fungal pathogens in molecular plant pathology. Mol Plant Pathol 13:414–430
3. Beddow JM, Pardey PG, Chai Y, Hurley TM, Kriticos DJ, Braun J-C, Park RF, Cuddy WS, Yonow T (2015) Research investment implications of shifts in the global geography of wheat stripe rust. Nat Plants 1:15132–15135
4. Ali S, Gladieux P, Leconte M, Gautier A, Justesen AF, Hovmoller MS, Enjalbert J, De Vallavieille-Pope C (2014) Origin, migration routes and worldwide population genetic structure of the wheat yellow rust pathogen *Puccinia striiformis* f. sp. *tritici*. PLoS Pathog 10:e1003903
5. Bahri B, Leconte M, Ouffroukh A, de Vallavieille-Pope C, Enjalbert J (2009) Geographic limits of a clonal population of wheat

yellow rust in the Mediterranean region. Mol Ecol 18:4165–4179

6. Thach T, Ali S, de Vallavieille-Pope C, Justesen AF, Hovmøller MS (2016) Worldwide population structure of the wheat rust fungus *Puccinia striiformis* in the past. Fungal Genet Biol 87:1–8
7. Ali S, Rodriguez-Algaba J, Thach T, Sørensen CK, Hansen JG, Lassen P, Nazari K, Hodson DP, Justesen AF, Hovmøller MS (2017) Yellow rust epidemics worldwide were caused by pathogen races from divergent genetic lineages. Front Plant Sci 8:1057. doi:10.3389/fpls.2017.01057
8. Walter S, Ali S, Kemen E, Nazari K, Bahri B, Enjalbert J, Brown JK, Sicheritz-Pontén T, Jones J, de Vallvieille-Pope C, Hovmøller MS, Justesen AF (2016) Molecular markers for tracking the origin and worldwide distribution of invasive strains of *Puccinia striiformis*. Ecol Evol 6:2790–2804
9. Hovmøller MS, Walter S, Bayles R, Hubbard A, Flath K, Sommerfeldt N, Leconte M, Rodriguez-Algaba J, Hansen JG, Lassen P, Justesen AF, Ali S, de Vallavieille-Pope C (2016) Replacement of the European wheat yellow rust population by new races from the centre of diversity in the near-Himalayan region. Plant Pathol 65:402–411
10. Ali S, Gladieux P, Rahman H, Saqib MS, Fiaz M, Ahmed H, Leconte M, Gautier A, Justesen AF, Hovmøller MS, Enjalbert J, de Vallavieille-Pope C (2014) Inferring the contribution of sexual reproduction, migration and off-season survival to the temporal maintenance of microbial populations: a case study on the wheat fungal pathogen *Puccinia striiformis* f. sp. *tritici*. Mol Ecol 23:603–617
11. Mboup M, Leconte M, Gautier A, Wan AM, Chen W, de Vallavieille-Pope C, Enjalbert J (2009) Evidence of genetic recombination in wheat yellow rust populations of a Chinese oversummering area. Fungal Genet Biol 46 (4):299–307
12. Ali S, Leconte M, Walker A-S, Enjalbert J, de Vallavieille-Pope C (2010) Reduction in the sex ability of worldwide clonal populations of *Puccinia striiformis* f. sp. *tritici*. Fungal Genet Biol 47:828–838
13. Ali S, Gautier A, Leconte M, Enjalbert J, de Vallavieille-Pope C (2011) A rapid genotyping method for an obligate fungal pathogen, *Puccinia striiformis* f. sp. *tritici*, based on DNA extraction from infected leaf and Multiplex PCR genotyping. BMC Res Notes 4:240
14. Rodriguez-Algaba J, Walter S, Sørensen CK, Hovmøller MS, Justesen AF (2014) Sexual structures and recombination of the wheat rust fungus *Puccinia striiformis* on Berberis vulgaris. Fungal Genet Biol 70:77–85
15. Enjalbert J, Duan X, Giraud T, Vautrin C, de Vallavieille-Pope C, Solignac M (2002) Isolation of twelve microsatellite loci, using an enrichment protocol, in the phytopathogenic fungus Puccinia striiformis f. sp. tritici. Mol Ecol Notes 2:563–565
16. Justesen AF, Ridoutb CJ, Hovmøller MS (2002) The recent history of *Puccinia striiformis* f. sp. *tritici* in Denmark as revealed by disease incidence and AFLP markers. Plant Pathol 51:13–23

Part II

Molecular Pathogenicity

Chapter 7

Computational Methods for Predicting Effectors in Rust Pathogens

Jana Sperschneider, Peter N. Dodds, Jennifer M. Taylor, and Sébastien Duplessis

Abstract

Lower costs and improved sequencing technologies have led to a large number of high-quality rust pathogen genomes and deeper characterization of gene expression profiles during early and late infection stages. However, the set of secreted proteins expressed during infection is too large for experimental investigations and contains not only effectors but also proteins that play a role in niche colonization or in fighting off competing microbes. Therefore, accurate computational prediction is essential for identifying high-priority rust effector candidates from secretomes.

Key words Rust, Effectors, Effector prediction, Diversifying selection

1 Introduction

Plants provide us with food, fiber, and biofuels, but the worldwide demands are threatened by plant pathogens such as rust fungi that can cause devastating crop losses [1]. A common feature of many pathogens is a reliance on effector proteins that are secreted from the pathogen to the host plant and alter host-cell structure and function, thereby facilitating infection [2]. Effectors can also be recognized by host immune receptors, triggering defence responses. In silico effector prediction is essential to prioritize candidates for experimental investigation, to understand how pathogens cause diseases and to enable the breeding of crops that are disease resistant [3]. However, for many agronomically important pathogens, including the rust fungi, only a handful of effectors have thus far been identified [4].

We have summarized common approaches for predicting rust effector candidates in Fig. 1. In the following, we make recommendations on the practical aspects of rust effector prediction. Note that in silico effector prediction is an active field of research.

Sambasivam Periyannan (ed.), *Wheat Rust Diseases: Methods and Protocols*, Methods in Molecular Biology, vol. 1659, DOI 10.1007/978-1-4939-7249-4_7, © Springer Science+Business Media LLC 2017

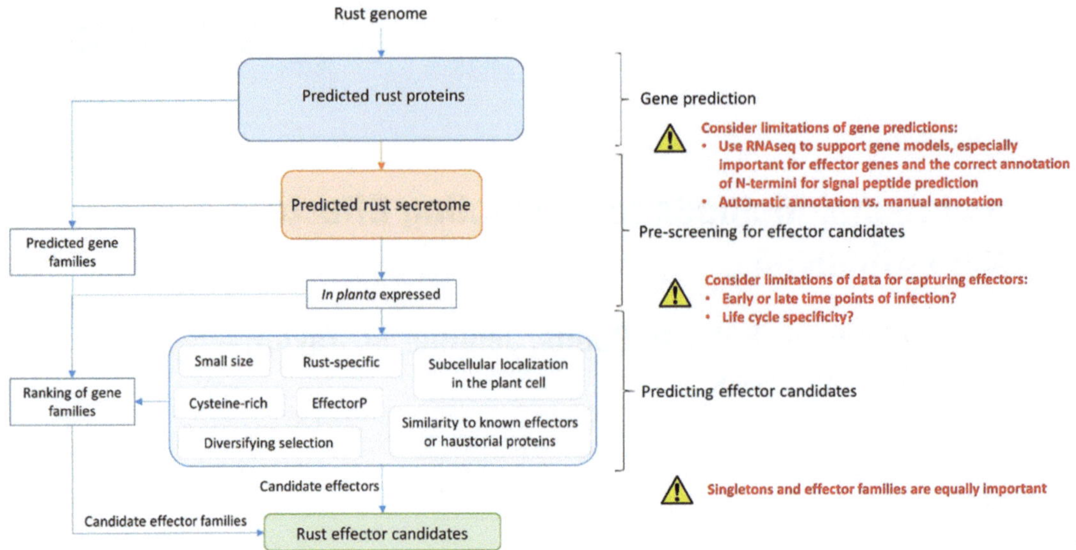

Fig. 1 A suggested workflow for rust effector prediction. Starting from the assembled rust genome, gene sets are predicted. Especially for effector gene prediction, using RNAseq data to support gene models is highly recommended [5, 6]. Effectors are both secreted and expressed during infection. Therefore, *in planta* RNAseq data can be used to filter the predicted secretome for those genes that are differentially expressed during infection. Several criteria associated with rust effectors can then be used to predict likely effector candidates, depending on the researcher's aims. For example, some apoplastic effectors have been associated with a small size and high cysteine content [7], whereas cytoplasmic effectors might carry subcellular localization signals or undergo diversifying selection. Gene families can be predicted either from the whole proteome or the secretome. Out of a gene family ranking based on typically observed effector features, candidate effector families can directly be derived and conserved sequence motifs can be highlighted from such families using tools such as MEME [8]. However, effectors occurring as singletons or in small families are equally important

Best practice procedures and tools are likely to change and improve with time, particularly as greater understanding of the functional aspects is derived from increasing numbers of validated effectors.

1.1 Identification of Secreted Rust Proteins

Rust effector proteins are secreted from the pathogen to the host and, therefore, signal peptide prediction is the first step in effector prediction. One of the challenges is the ability to distinguish between signal peptides and N-terminal transmembrane regions, which are both hydrophobic. Before SignalP 4 was published in 2011 [9], the most accurate tools for eukaryotes were reported as SignalP 2.0, SignalP 3.0, and TargetP [10]. While SignalP 4 has been evaluated as more accurate than older versions of SignalP for discriminating transmembrane regions from signal peptides, it has lower sensitivity on cleavage site prediction and for signal peptide prediction when no transmembrane domains are present. A recent benchmark including SignalP 4 found that the neural network predictors of SignalP 2 and 3, as well as TargetP were the most sensitive tools for fungal effector secretion prediction [11].

For example, most variants of the AvrP4 effector from *Melampsora lini* are not predicted to be secreted by SignalP 4, but are predicted to be secreted by earlier versions of SignalP. The use of the mammalian-trained SecretomeP for predicting nonclassically secreted proteins is not recommended for eukaryotes due to a very high false positive rate [12].

Taken together, we currently recommend the following pipeline for predicting rust secretomes for subsequent effector identification:

1. Predict signal peptides using an earlier version of SignalP (neural network predictors 2.0 or 3.0) and TargetP [13–15] for signal peptide prediction with increased sensitivity.
2. Remove secreted proteins with predicted transmembrane domains using TMHMM [16] or Phobius [17].

It is important to keep in mind that signal peptide prediction is dependent on the quality of gene model predictions. N-termini of effector genes can be misannotated and consequently, these genes will not be called as secreted by tools such as SignalP. Improved accuracy in gene models through the incorporation of RNAseq data, more complete genome sequence or manual correction and curation of gene models is likely to lead to improved signal peptide prediction. Similarly, predicted gene families from the whole proteome can reveal misannotated N-termini by comparing the structure and exon/intron organization of genes falling in a given family (*see* details in Duplessis et al. [18]).

1.2 Prediction of Small, Cysteine-rich Rust Effector Candidates from Secretomes

Rust secretomes tend to be large, with over a thousand predicted secreted proteins [11], and thus need to be mined for likely effector candidates for functional validation. Fungal effector proteins have commonly been predicted from secretomes using the criteria of small size and a cysteine-rich sequence [7]. Several recent studies have cautioned against these requirements in effector prediction, as many cytoplasmic effectors have been found to be relatively large (e.g., some *M. lini* AvrM variants are larger than 300 aas), and contain no or few cysteines in their sequences (*M. lini* AvrM, AvrL567, AvrM14 and *Puccinia graminis* f. sp. *tritici* PGTAUSPE-10-1 have only one cysteine in their sequence) [4, 19–21]. While the criteria of small size and cysteine-rich can be appropriate for some rust effectors, especially apoplastic ones, there are exceptions. For example, the *M. lini* AvrP123 family proteins are small and cysteine-rich, yet intracellular [22]. Therefore, additional methods for prioritizing effector candidates in rust secretomes are essential.

1.3 Analysis of Effector Gene Families in Rust Fungi

Sequencing of rust genomes has revealed the presence of expanded gene families, with many containing secreted proteins [18]. Such family organization can be determined using gene family prediction, e.g., using Markov Clustering (MLC; Enright et al. [23]), from the whole predicted proteome or directly from the predicted secretome. Clusters of paralogous genes derived from this analysis can be used for further diversifying selection tests (*see* Subheading 1.4). Beyond, gene family prediction can be performed at a broader scale considering close-relative rust species or other fungi with available genome data to reveal contraction and expansion in given families. Specific conserved motifs can also be found from protein conservation profiles in families [24–26]. Such motifs maybe useful for functional analysis and valuable when more knowledge will be gained on rust effectors by structural biology.

To rank gene families depending on their likelihood of containing effectors, Saunders et al. [25] developed a tribe clustering method, which includes the following steps:

1. Two or more rust secretomes are clustered into "protein tribes" based on sequence similarity using TribeMCL [23]. This returns protein tribes or families that contain either more than two members or only two members (protein pairs). Proteins that share no significant sequence similarity with another one are classed into singletons. Note that secretomes can also be clustered with predicted non-secreted proteins to identify potentially misannotated N-termini of *bona fide* secreted proteins.
2. The protein tribes are then annotated using a predefined list of effector features. This can include *in planta* expression data, sequence similarity to other rust effectors, the presence of effector motifs, cysteine content or protein size [25, 27]. Note that so far rust effectors have not been found to occur in specific genomic location, e.g., in repeat-rich regions or in effector gene clusters.
3. Protein tribes can be scored according to their likelihood of containing effectors. Each effector feature is associated with an *e*-value that corresponds to the likelihood of obtaining at least the same number of proteins with the given property by chance. These *e*-values can then be log-converted into a score. Finally, a combined score is calculated as the sum of scores associated to each effector feature. Tribes are then ranked according to their combined score to prioritize effector candidates for functional validation.

The tribe ranking method can be very effective for finding rust effectors that occur in large, expanded families across different rust pathogens and that are enriched for known effector features. However, some of the known *M. lini* effectors ranked fairly low in the tribe scoring (e.g., AvrP4 and AvrP123 are ranked as 156 out of 940; Nemri et al. [27]). It is important to also investigate singleton rust proteins as these are just as likely to constitute *bona fide* effectors. Therefore, approaches complementary to the analysis of gene families are needed for identifying rust effectors.

1.4 Identification of Rapidly Evolving Rust Effector Candidates

Effector proteins may be recognized by intracellular immune receptors in the host plant leading to resistance responses that prevent pathogen spread to other cells [28]. Effectors that are the target of such host receptors are under strong selection to escape recognition and often display signatures of diversifying selection that can be identified when sufficient sequence data is available. For instance, the alleles of the AvrL567, AvrP4, and AvrP123 effectors in *M. lini* are highly diverse and show elevated rates of nonsynonymous substitutions [29, 30]. Diversifying selection analysis is a powerful way to prioritize effector candidates, especially for those effectors that interact with resistance genes [24, 31].

The relative rates of nonsynonymous (d_N) and synonymous substitutions (d_S) in genes is commonly used to assess selection between species, whereas nonsynonymous and synonymous polymorphisms (P_N and P_S) are used similarly for comparison within species [32]. A d_N/d_S ratio <1 indicates purifying selection and functional conservation, $d_N/d_S = 1$ is consistent with neutral evolution, and $d_N/d_S > 1$ is indicative of diversifying selection or potential functional divergence. The d_N/d_S measure was originally developed to analyse distantly diverged genetic sequences, and lacks power for samples drawn from a single population. However, for microbes the distinction can be ambiguous [33] and the power of the d_N/d_S ratio to detect selection depends on the time-scale over which for example different Avr alleles have evolved and if they can be considered as competing genotypes [33].

1.5 Other Evidence for Effector Function: EffectorP and Subcellular Localization Prediction

While diversifying selection analysis is a powerful approach for prioritizing rust effector candidates, it can only be applied to genes for which sufficient sequence data is available. A machine learning approach called EffectorP was recently developed which has been trained on fungal effectors and can be used to predict rust effectors from secretomes [34]. Predicted effectors are assigned a probability of their likelihood of being effectors. EffectorP correctly predicts AvrL567, AvrP123, AvrP4, RTP1, AvrM14, and AvrL2 [21] as effectors, but predicts PGTAUSPE-10-1 and AvrM as non-effectors [34].

Cytoplasmic effectors are operating inside the plant cell and many have been found to exploit the plant machinery to enter specific subcellular compartments. Several high-confidence rust effector candidates have recently been found to mimic chloroplast transit peptides to enter plant chloroplasts and to carry nuclear localization signals (NLSs) to enter plant nuclei [35–37]. Subcellular localization in the plant cell can be predicted by using plant-trained classifiers such as WoLF PSORT [38] or ChloroP [39] on mature effector sequences (without the signal peptide). However, the variable lengths of signal peptides and pro-domains as well as the rapid evolution of effectors pose challenges to plant-based predictors, which assume that potential transit peptides start at the first residue after the signal peptide and often rely on homology-based information. A recent tool called LOCALIZER has been developed to predict effector localization to chloroplasts, mitochondria and nuclei and has been shown to give higher accuracy than plant-based predictors [40]. In particular, it predicts a nuclear localization signal for the rust effector RTP1 from *Uromyces fabae* (Uf-RTP1), which localizes to the plant nucleus, but not for its homolog from *U. striatus* (Us-RTP1) which has been found only in the host cytoplasm and is barely visible in nuclei [41]. LOCALIZER and WoLF PSORT predict chloroplast localization for the rust effector candidates CTP1, CTP2, and CTP3 which mimic chloroplast transit peptides [35], whereas ChloroP only predicts a transit peptide for CTP1 [40].

2 Materials

Many tools for effector prediction provide web servers, where researchers can submit their sequences of interest and get results back instantly. However, for reproducible workflows the recommended procedure is the use of command line tools, Python or Perl scripts or the Galaxy web platform and the release of these workflows and scripts alongside publication.

3 Methods

3.1 Identification of Secreted Rust Proteins

For secretome predictions, we refer to two comprehensive and practical descriptions for reproducible workflows using the command line or Galaxy published by Cock and Pritchard [42] and Reid and Jones [43].

3.2 Prediction of Small, Cysteine-rich Rust Effector Candidates from Secretomes

To filter a FASTA file of secreted proteins for those that are of a small size, say <300 aas and ≥4 cysteines, a Python script can be used. Note that this requires an installation of BioPython to read and write FASTA files.

```
from Bio import SeqIO
small_cys_proteins = [] # Setup an empty list for the output
# Make the sequences uppercase, for finding cysteines later
records = (rec.upper() for rec in SeqIO.parse("secretome.fasta", "fasta"))
# Go through each sequence and record the small, cysteine-rich ones
for record in records:
   if len(record.seq) < 300 and record.seq.count('C') >= 4.0:
    small_cys_proteins.append(record) # Add this record to our list
print("Found %i small, cysteine-rich proteins" % len(small_cys_proteins))
# Write the small, cysteine-rich proteins to new FASTA file
with open("small_cys_seqs.fasta", "w") as output_handle:
   SeqIO.write(small_cys_proteins, output_handle, "fasta")
```

3.3 Analysis of Effector Gene Families in Rust Fungi

The procedure for classifying effectors family with MCL detailed in Enright et al. [23] is based on all-against-all sequence similarities. The MCL algorithm is available from the authors (http://micans.org/mcl/). We stress that the inflation parameter value, which controls the "granularity/tightness" of clusters is critical to accurate prediction, particularly with the low sequence similarity noticed in the C-termini of candidate effector families. Several runs with different inflation rates may be used, depending also if intraspecific or interspecific classifications are regarded. Manual annotation of a few families of conserved candidate effectors in rust fungi (e.g., RTP1, some haustorially expressed secreted proteins) can be used as a guide to assess the quality of the family classification.

The selection of the most promising effector families is usually done manually from simple table sorting upon the user's selection criteria (e.g., see pipeline described in Petre et al. [37]). Candidates can also be selected using a tribe ranking approach as described in Saunders et al. [25] using the Multiple Experiment Viewer program MeV4 [44]. In both cases, we stress the importance of the weight applied to the different effector features considered for analysis. One might consider a given feature of prior importance over others, depending on the type of effector that is under investigation.

3.4 Identification of Rapidly Evolving Rust Effector Candidates

CodeML from the PAML package is a well-established software to calculate diversifying selection [45], however it can be difficult to use especially for researchers with limited exposure to the command line. One recommended, accessible way for performing a CodeML diversifying selection analysis is the ETE Toolkit and the ete evol module [46]. The ETE Toolkit can also be used to assemble a reproducible workflow, from performing a multiple sequence alignment, masking gapped alignment columns and calculating phylogenetic trees from the alignment.

Install the ETE toolkit following the instructions (http://etetoolkit.org/). To investigate d_N/d_S ratios for more than two sequences, a multiple nucleotide (CDS) alignment and a phylogenetic tree of the sequences under investigation are required. To calculate the CDS alignment and phylogenetic tree, it can be beneficial to use the corresponding protein sequences, as this can give a more accurate alignment than using the nucleotide sequences. The ETE toolkit can be used for alignment and tree calculations. For example, a phylogenetic tree can be built from unaligned protein sequences using Clustal Omega [47] and RAxML [48] with a simple command like this:

ete3 build -a seqs.fasta -w standard_raxml_bootstrap -o mytree/.

Given the protein alignment, the gaps need to be mapped to the nucleotide (CDS) alignment for input into CodeML. This can be done for example using PyCogent (http://pycogent.org/examples/align_codons_to_protein.html). Given the nucleotide alignment and phylogenetic tree, a test for diversifying selection acting on specific residues can be applied if more than three sequences are used [49]. Site-specific diversifying selection prediction can be more powerful for effectors than pairwise d_N/d_S ratios from entire genes as it allows varying d_N/d_S ratios across codons [31]. Two likelihood ratio tests of site-specific diversifying selection are commonly used in CodeML: model M1 (neutral) to model M2 (selection) and model M7 (beta) to M8 (beta&ω) [45]. A full example is provided here: http://etetoolkit.org/cookbook/ete_evol_hiv-env_site.ipynb.

If only two sequences are available, a pairwise d_N/d_S ratio test can be used and is available through the yn00 module in PAML. BioPython offers a simpler interface for running the yn00 than PAML itself (http://biopython.org/wiki/PAML). For polymorphism data, the McDonald-Kreitman test is appropriate for assessing adaptive evolution and is available through web servers such as http://mkt.uab.es/mkt/MKT.asp

3.5 Other Evidence for Effector Function: EffectorP and Subcellular Localization Prediction

EffectorP is available as both a stand-alone Python script (http://effectorp.csiro.au/software.html) as well as a web server (http://effectorp.csiro.au/). For subcellular localization prediction of effectors in the plant cell, we first need to obtain the mature effectors sequences (without signal peptides). The location of the cleavage site can be found in the SignalP output table.

Subcellular localization prediction tools such as ChloroP, WoLF PSORT, or LOCALIZER are available as both stand-alone scripts as well as web servers (http://www.cbs.dtu.dk/services/ChloroP/; http://www.genscript.com/wolf-psort.html; http://localizer.csiro.au/).

4 Notes

Effector prediction and its associated prediction methods (gene prediction, signal peptide prediction, localization prediction, and gene family and orthology prediction) are an active area of research and the tools mentioned here are likely to improve in the future. Improvements in effector prediction as well as in genome sequencing technologies and assemblies will reveal more insight into rust effectors. While the general workflows and pitfalls described will stay relevant, readers are encouraged to consider future developments in the area beyond the methods described here.

References

1. Leonard KJ, Szabo LJ (2005) Stem rust of small grains and grasses caused by *Puccinia graminis*. Mol Plant Pathol 6(2):99–111
2. Kamoun S (2006) A catalogue of the effector secretome of plant pathogenic oomycetes. Annu Rev Phytopathol 44:41–60

3. Figueroa M, Upadhyaya NM, Sperschneider J, Park RF, Szabo LJ, Steffenson B, Ellis JG, Dodds PN (2016) Changing the game: using integrative genomics to probe virulence mechanisms of the stem rust pathogen *Puccinia graminis* f. sp. *tritici*. Front Plant Sci 7:205
4. Petre B, Joly DL, Duplessis S (2014) Effector proteins of rust fungi. Front Plant Sci 5:416
5. Testa AC, Hane JK, Ellwood SR, Oliver RP (2015) CodingQuarry: highly accurate hidden Markov model gene prediction in fungal genomes using RNA-seq transcripts. BMC Genomics 16:170
6. Hoff KJ, Lange S, Lomsadze A, Borodovsky M, Stanke M (2016) BRAKER1: unsupervised RNA-Seq-based genome annotation with GeneMark-ET and AUGUSTUS. Bioinformatics 32(5):767–769
7. Stergiopoulos I, de Wit PJ (2009) Fungal effector proteins. Annu Rev Phytopathol 47:233–263
8. Bailey TL, Boden M, Buske FA, Frith M, Grant CE, Clementi L, Ren J, Li WW, Noble WS (2009) MEME SUITE: tools for motif discovery and searching. Nucleic Acids Res 37(Web Server issue):W202–W208
9. Petersen TN, Brunak S, von Heijne G, Nielsen H (2011) SignalP 4.0: discriminating signal peptides from transmembrane regions. Nat Methods 8(10):785–786
10. Klee EW, Ellis LB (2005) Evaluating eukaryotic secreted protein prediction. BMC Bioinformatics 6:256
11. Sperschneider J, Williams AH, Hane JK, Singh KB, Taylor JM (2015) Evaluation of secretion prediction highlights differing approaches needed for Oomycete and fungal effectors. Front Plant Sci 6:1168
12. Lonsdale A, Davis MJ, Doblin MS, Bacic A (2016) Better than nothing? Limitations of the prediction tool secretomeP in the search for Leaderless Secretory Proteins (LSPs) in plants. Front Plant Sci 7:1451
13. Nielsen H, Engelbrecht J, Brunak S, von Heijne G (1997) Identification of prokaryotic and eukaryotic signal peptides and prediction of their cleavage sites. Protein Eng 10(1):1–6
14. Bendtsen JD, Nielsen H, von Heijne G, Brunak S (2004) Improved prediction of signal peptides: SignalP 3.0. J Mol Biol 340 (4):783–795
15. Emanuelsson O, Nielsen H, Brunak S, von Heijne G (2000) Predicting subcellular localization of proteins based on their N-terminal amino acid sequence. J Mol Biol 300 (4):1005–1016
16. Krogh A, Larsson B, von Heijne G, Sonnhammer EL (2001) Predicting transmembrane protein topology with a hidden Markov model: application to complete genomes. J Mol Biol 305(3):567–580
17. Kall L, Krogh A, Sonnhammer EL (2004) A combined transmembrane topology and signal peptide prediction method. J Mol Biol 338 (5):1027–1036
18. Duplessis S, Cuomo CA, Lin YC, Aerts A, Tisserant E, Veneault-Fourrey C, Joly DL, Hacquard S, Amselem J, Cantarel BL, Chiu R, Coutinho PM, Feau N, Field M, Frey P, Gelhaye E, Goldberg J, Grabherr MG, Kodira CD, Kohler A, Kues U, Lindquist EA, Lucas SM, Mago R, Mauceli E, Morin E, Murat C, Pangilinan JL, Park R, Pearson M, Quesneville H, Rouhier N, Sakthikumar S, Salamov AA, Schmutz J, Selles B, Shapiro H, Tanguay P, Tuskan GA, Henrissat B, Van de Peer Y, Rouze P, Ellis JG, Dodds PN, Schein JE, Zhong S, Hamelin RC, Grigoriev IV, Szabo LJ, Martin F (2011) Obligate biotrophy features unraveled by the genomic analysis of rust fungi. Proc Natl Acad Sci U S A 108 (22):9166–9171
19. Lorrain C, Hecker A, Duplessis S (2015) Effector-mining in the poplar rust fungus *Melampsora larici-populina* secretome. Front Plant Sci 6:1051
20. Sperschneider J, Dodds PN, Gardiner DM, Manners JM, Singh KB, Taylor JM (2015) Advances and challenges in computational prediction of effectors from plant pathogenic fungi. PLoS Pathog 11(5):e1004806
21. Anderson C, Khan MA, Catanzariti AM, Jack CA, Nemri A, Lawrence GJ, Upadhyaya NM, Hardham AR, Ellis JG, Dodds PN, Jones DA (2016) Genome analysis and avirulence gene cloning using a high-density RADseq linkage map of the flax rust fungus, *Melampsora lini*. BMC Genomics 17:667
22. Catanzariti AM, Dodds PN, Lawrence GJ, Ayliffe MA, Ellis JG (2006) Haustorially expressed secreted proteins from flax rust are highly enriched for avirulence elicitors. Plant Cell 18(1):243–256
23. Enright AJ, Van Dongen S, Ouzounis CA (2002) An efficient algorithm for large-scale detection of protein families. Nucleic Acids Res 30(7):1575–1584
24. Hacquard S, Joly DL, Lin YC, Tisserant E, Feau N, Delaruelle C, Legue V, Kohler A, Tanguay P, Petre B, Frey P, Van de Peer Y, Rouze P, Martin F, Hamelin RC, Duplessis S (2012) A comprehensive analysis of genes encoding small secreted proteins identifies candidate effectors in *Melampsora larici-populina* (poplar leaf rust). Mol Plant-Microbe Interact 25 (3):279–293
25. Saunders DG, Win J, Cano LM, Szabo LJ, Kamoun S, Raffaele S (2012) Using hierarchical clustering of secreted protein families to

classify and rank candidate effectors of rust fungi. PLoS One 7(1):e29847

26. Link TI, Lang P, Scheffler BE, Duke MV, Graham MA, Cooper B, Tucker ML, van de Mortel M, Voegele RT, Mendgen K, Baum TJ, Whitham SA (2014) The haustorial transcriptomes of Uromyces appendiculatus and *Phakopsora pachyrhizi* and their candidate effector families. Mol Plant Pathol 15(4):379–393
27. Nemri A, Saunders DGO, Anderson C, Upadhyaya N, Win J, Lawrence GJ, Jones DA, Kamoun S, Ellis JG, Dodds PN (2014) The genome sequence and effector complement of the flax rust pathogen *Melampsora lini*. Front Plant Sci 5:98
28. Dodds PN, Rathjen JP (2010) Plant immunity: towards an integrated view of plant-pathogen interactions. Nat Rev Genet 11(8):539–548
29. Dodds PN, Lawrence GJ, Catanzariti AM, Teh T, Wang CI, Ayliffe MA, Kobe B, Ellis JG (2006) Direct protein interaction underlies gene-for-gene specificity and coevolution of the flax resistance genes and flax rust avirulence genes. Proc Natl Acad Sci U S A 103 (23):8888–8893
30. Barrett LG, Thrall PH, Dodds PN, van der Merwe M, Linde CC, Lawrence GJ, Burdon JJ (2009) Diversity and evolution of effector loci in natural populations of the plant pathogen *Melampsora lini*. Mol Biol Evol 26 (11):2499–2513
31. Sperschneider J, Ying H, Dodds P, Gardiner D, Upadhyaya NM, Singh K, Manners JM, Taylor J (2014) Diversifying selection in the wheat stem rust fungus acts predominantly on pathogen-associated gene families and reveals candidate effectors. Front Plant Sci 5:372
32. Stukenbrock EH, Bataillon T (2012) A population genomics perspective on the emergence and adaptation of new plant pathogens in agroecosystems. PLoS Pathog 8(9):e1002893
33. Kryazhimskiy S, Plotkin JB (2008) The population genetics of dN/dS. PLoS Gen 4(12): e1000304
34. Sperschneider J, Gardiner DM, Dodds PN, Tini F, Covarelli L, Singh KB, Manners JM, Taylor JM (2016) EffectorP: predicting fungal effector proteins from secretomes using machine learning. New Phytol 210 (2):743–761
35. Petre B, Lorrain C, Saunders DG, Win J, Sklenar J, Duplessis S, Kamoun S (2015) Rust fungal effectors mimic host transit peptides to translocate into chloroplasts. Cell Microbiol 18:453–465
36. Petre B, Saunders DG, Sklenar J, Lorrain C, Krasileva KV, Win J, Duplessis S, Kamoun S (2016) Heterologous expression screens in *Nicotiana benthamiana* identify a candidate effector of the wheat yellow rust pathogen that associates with processing bodies. PLoS One 11(2):e0149035
37. Petre B, Saunders DG, Sklenar J, Lorrain C, Win J, Duplessis S, Kamoun S (2015) Candidate effector proteins of the rust pathogen *Melampsora larici-populina* target diverse plant cell compartments. Mol Plant-Microbe Interact 28 (6):689–700
38. Horton P, Park KJ, Obayashi T, Fujita N, Harada H, Adams-Collier CJ, Nakai K (2007) WoLF PSORT: protein localization predictor. Nucleic Acids Res 35(Web Server issue): W585–W587
39. Emanuelsson O, Nielsen H, von Heijne G (1999) ChloroP, a neural network-based method for predicting chloroplast transit peptides and their cleavage sites. Protein Sci 8 (5):978–984
40. Sperschneider J, Catanzariti A-M, DeBoer K, Petre B, Gardiner DM, Singh KB, Dodds PN, Taylor JM (2017) LOCALIZER: subcellular localization prediction of both plant and effector proteins in the plant cell. Sci Rep 7:44598
41. Kemen E, Kemen AC, Rafiqi M, Hempel U, Mendgen K, Hahn M, Voegele RT (2005) Identification of a protein from rust fungi transferred from haustoria into infected plant cells. Mol Plant-Microbe Interact 18 (11):1130–1139
42. Cock PJ, Pritchard L (2014) Galaxy as a platform for identifying candidate pathogen effectors. Methods Mol Biol 1127:3–15
43. Reid AJ, Jones JT (2014) Bioinformatic analysis of expression data to identify effector candidates. Methods Mol Biol 1127:17–27
44. Saeed AI, Sharov V, White J, Li J, Liang W, Bhagabati N, Braisted J, Klapa M, Currier T, Thiagarajan M, Sturn A, Snuffin M, Rezantsev A, Popov D, Ryltsov A, Kostukovich E, Borisovsky I, Liu Z, Vinsavich A, Trush V, Quackenbush J (2003) TM4: a free, open-source system for microarray data management and analysis. BioTechniques 34(2):374–378
45. Yang Z (2007) PAML 4: phylogenetic analysis by maximum likelihood. Mol Biol Evol 24 (8):1586–1591

46. Huerta-Cepas J, Serra F, Bork P (2016) ETE 3: Reconstruction, analysis, and visualization of Phylogenomic data. Mol Biol Evol 33 (6):1635–1638
47. Sievers F, Higgins DG (2014) Clustal Omega, accurate alignment of very large numbers of sequences. Methods Mol Biol 1079:105–116
48. Stamatakis A (2014) RAxML version 8: a tool for phylogenetic analysis and post-analysis of large phylogenies. Bioinformatics 30 (9):1312–1313
49. Yang Z, Wong WS, Nielsen R (2005) Bayes empirical bayes inference of amino acid sites under positive selection. Mol Biol Evol 22 (4):1107–1118

Chapter 8

Protein–Protein Interaction Assays with Effector–GFP Fusions in *Nicotiana benthamiana*

Benjamin Petre, Joe Win, Frank L.H. Menke, and Sophien Kamoun

Abstract

Plant parasites secrete proteins known as effectors into host tissues to manipulate host cell structures and functions. One of the major goals in effector biology is to determine the host cell compartments and the protein complexes in which effectors accumulate. Here, we describe a five-step pipeline that we routinely use in our lab to achieve this goal, which consists of (1) Golden Gate assembly of pathogen effector–green fluorescent protein (GFP) fusions into binary vectors, (2) *Agrobacterium*-mediated heterologous protein expression in *Nicotiana benthamiana* leaf cells, (3) laser-scanning confocal microscopy assay, (4) anti-GFP coimmunoprecipitation–liquid chromatography–tandem mass spectrometry (coIP/MS) assay, and (5) anti-GFP western blotting. This pipeline is suitable for rapid, cost-effective, and medium-throughput screening of pathogen effectors *in planta*.

Key words Agroinfiltration, Live-cell imaging, Affinity chromatography, DNA assembly, Proteomics

1 Introduction

Over the last decade, postgenomic analyses of eukaryotic plant pathogens—such as nematodes, fungi, oomycetes, and aphids—revealed hundreds of effector proteins [1–4]. Although effectors are encoded by pathogen genomes, they are operationally plant proteins, i.e., they function and have a phenotypic expression in plant tissues [5]. One challenge in effector biology is to study effectors *in planta* [6]. In many plant species this task is challenging due to the lack of transient transformation method. To circumvent this obstacle, scientists often use the model plant *Nicotiana benthamiana* primarily because it enables transient expression of effectors in leaf cells [3].

N. benthamiana is a dicot plant used as a model in plant biology. Notably, the infiltration of leaves with solutions of *Agrobacterium tumefaciens* carrying a binary vector—the so-called *Agrobacterium*-mediated transient expression or agroinfiltration assay—allows rapid expression of proteins in leaf cells [7, 8].

Sambasivam Periyannan (ed.), *Wheat Rust Diseases: Methods and Protocols*, Methods in Molecular Biology, vol. 1659, DOI 10.1007/978-1-4939-7249-4_8, © Springer Science+Business Media LLC 2017

If the protein expressed is fused to a fluorescent protein tag, such as the green fluorescent protein (GFP) for instance, it is then possible to combine two different assays. Firstly, a live-cell imaging assay—by using a laser-scanning confocal microscope—can be carried out to identify the subcellular compartment in which the effector–GFP fusion accumulates [9]. Secondly, a protein–protein interaction assay—by using anti-GFP coimmunoprecipitation–liquid chromatography–tandem mass spectrometry (coIP-LC/MS-MS or coIP/MS)—can be performed to identify the plant protein complexes with which the effector–GFP fusion associates [10–12].

Here, we detail the five-step pipeline that we routinely use in our lab to identify the leaf cell compartments and protein complexes in which effectors accumulate. This pipeline is aimed at achieving fast-forward screening of medium-sized effector sets in a cost-effective manner.

2 Materials

2.1 Golden Gate DNA Assembly

1. *Escherichia coli* subcloning efficiency DH5α competent cells. Store 50 μL aliquots at −80 °C.
2. Digestion/Ligation mix 1: 1 μL BbsI restriction enzyme, 1 μL T4 DNA ligase, 2 μL Bovine Serum Albumin (1 mg/mL stock solution), 4 μL T4 Ligase buffer, 1 μL Golden Gate Level 0 acceptor vector pICSL01005 (50 ng/μL in ddHOH stock solution), 6 μL ddHOH. Prepare right before use.
3. Digestion/Ligation mix 2: 1 μL BsaI-High fidelity restriction enzyme, 1 μL T4 DNA ligase, 2 μL BSA, 4 μL T4 Ligase buffer, 1 μL Golden Gate Level 1 binary acceptor vector (CaMV 35S promoter, OCS terminator, 50 ng/μL in ddHOH stock solution), 1 μL Golden Gate C-terminal GFP tag module vector pICSL50008 (50 ng/μL in ddHOH stock solution), 9 μL ddHOH. Prepare right before use.
4. Blue/White selection mix: 1/1 (v/v) 100 mM Isopropyl β-D-1-thiogalactopyranoside (IPTG)/20 mg/mL 5-bromo-4-chloro-3-indolyl-β-D-Galactopyranoside (X-Gal) in DiMethyl-Formamide (DMF). Prepare right before use.
5. Luria–Bertani (LB) growth medium: 10 g/L Bacto tryptone, 5 g/L yeast extract, 10 g/L NaCl. Adjust pH to 7.0 with NaOH. Add 20 g/L Agar for solid medium. Autoclave and store at room temperature up to a month. Supplement with appropriate combination of antibiotics at the following final concentration: 100 μg/mL Carbenicillin (1000× stock solution), 50 μg/mL Kanamycin (1000× stock solution), 100 μg/mL Spectinomycin (1000× stock solution in 50% Dimethyl sulfoxide [DMSO]), 100 μg/mL Rifampicin (100× stock solution in Methanol).

2.2 Vector Insertion into A. tumefaciens and Infiltration of N. benthamiana Leaves

1. *A. tumefaciens* electrocompetent strain GV3101 (pMP90). Store 50 μL aliquots at −80 °C (*see* **Note 1**).
2. *N. benthamiana* 3- to 5-week-old plants grown at 22 °C in a glasshouse with 16 h day and 8 h night cycles.
3. MicroPulser Electroporator.
4. 2 mm gap electroporation cuvette.
5. Agroinfiltration buffer: 10 mM $MgCl_2$ (100× stock solution), 150 μm Acetosyringone (3333× stock solution in DMSO).

2.3 Laser-Scanning Confocal Microscopy

1. Leica DM6000B/TCS SP5 laser-scanning confocal microscope equipped with a 488 nm laser line, a 10× air, and a 63× water-immersion objectives.
2. Super Premium 1.2 mm microscope slides.
3. 17 μm-thick 22 × 50 mm cover glass.

2.4 Protein Isolation and Anti-GFP Immunoprecipitation

1. Ultrasonic cleaner.
2. Miracloth.
3. 0.22 μm syringe filter.
4. 5 mL syringe.
5. 30 mL ultracentrifuge tube.
6. GFP_Trap_A beads. Store at 4 °C.
7. Immunoprecipitation buffer: 10% (v/v) glycerol, 25 mM Tris–HCl pH 7.5 (40× stock solution), 1 mM Ethylenediaminetetraacetic (EDTA, 500× stock solution), 150 mM sodium chloride (NaCl, 33× stock solution), 0.1% Tween 20. Store at 4 °C up to a week.
8. Protein isolation buffer: Immunoprecipitation buffer, 2% (w/v) Polyvinylpyrrolidone (PVPP), 10 mM dithiothreitol (DTT, 100× stock solution), protease inhibitor cocktail (100× stock solution). Prepare right before use.
9. Laemmli buffer: 0.5 M Tris–HCl pH 6.8, 50 mM DTT, 2% [w/v] sodium dodecylsulfate [SDS], 20% glycerol, 0.0001% [w/v] bromophenol blue. Store at 4 °C up to a month (*see* **Note 2**).

2.5 Western Blotting

1. ImageQuant LAS 4000 luminescent imager.
2. Trans-Blot Turbo transfer machine.
3. Trans-Blot Turbo Mini PVDF Transfer Pack.
4. A4 transparency film.
5. Single step GFP (B2): sc-9996 horseradish peroxidase (HRP)-conjugated antibody.

6. Tris Buffer Saline (TBS): 24.2 g/L Tris, 80 g/L NaCl, adjust pH to 7.6 with HCl (10×).
7. TBS-T: TBS (1×), 0.1% Tween 20.
8. Blocking buffer: 3% [w/v] BSA in TBS-T.
9. Probing solution: 1/5000 single step GFP (B2): sc-9996 horseradish peroxidase (HRP)-conjugated antibody in TBS-T. Prepare right before use (*see* **Note 3**).
10. Pierce ECL Western Blotting substrate: 1/1 [v/v] Luminol/Peroxidase.
11. SuperSignal West Femto Maximum Sensitivity Substrate: 1/1 [v/v] Luminol/Peroxidase.
12. Revelation buffer: 19/1 [v/v] Pierce ECL Western Blotting substrate/SuperSignal West Femto Maximum Sensitivity Substrate.
13. Ponceau S solution: 0.1% [w/v] Ponceau S, 5% [v/v] Acetic Acid.

2.6 Protein Separation by SDS-PAGE, Gel Excision, and Trypsin Digestion

1. Mini-Protean electrophoresis system.
2. EZ-2 Genevac evaporator.
3. TGX Precast polyacrylamide gels.
4. Prestained Plus protein ladder.
5. Instant Blue.
6. Protein low binding 1.5 mL centrifuge tubes.
7. 10× SDS-PAGE running buffer: 30 g/L Tris base, 144 g/L glycine, 10 g/L SDS.
8. ABC buffer: 50 mM Ammonium Bicarbonate in ultrapure water.
9. Gel destaining solution: 1/1 [v/v] ABC buffer/100% Acetonitrile (ACN).
10. Reduction solution: ABC buffer, 10 mM DDT.
11. Alkylation solution: ABC buffer, 55 mM Chloroacetamide (CAM).
12. Trypsin buffer: 100 ng/μL Trypsin, 5% [v/v] ACN, 50% [v/v] ABC buffer.
13. Peptide extraction buffer: 5% [v/v] formic acid (FA), 45% [v/v] ultrapure water, 50% [v/v] ACN.

2.7 LC-MS/MS

1. Orbitrap Fusion trihybrid mass spectrometer in positive ion mode.
2. Nanoflow-UHPLC system Dionex Ultimate 3000.

3. Reverse phase trap column Acclaim PepMap, C18 5 μm, 100 μm × 2 cm connected to an analytical column Acclaim PepMap 100, C18 3 μm, 75 μm × 50 cm.
4. Nano-electrospray ion source with ID 0.02 mm fused silica emitter.
5. Mobile phase A: 3% ACN, 0.1% FA.
6. Mobile phase B: 80% ACN, 0.1% FA.

2.8 In Silico Data Analysis

1. Fiji (https://fiji.sc/).
2. Excel.
3. Scaffold v.4 (Proteome Software).
4. Mascot (Matrix Science).
5. Perl (https://www.perl.org/).
6. Blastclust program from Blast standalone program (http://blast.ncbi.nlm.nih.gov/).
7. TextWrangler (Bare Bones Softwares).
8. MSConvert package (Matrix Science).

3 Methods

We obtain the coding sequence of effectors by PCR cloning or by gene synthesis with codon optimization for expression in *N. benthamiana* and removal of BbsI and/or BsaI restriction sites if necessary. During this process, we replace the sequence coding the predicted signal peptide by the following nucleotides: CACCGAAGACACAATG. GAAGAC is the BbsI restriction site, AATG is the overhang for Golden Gate assembly with a promoter, the last three nucleotides (ATG) are the start codon. We also replace the stop codon by the following nucleotides: ggTTCGCCGTCTTCGTAG. GTCTTC is the BbsI restriction site, TTCG is the overhang for Golden Gate assembly with the coding sequence of a C-terminal tag. For more information about the Golden Gate DNA assembly method in plant biology, we refer readers to these recent reviews [13, 14].

3.1 DNA Assembly into Golden Gate Vectors

1. Mix 5 μL of purified PCR products or 1 μL of plasmid with a synthesized DNA fragment with the Digestion/Ligation mix 1. Incubate in a thermocycler with the following program: 20 × (4 min at 37 °C, 4 min at 16 °C), 10 min at 50 °C, 10 min at 80 °C.
2. Thaw a 50 μL-aliquot of *E. coli* DH5 α thermocompetent cells on ice for 5 min and add 5 μL of digestion/ligation reaction product from **step 1**. Keep on ice for 30 min.

3. Insert plasmids into *E. coli* cells by incubating at 42 °C for 20–30 s in a waterbath.
4. Keep the bacteria on ice for 1 min, then add 500 μL of LB liquid medium preheated at 37 °C and incubate at 37 °C for 90 min.
5. Spread 100 μL of Blue/White selection mix on a plate with LB-Agar supplemented with Spectinomycin. Keep for 1 h at 37 °C.
6. Plate 200 μL of bacteria and keep at 37 °C for 16–24 h.
7. Select 3–5 white bacterial colonies and verify the presence of the recombinant vector by colony PCR (*see* **Note 4**).
8. Grow bacteria from one colony PCR-positive colony in 10 mL liquid LB supplemented with Spectinomycin at 37 °C in a shaking incubator for 16–24 h.
9. Purify the plasmid, adjust it to 50 ng/μL in ddHOH, and store at −20 °C for further use or proceed directly to next step.
10. Mix 1 μL of plasmid with the Digestion/Ligation mix 2. Incubate in a thermocycler with the program described at **step 1**.
11. Repeat **steps 2–9** but replace Spectinomycin by Kanamycin.

3.2 Vector Insertion into A. tumefaciens Strains and Infiltration of N. benthamiana Leaves

1. Thaw a 50 μL-aliquot of *A. tumefaciens* competent cells on ice.
2. Add 0.2 μL of plasmid (*see* Subheading 3.1, **step 11**) to the cells into a 2 mm-gap electroporation cuvette.
3. Electroporate with a micropulser with the following setting: capacitance of 25 μF, voltage of 2.4 kV, resistance of 200 Ω (*see* **Note 5**).
4. Add 0.5 mL of liquid LB medium preheated at 28 °C and incubate at 28 °C for 1 h.
5. Plate 200 μL of bacteria on LB-Agar medium supplemented with Rifampicin and Kanamycin. Incubate at 28 °C for 36–48 h.
6. Select 3–5 colonies and verify the presence of the recombinant vector by colony PCR (*see* **Note 5**).
7. Grow bacteria from one PCR-positive colony in 10 mL liquid LB medium supplemented with Rifampicin and Kanamycin at 37 °C in a shaking incubator for 16–24 h.
8. Mix 1 mL of bacteria with 500 μL of 60% glycerol in a 2 mL centrifuge tube. Invert five times and store at −80 °C. Centrifuge the remaining 9 mL of culture at 4000 × *g* for 10 min (*see* **Note 6**).
9. Resuspend the bacterial pellet in 10 mL Agroinfiltration buffer and adjust to OD_{600} of 0.1–0.4. Keep on ice for 1 h to activate the bacteria.

10. Infiltrate 2–4 leaves from ranks 3–5 (starting from top) of three to five week-old *N. benthamiana* plants with a 1-mL syringe without needle (*see* **Note 7**).
11. Harvest the leaves 2–3 days after infiltration and transport/ keep them in a plate with high humidity (*see* **Note 8**).
12. Cut 2–3 leaf stripes of approximately 2 mm by 10 mm per leaf for immediate use for confocal microscopy (*see* Subheading 3.3).
13. Snap-freeze the leaf in liquid nitrogen, and store at −80 °C or proceed further immediately (*see* Subheadings 3.4 and 3.5) (*see* **Note 9**).

3.3 Laser-Scanning Confocal Microscopy

1. Mount a leaf stripe (*see* Subheading 3.2, **step 12**) in ddHOH between slide and cover glass, with the lower epidermis toward the objective. Remove air bubbles from the mounting by gently tapping the cover glass.
2. Place the mounting on the microscope and set up the microscope with the following parameters: laser line: 488 nm at 15% power; receptor 1: collection from 505 to 530 nm, 100% gain (GFP fluorescence); receptor 2: collection from 680 to 700 nm, 50% gain (chlorophyll autofluorescence); receptor 3: bright field; scanning frequency: 400 Hz; image resolution: 1024 × 512 pixels; line average: 4 (*see* **Note 10**).
3. Adjust the focus to epidermal cells, and screen the leaf using the 10× objective to select a region of interest with pavement cells that show detectable accumulation of the GFP signal and no sign of stress (*see* **Note 11**).
4. Use the 63× water-immersion objective to perform high magnification imaging (*see* **Note 12**).
5. Repeat **steps 1–4** with leaf stripes from other leaves.
6. Save data in .lif format.
7. Use the software Fiji to read the .lif file, perform post-treatment as needed, and export final images as .png or .tif files.

3.4 Anti-GFP Immunoblotting

1. Grind leaf tissues (*see* Subheading 3.2, **step 13**) into powder in liquid nitrogen using mortar and pestle. Transfer the leaf powder into a 50-mL centrifuge tube.
2. Use a 1 mL pipet tip with the narrow extremity bent on approx. 0.5 cm to pick up a small amount of leaf powder. Resuspend the powder in 100 μL Laemmli buffer in a 1.5 mL centrifuge tube and immediately incubate at 95 °C under vigorous agitation for 15 min. Keep the 50 mL tube with the remaining powder at −80 °C up to a week or proceed further directly (*see* Subheading 3.5).
3. During **step 2**, set up a precast gel in the Mini Protean device following manufacturer's instructions. Poor 1 L of SDS-PAGE

running buffer in the upper and lower chamber up to the indicated levels 5 min before starting the electrophoresis.

4. Centrifuge at 15,000 × *g* for 5 min.
5. Transfer the supernatant in a new 1.5 mL centrifuge tube. Add 100 μL of Laemmli and incubate at 95 °C for 10 min.
6. Centrifuge at 15,000 × *g* for 5 min. Transfer the supernatant in a new tube.
7. Load 10 μL of protein mixture on the gel. Load 5 μL of Prestained Page ruler in the wells at each extremity of the sample(s) (*see* **Note 13**).
8. Start the electrophoresis at 120 V for 5 min, then increase to 160 V. Stop the electrophoresis when the migration front reaches the bottom of the gel.
9. Disassemble the cassette and incubate the gel two min in ddHOH.
10. Transfer the proteins to the PVDF membrane using the Trans-Blot Turbo machine and kit following manufacturer's instructions (*see* **Note 14**).
11. Incubate the membrane in TBS-T for 2 min.
12. Block the membrane in 15 mL of Blocking Buffer for 1 h under gentle rotating agitation.
13. Incubate the membrane in 15 mL of Probing Solution 1 h under gentle rotating agitation.
14. Wash the membrane five times with 15 mL of TBS-T for 1 min. Then wash the membrane with 15 mL of TBS for 1 min.
15. Cover the membrane with the 2 mL of Revelation Solution for 3 min.
16. Remove the excess of solution and place the membrane between two transparency films. Position the montage in the ImageQuant LAS 4000 luminescent imager, with the side of the membrane facing the camera objective.
17. Expose the membrane for 30 s using the following settings: chemiluminescence, tray position 1, precision, high sensitivity (*see* **Note 15**).
18. Save data as a .gel file. Adjust image brightness/contrast and save the image as an 8-bit .tif file.
19. Wash the membrane one minute with ddHOH. Stain the proteins by incubating the membrane in Ponceau S solution for 30 min. Wash the membrane with ddHOH for 1 min. Place the membrane between two transparency films and scan it with a standard scanner (*see* **Note 16**).

3.5 Protein Isolation and Anti-GFP Immunoprecipitation (IP)

1. Grind leaf tissues (*see* Subheading 3.2, **step 13**) into powder in liquid nitrogen using a mortar and a pestle. Transfer the leaf powder into a 50-mL centrifuge tube.
2. Weight leaf powder and resuspend it into 300% [v/w] ice-cold Protein Isolation Buffer. Vortex and shake vigorously for 30 s or until the powder is completely thawed and heterogeneously in solution (*see* **Note 17**).
3. Sonicate the samples at 4 °C for 15 min in a waterbath sonicator, using maximal sonication parameters.
4. Centrifuge at 5000 × *g* at 4 °C for 20 min. Filter the supernatant through a four-layered piece of Miracloth and transfer it into a 30-mL ultra-centrifuge tube.
5. Centrifuge at 50,000 × *g* at 4 °C for 90 min. Collect 10 mL of the filtered solution into a 15-mL centrifuge tube.
6. During step 5, pipet GFP_trap beads with a 1 mL tip with a cut extremity into a 1.5 mL tube. Use 30 μL of beads per sample.
7. Equilibrate the GFP_trap beads into IP buffer by adding 1 mL of IP Buffer, inverting for 1 min, centrifugating at 800 × *g* for 1 min, and discarding the supernatant. Repeat two more times. Keep the beads into 1 mL of IP buffer before use.
8. Mix the equivalent of the initial 30 μL of beads from **step 6** with the 10 mL of protein solution from **step 5**. Incubate at 4 °C with gentle inversion for 30 min (*see* **Note 18**).
9. Centrifuge at 800 × *g* at 4 °C for 5 min. Discard the supernatant without disturbing the pellet, resuspend the pellet into 1 mL of IP Buffer and transfer to a new 1.5 mL centrifuge tube.
10. Centrifuge at 800 × *g* for 30 s. Discard supernatant without disturbing the pellet and resuspend the pellet into 1 mL of IP Buffer.
11. Repeat **step 10** four more times.
12. Centrifuge at 800 × *g* for 30 s. Discard supernatant without disturbing the pellet and resuspend the pellet into 200 μL of IP buffer.
13. Centrifuge at 800 × *g* for 30 s. Discard the supernatant. Add 50 μL of Laemmli buffer to the beads and incubate at 70 °C for 15 min in a heating block under vigorous agitation.
14. Centrifuge at 800 × *g* for 1 min. Transfer the supernatant into a new 1.5 mL centrifuge tube.
15. Centrifuge at 15,000 × *g* for 5 min. Transfer the supernatant into a 1.5 mL centrifuge tube and proceed further directly (*see* Subheading 3.6) or keep at −20 °C.

3.6 Protein Separation by SDS-PAGE, Gel Excision, and Trypsin Digestion

1. Set up a precast gel in the Mini Protean device following manufacturer's instructions.
2. Poor 1 L of SDS-PAGE running buffer in the upper and lower chamber up to the indicated levels 5 min before starting the electrophoresis.
3. Load 15 μL of protein solution (*see* Subheading 3.5, **step 15**) in one well. Load 5 μL of Prestained Page ruler in the wells at each extremity of the sample(s).
4. Start the electrophoresis at 120 V for 5 min, then increase to 160 V. Stop the electrophoresis when the migrating front is approximately 3 cm away from the wells.
5. Disassemble the cassette and incubate the gel in ddHOH for 2 min.
6. Stain the gel in 20 mL of Instant Blue for 30 min under gentle rotating agitation.
7. Wash the gel with ddHOH for 1 min, then incubate the gel in 20 mL of 10% EtOH overnight to ensure destaining.
8. Transfer the gel to a fresh 10% EtOH solution. Cut up to five gel slices and store them in 10% EtOH in 1.5 mL protein low binding centrifuge tubes. Cut small gel slices for each major protein band, and larger slices for gel sections that do not show a band signal. Gel slices can be stored at −20 °C.
9. Incubate gel slices in destaining Solution for 2 × 30 min.
10. Destain gel slices in the Destaining Solution at 25 °C for 30 min under vigorous shaking. Repeat until gel pieces are completely colorless (*see* **Note 19**).
11. Dry the gel pieces in 100% ACN for 10 min and remove supernatant.
12. Reduce Cysteine residues by incubating the gel pieces in the Reducing Solution at 25 °C for 30 min with gentle agitation. Make sure that gel pieces are well-covered by the solution. Then remove supernatant.
13. Alkylate Cysteine residues by incubating the gel pieces in the Alkylation Solution in the dark at room temperature for 30 min. Make sure gel pieces are well-covered by the solution.
14. Wash the gel pieces in the destaining Solution for 2 × 10 min.
15. Dry the gel pieces in 100% ACN for 10 min.
16. Cover the gel pieces with the Trypsin Solution. When the gel pieces are fully rehydrated (i.e., they are transparent), cover them with ABC Buffer and incubate overnight at 37 °C.
17. Add one volume of Peptide Extraction Buffer, vortex for 10 s and sonicate for 10 min.

18. Transfer the supernatant to a 1.5 mL protein low binding centrifuge tube.
19. Cover the gel pieces from **step 17** with Peptide Extraction Buffer, vortex for 10 s and sonicate for 10 min.
20. Transfer supernatant to the 1.5 mL tube from **step 18**.
21. Evaporate the peptide solution at 30 °C in an evaporator with the HPLC setting until all the liquid is evaporated.

3.7 LC-MS/MS and Peptide Search

1. Trap peptides to the reverse phase trap column.
2. Elute peptides using a 3–30% ACN gradient over 50 min, followed by a 6 min gradient of 30–80% ACN at a flow rate of 300 nL/min at 40 °C.
3. Operate the mass spectrometer in positive ion mode, apply a spray voltage of +2200 V, with transfer capillary temperature set to 275 °C. Use a scan resolution of 120,000 at 400 *m/z*, range 300–1800 *m/z*, automatic gain control set 2e5, and maximum inject time to 50 ms. In the linear ion trap, use data dependent acquisition method with "top speed" and "most intense ion" settings to trigger MS/MS spectra. Use the Universal Method (above 100 counts, rapid scan rate, maximum inject time to 500 ms) to set the threshold for collision induced dissociation (CID) and HCD. Set dynamic exclusion to 30 s. Allow charge state between 2+ and 7+ to be selected for MS/MS fragmentation.
4. Prepare peak lists in .mgf format from raw data using the MSConvert package.
5. Search peak lists on Mascot server against a in-house *N. benthamiana* database and a separate in-house constructs database and an in-house contaminants database. Allow in the search tryptic peptides with up to two possible miscleavage and charge states +2, +3, +4. Include the following modifications in the search: oxidized methionine as variable modification and carbamidomethylated cysteine as static modification. Search data with a monoisotopic precursor and fragment ions mass tolerance of 10 ppm and 0.6 Da, respectively.
6. Combine Mascot results in Scaffold and export in Excel.

3.8 Protein Merging and Removal of Contaminants

1. Extract the sequences of putative interactors ("Accession Number" column) reported in the Excel spreadsheet exported from Scaffold (*see* Subheading 3.7, **step 6**) into a .fasta file.
2. Cluster sequences that have at least 80% identity over 80% of their length using the blastclust program. The command line to use is as follows: blastclust -i infile -o outfile -L 0.8 -S 80 -e F. The output of this program is a text file containing each cluster per line of protein identifiers separated by spaces.

3. Select a representative sequence (usually the longest sequence) for each cluster and create a "lookup table" in the spreadsheet to be used in next step.
4. Replace all the interactors belonging to a cluster with just one representative sequence and its description using "vlookup" function of the spreadsheet program and the lookup table created above.
5. Consolidate the interactors by adding the peptide hits for sequences in the same group.
6. Remove any rows containing the word "Decoy" in the spreadsheet. These are added by spectral search programs as internal controls.

4 Notes

1. To prepare competent cells, keep a liquid culture (OD_{600} at 0.5–0.7) 30 min on ice, then wash twice with ice-cold 10% glycerol by centrifugation (3000 × *g* for 15 min). Prepare 50 μL aliquots in 10% glycerol and store at −80 °C.
2. Do not try to weight the bromophenol blue powder. Rather, scratch it with a 1 mL pipet tip and dip it in the Laemmli buffer, which should instantly become blue.
3. Spin the antibody solution before using it to pellet precipitates. Also, it is possible to collect the antibody solution after use and keep it at −20 °C, to be reused up to two times.
4. Standard reactions result in 10–200 colonies, 80–100% being white. A lower rate of white colonies indicates a low efficiency of the digestion/ligation reaction. Over 95% of the white colonies are usually positive for the colony PCR screening. Primers for the colony PCR are plCSL01005_For: GTCTCATGAGCGGATACATATTTGAATG and plCSL01005_Rev: CGTTATCCCCTGATTCTGTGGATAAC (primers amplify 350 nt in addition of the effector coding sequence), and pICH86988_For: GGACACGCTCGAGTATAAGAGCTC and pICH86988_Rev: GGATCTGAGCTACACATGCTCAGG (primers amplify 190 nt in addition of the effector-GFP coding sequence).
5. Electroporation time is usually between 5 and 6 ms. Smaller time will decrease the electroporation efficiency.
6. Use the glycerol stock to start a fresh liquid culture or plate when needed. If using a plate culture, it is recommended to wash once the bacteria by resuspension–centrifugation in Agroinfiltration buffer before proceeding to leaf infiltration, in order to remove the excess of antibiotics.

7. Perform this task during the light cycle to ensure maximal opening of the stomata and optimal infiltration.
8. Achieve high humidity by placing a humid piece of paper roll at the bottom of the plate.
9. When using large leaves, cut out petiole and main nerves before snap-freezing.
10. All settings must be fine-tuned for each sample. As a general recommendation, keep the gain and the laser power as low as possible, the scanning frequency as fast as possible, and the windows for fluorescence collection as narrow as possible.
11. Sign of stress includes autofluorescence and packing of chloroplasts, irregular cell shapes, and bright artefactual light signal. Avoid regions near the edge of the sample. Guard cells are not transformed in agroinfiltration assays and should therefore show no fluorescence.
12. We recommend performing z-stack imaging as often as possible as they allow to better appreciate the tridimensional context of the sample.
13. Always keep the loaded volume as low as possible to avoid migration artifacts and cross-contamination between wells.
14. Make sure to select the transfer mode according to the size of the protein of interest. Disassemble the stack as soon as the transfer finishes.
15. Adjust exposure time from 10 s to 1 h according to the intensity of the signal.
16. Use the intensity of the band signal at 55 kDa to control the equal loading and transfer of the proteins.
17. The more powder you use the longer it takes to thaw it. Multiple steps of 30 s vortexing might be necessary. Keep the tubes on ice 1 min between each vortexing.
18. Resuspend the beads to homogeneity by gently tapping the tube before use.
19. If gel pieces still show some coloration, increase temperature to 55 °C.

Acknowledgments

Research at The Sainsbury Laboratory (TSL) is supported by the Gatsby Charitable Foundation, the European Research Council (ERC), and the Biotechnology and Biological Sciences Research Council (BBSRC). We thank the Kamoun, Menke (proteomics), and the synthetic biology (Synbio) groups at TSL for continuous support and fruitful discussions. We also thank Dr. Thysis Buck for insightful discussions on protein–protein interaction assays.

References

1. Rehman S, Gupta VK, Goyal AK (2016) Identification and functional analysis of secreted effectors from phytoparasitic nematodes. BMC Microbiol 16:48
2. Lo Presti L, Lanver D, Schweizer G, Tanaka S, Liang L, Tollot M, Zuccaro A, Reissmann S, Kahmann R (2015) Fungal effectors and plant susceptibility. Annu Rev Plant Biol 66:513–545
3. Pais M, Win J, Yoshida K, Etherington GJ, Cano LM, Raffaele S, Banfield MJ, Jones A, Kamoun S, Saunders DG (2013) From pathogen genomes to host plant processes: the power of plant parasitic oomycetes. Genome Biol 14:211
4. Rodriguez PA, Bos JI (2013) Toward understanding the role of aphid effectors in plant infestation. Mol Plant-Microbe Interact 26:25–30
5. Win J, Chaparro-Garcia A, Belhaj K, Saunders DG, Yoshida K, Dong S, Schornack S, Zipfel C, Robatzek S, Hogenhout SA, Kamoun S (2012) Effector biology of plant-associated organisms: concepts and perspectives. Cold Spring Harb Symp Quant Biol 77:235–247
6. Petre B, Joly DL, Duplessis S (2014) Effector proteins of rust fungi. Front Plant Sci 5:416
7. Bombarely A, Rosli HG, Vrebalov J, Moffett P, Mueller LA, Martin GB (2012) A draft genome sequence of *Nicotiana benthamiana* to enhance molecular plant-microbe biology research. Mol Plant-Microbe Interact 25:1523–1530
8. Goodin MM, Zaitlin D, Naidu RA, Lommel SA (2008) *Nicotiana benthamiana*: its history and future as a model for plant-pathogen interactions. Mol Plant-Microbe Interact 21:1015–1026
9. Schornack S, van Damme M, Bozkurt TO, Cano LM, Smoker M, Thines M, Gaulin E, Kamoun S, Huitema E (2010) Ancient class of translocated oomycete effectors targets the host nucleus. Proc Natl Acad Sci U S A 107:17421–17326
10. Petre B, Saunders DG, Sklenar J, Lorrain C, Krasileva KV, Win J, Duplessis S, Kamoun S (2016) Heterologous expression screens in *Nicotiana benthamiana* identify a candidate effector of the wheat yellow rust pathogen that associates with processing bodies. PLoS One 11:e0149035
11. Petre B, Saunders DG, Sklenar J, Lorrain C, Win J, Duplessis S, Kamoun S (2015) Candidate effector proteins of the rust pathogen *Melampsora larici-populina* target diverse plant cell compartments. Mol Plant-Microbe Interact 28:689–700
12. Win J, Kamoun S, Jones AM (2011) Purification of effector-target protein complexes via transient expression in *Nicotiana benthamiana*. Methods Mol Biol 712:1811–1894
13. Patron NJ (2016) Blueprints for green biotech: development and application of standards for plant synthetic biology. Biochem Soc Trans 44:702–708
14. Patron NJ, Orzaez D, Marillonnet S, Warzecha H, Matthewman C, Youles M, Raitskin O, Leveau A, Farré G, Rogers C, Smith A, Hibberd J, Webb AA, Locke J, Schornack S, Ajioka J, Baulcombe DC, Zipfel C, Kamoun S, Jones JD, Kuhn H, Robatzek S, Van Esse HP, Sanders D, Oldroyd G, Martin C, Field R, O'Connor S, Fox S, Wulff B, Miller B, Breakspear A, Radhakrishnan G, Delaux PM, Loqué D, Granell A, Tissier A, Shih P, Brutnell TP, Quick WP, Rischer H, Fraser PD, Aharoni A, Raines C, South PF, Ané JM, Hamberger BR, Langdale J, Stougaard J, Bouwmeester H, Udvardi M, Murray JA, Ntoukakis V, Schäfer P, Denby K, Edwards KJ, Osbourn A, Haseloff J (2015) Standards for plant synthetic biology: a common syntax for exchange of DNA parts. New Phytol 208:13–19

Chapter 9

Proteome Profiling by 2D–Liquid Chromatography Method for Wheat–Rust Interaction

Semra Hasançebi

Abstract

Wheat–rust interactions are extremely complex biological processes which are accompanied with the defense/attack responses to survive and overcome pathogen attack or plant defense. Understanding of molecular mechanism of these interactions is a promising way to develop sustainable combat. Therefore, many studies have been performed to reveal the active host and pathogen-derived genes and their products during the infection or defense using different approaches for many decades. Particularly proteomics technology and proteome profiling which is a large scale analysis of a protein mixture to reveal differently expressed proteins under a certain conditions has become a very important tool for providing real insights into the extremely complex interactions. Moreover, this type of research has the potential to explore target proteins/genes such as effectors that can be used in disease management strategies. Hence, in this chapter we describe the proteome profiling protocols by using 2D–LC system.

Key words Proteomics, Proteome profiling, Plant–pathogen interaction, Defense, 2D–LC, PF2D, Leaf protein extraction, Wheat, Stripe rust

1 Introduction

One of the most important problems in our life has been feeding the ever increasing world's population. Agriculture and agricultural production is the main source of our life. Therefore, agricultural problems and yield losses can affect each person, and development of fast, reliable, and sustainable solutions for them is of vital importance. Diseases caused by phytopathogens are one of the most important factors that can bring about yield losses on agricultural products. To develop effective combating strategies, we need to understand the plant and pathogen communication and following cellular events at the molecular level. Therefore, research on the host–pathogen relationships has become one of the most interesting and rapidly advancing fields in plant science. These interactions include the most complex biological proccesses that require power-

Sambasivam Periyannan (ed.), *Wheat Rust Diseases: Methods and Protocols*, Methods in Molecular Biology, vol. 1659,
DOI 10.1007/978-1-4939-7249-4_9, © Springer Science+Business Media LLC 2017

ful and fast transcriptional and posttranscriptional changes in numerous genes belonging to both the plant and pathogen. This complex process evolutionarily continues on both sides to overcome defense and attack strategies of each other. In general, molecular studies about the plant–pathogen interactions are more focused on plant resistance than susceptibility, as it is thought from the resistance perspective. However, disease and resistance are a cluster of interaction responses, and pathogenic genes and gene products are also important as much as the host-derived counterparts.

Proteins are main actors of deep and extremely complex molecular world that conduct all biological processes in living organisms. Therefore, investigating the changes among proteome profiles from one situation to another in an organism is more informative because, unlike transcripts, these proteins participate directly in all cellular events. As a natural consequence of this, proteomics technology has become a very important tool and is rapidly developing for providing real insights into complex bioprocesses in cells and helping us understand how changes in gene expression become a cell response by completing the genomic and transcriptomic data. Thus, molecular studies in plant science, particularly those of plant–pathogen interactions, have recently moved to proteomics area.

The success of proteomics approach is based on high-resolution separation of complex protein mixtures and its reproducibility [1]. Two-dimensional gel electrophoresis (2D–PAGE) is routinely used in many laboratories for this purpose. Alternative separation methods such as 2D liquid chromatography (2D–LC) systems have been developed because of the disadvantages of 2D–PAGE that include application difficulties, low reproducibility, and inability to separate hydrophobic proteins. Here we describe protocols based on Proteome-Lab™ PF2D that is one of the 2D–LC systems. The system separates the protein mixtures in the first dimension according to their p*I* using chromatofocusing column, followed by a fractionation according to hydrophobicity, using reversed phase chromatography in the second dimension [2]. Additionally, it has automation for fractionation processes, which help separation of a protein mixture into several hundred fractions, and thus a large number of samples can be fractionated in a short time. Moreover, 2D–LC allows the determination of a large set of proteins and novel protein discovery [3].

In this chapter, detailed protocols for proteome profiling of wheat leaves are presented.

2 Materials

2.1 Total Protein Extraction

1. **PVPP:** 0.1 g PVPP for 1 g tissue.
2. **Mg/NP-40 extraction buffer** (100 mL): 0.5 M Tris–HCl, pH 8.3, 2% NP-40 (igepal), 20 mM $MgCl_2$, 2% Beta-mercaptoethanol, and 1 mM PMSF are weighed and dissolved in 50 mL ultrapure water one by one, and the total volume is filled up to 100 mL. The buffer solution is aliquoted into separate 2.0 mL vials and stored at −20 °C.
3. **50% PEG4000** (100 mL): 50 g PEG4000 is dissolved in 80 mL ultrapure water. After PEG dissolves completely, its total volume is filled up to 100 mL with ultrapure water, and it is stored in room temperature.
4. **Protease inhibitor cocktail** (1 mL): Protease inhibitor is dissolved in 1 mL ultrapure water. The prepared protease inhibitor is divided into 100 μL aliquots and stored at −20 °C. Right before the extracted proteins are dissolved, they are added to the solubilization buffer in proportion of 50 μL for 1 g tissue.
5. **Solubilization Buffer** (50 mL): 20 mL of 12.5% glycerol is added to a 100-mL beaker. Add the following components in order and keep stirring until it dissolves:

   62.5 mM Tris–HCl, pH 7.8.

   Slowly add 2.5 M Thiourea, stirring until dissolved.

   Slowly add 7.5 M Urea, stirring until dissolved.

   Slowly add 1.25 g of n-octylglucoside, FW 292.4, stirring until dissolved.

   6.25 mM TCEP [Tris-(2-Carboxyethyl) phosphine, hydrochlo ride], stirring until dissolved.

   Add 2.5% Octyl β-D-glucopranoside (OG), stir until dissolved.

   Transfer entire solution to a 50-mL volumetric flask and add 12.5% glycerol to volume.

   Aliquot the buffer solution into separate 2.0 mL vials and store at −20 °C.

2.2 First Dimensional Separation by PF2D

1. **Start Buffer (SB)** (1 L): 6 M urea, 25 mM Bis-Tris and 0.2% Oktil-Beta-D-glucopyranoside are dissolved in ultrapure water (pH 8.5, where pH is adjusted by ammonium hydroxide). It is wrapped in aluminum foil and stored at 2–8 °C.
2. **Eluent Buffer (EB)** (1 L): 6 M Urea, 10–12.5% v/v Polybuffer and 0.2% OG are dissolved in ultrapure water (pH 4.0, where pH is adjusted by iminodiacetic acid).
3. **High Ionic Strength Solution** (HISS) (1 L): 1 M NaCl is dissolved in 800 mL of ultrapure water. Its volume is filled up

to 1 L with ultrapure water after complete dissolution. The prepared buffer HISS (including SB and EB above) are sieved through a cellulose acetate membrane filter with pore diameter of 0.45 μm, and stored at 4 °C in a glass bottle wrapped in aluminum foil. All buffers must be degassed for 5 min in ultrasonic bath right before usage.

4. **Water**: suitable for HPLC.
5. **HPCF** (High Performance Chromatofocusing Fractionation) **column**.
6. **PD-10 SephadexTM G-25 gel filtration column**.

2.3 Second Dimensional Separation by PF2D

1. **Buffer A** (0.1% Trifluoracetic acid (TFA) in water): 1 mL TFA is added to 999 mL HPLC grade water and stirred intensively. TFA is added right before usage.
2. **Buffer B** (0.08% TFA in acetonitrile): 800 μL TFA is added to 999.2 L HPLC grade acetonitrile and stirred thoroughly. TFA must be added right before usage. All buffers are degassed for 5 min in ultrasonic bath.
3. **HPRP** (high performance reverse phase) **column**.

2.4 Trypsin Digestion

1. **50 mM NH_4HCO_3**: 40 mg NH_4HCO_3 is dissolved in 7 mL HPLC grade water and the volume is filled up to 10 mL.
2. **100 mM DDT:** 15.5 mg Dithiothreitol is dissolved in 1 mL 50 mM NH_4HCO_3.
3. **200 mM Iodoacetamide**: 56 mg Iodoacetamide is dissolved in 1.5 mL 50 mM NH_4HCO_3.
4. **Trypsin:** Commercially supplied trypsin is 20 μg in each vial. Prepare a solution by adding 1 mL 50 mM NH_4HCO_3 to vial. Mix the vial briefly to ensure the trypsin is dissolved. The final concentration of trypsin is 20 μg/mL.
5. **ZipTip with 0.2 μL C18 resin**.

3 Methods

3.1 Plant Material

Minimum 5 g wheat leaves are harvested from *Puccinia striiformis* inoculated and mock-inoculated (control) plants at different time points and the leaves is stored in a − 80 °C freezer until protein extraction (*see* **Notes 1** and **2**).

3.2 Total Protein Extraction

1. 0.3 g PVPP in powder form is added onto 3 g leaf tissue from infected and control plant samples stored in the −80 °C freezer.
2. Leaf tissues are ground in liquid nitrogen using mortar and pestle or grinder until they take fine powder form.

3. The powdered leaf tissues are transferred to 50 mL centrifuge tubes, and 30 mL Mg/NP-40 extraction buffer cooled in ice is added on them. The mixture is stirred until it becomes homogenous and left in ice for 15 min.
4. The mixture is centrifuged at 12,000 × *g* and 4 °C for 15 min and the supernatant is transferred to a new tube.
5. PEG4000 (50%) solution is added onto the supernatant for a final concentration of 15% and left in ice for 30 min. During the waiting period, the mixture is vortexed every 10 min. In this step, Rubisco proteins are removed from the total protein extraction by precipitating.
6. At the end of the duration, it is centrifuged at 15,000 × *g* and 4 °C for 15 min. In this stage, Rubisco proteins are precipitated and separated from the total protein extract. The supernatant is taken into a new tube.
7. It is left at −20 °C for overnight by adding 4 volumes of acetone on it. In this stage, removal of metabolites soluble in acetone is achieved.
8. The next day, it is centrifuged at 12,000 × *g* and 4 °C for 20 min and the supernatant is disposed of.
9. The protein precipitate is washed. For this, 10 mL cold acetone is added on the pellet, vortexed, and left at −20 °C for at least 30 min.
10. It is centrifuged at 5000 × *g*, 4 °C for 10 min and the fluid at the top is removed.
11. The **steps 9** and **10** are repeated 5–6 times until the pellet is cleaned and the supernatant becomes colorless.
12. The fluid at the top is removed after the last wash and the protein precipitate is left for drying under fume hood in room temperature until the smell of acetone disappears (for about 30 min).
13. The pellet is dissolved by adding 3 mL Solubilization Buffer and 150 μL protease inhibitor cocktail. To increase dissolution, the mixture is sonicated five times at 7 W for 5 s each time.
14. It is centrifuged at 30,000 × *g* and 22 °C for 30 min.
15. Later, the dissolved protein mixture is transferred to ultracentrifuge tubes, centrifuged for 1 h at 100,000 × *g* and 20 °C, and molecules with low molecular weight are separated from the protein mixture by precipitation.
16. The protein mixture (supernatant) is taken into a clean LoBind tube and 100 mL of it is separated for protein concentration determination and one-dimensional SDS-PAGE analysis. The remaining amount is divided into tubes to achieve a

distribution of 500 μg/500 μL, and stored at −80 °C until 2D–LC separation.

17. The amount of protein is determined with the Bradford method.
18. The qualitative analysis of extracted total protein samples are achieved with the one-dimensional classical SDS-PAGE.

3.3 First Dimension Separation of Total Proteins by PF2D System

The PF2D system is a 2D–LC system which consists of two modules that separate proteins based on their different characteristics. The first dimension separation module carries a HPCF column which separates proteins based on their isoelectric points (p*I*). In order to separate proteins based on their p*I*, a pH gradient starting from 8.5 and reaching 4.0 is established in the HPCF column. The pH gradient is established using "Start Buffer" (SB, pH 8.5) and "Eluent Buffer" (EB, pH 4.0). The proteins passing through the column along the pH gradient via the mobile phase are sent out of the column if their p*I* is higher than the pH in the column, stuck on the column if the values are equal, and kept in the internal surface of the column if p*I* is lower (Fig. 1). While the proteins leaving the column are passing through the UV-1 detector, their absorbance values at 280 nm are measured, and then proteins are fractionated into the wells of a 2.2 mL 96-deep well plate. The preparation stages and all operations for the device required for healthy operation of the PF2D system are achieved as follows.

3.3.1 Preparation of Glass Materials

All glass materials to be used in the operation of the PF2D system are subjected to a special washing process so that they do not

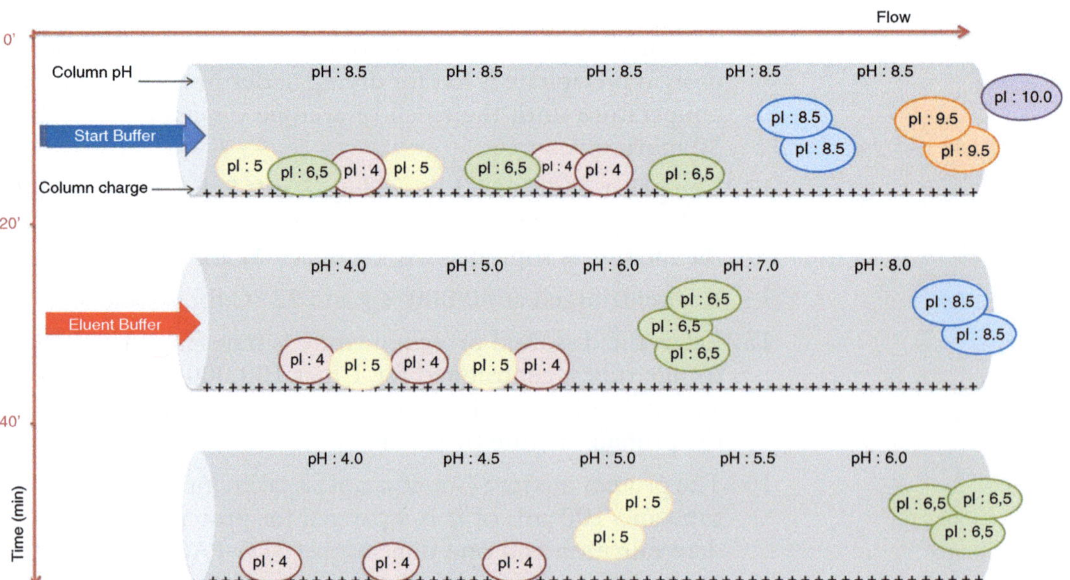

Fig. 1 Separation of proteins in the first dimension column based on their p*I*

contain any oil and protein contaminants. For this, all glass materials are washed with detergent, rinsed, and dried after passing through distilled water. In the next step, all materials are washed in order with HPLC grade water, 2-propanol, dichloromethane, and hexane. Then they are passed through dichloromethane and 2-propanol again, and finally dried after rinsing with HPLC grade water.

3.3.2 PF2D System Operation

It is crucial to establish a healthy and reproducible pH gradient in first dimension separation of proteins. Therefore, Start (SB) and Eluent (EB) buffers are prepared fresh right before separation of each time point samples. The same SB and EB should be used for all samples belongs to three biological replicates in one time point. Total proteins are separated as the following:

1. Right before starting, the first dimension HPCF column, 50 mL SB and 50 mL EB, 500 mL HISS and 1000 mL HPLC grade water are brought to room temperature.
2. PF2D operation program (32 Karat™ Software) is run by turning on the computer and all modules connected to the system.
3. All buffers are incubated in ultrasonic water bath for 5 min to get rid of air bubbles. The buffers are put in their place on the PF2D system (*see* **Note 3**).
4. The pH probe provided with the system is checked with standards of pH 4.0, 7.0 and 10.0. It is calibrated if needed.
5. The pH values of SB and EB are checked using the same probe. pH should be 8.5 for SB and 4.0 for EB. pH adjustment is made with NaOH and IAA if needed.
6. The same pH probe is mounted on its place on the PF2D system, and measurements are made with the same probe along the pH gradient.
7. Firstly, water is passed through the system for 10 min at a flow rate of 0.2 mL/min, and it is ensured that all passages of the system are clear.
8. First dimension HPCF column is put in its place.
9. Flush the HPCF column with 100% distilled water for 45 min at 0.2 mL/min.
10. Then equilibrate column with 100% SB for 130 min at 0.2 mL/min.

3.3.3 Preparation of Total Protein Samples

The samples to be loaded on the PF2D device should be purified from salts and taken in to SB (*see* **Note 4**). For this:

1. The total protein extract (~4 mg) stored at −80 °C is brought to room temperature.

2. Volume of the samples is filled up to 2.5 mL with SB.
3. They are loaded onto the PD-10 column and passed through the column completely.
4. 3.5 mL SB is added to the PD-10 column, the first couple of drops are discarded, and the remaining amount is collected in a clean tube.
5. The protein concentration in the collected sample is determined by micro BCA Protein Assay.

3.3.4 First Dimension Separation

1. Before loading the sample, the fraction collector is cooled down to 10 °C and the 96-deep-well plate is placed and made ready.
2. About 3 mg of the protein sample prepared in the above Subheading 3.3.3 is loaded onto the system with Hamilton injector through the injection segment of the PF2D system, so the maximum volume will be 3 mL. Attention should be paid to avoid providing the system with air bubbles during the sample loading process.
3. The program where the first dimension separation steps are located in the system is selected and the process is given start command.
4. In the first 40 min, only SB is passed through the column and the pH value is kept at 8.5. In this period, proteins with pI greater than 8.5 are separated and fractioned based on time (90 s intervals).
5. Later, EB flow is slowly started and the pH inside the column starts to drop. The proteins leaving the column are fractioned by 96-deep-well plate for each 0.3 pH interval.
6. Along the gradient, proteins with p*I* values greater than the in-column pH are taken out of the column (Fig. 1).
7. Proteins leaving the column pass through UV-1 detector before being collected on the 96-deep-well plate. This detector measures the absorbance values of proteins at 280 nm and the first dimension chromatogram is established.
8. When the SB flow end, 100% of the flow from the column is EB and the in-column pH drops to 4.0. During this 20-min process, separation and fractionation of acidic proteins are achieved based on time (90 s intervals).
9. After the completion of the pH gradient, the column is washed with HISS solution. This way, proteins stuck on the column are separated from the column and the column is cleaned.
10. In the last stage, the column is cleaned completely by passing water through it for 45 min and removed from the system.

11. Fractions should be stored at −20 °C if the second dimension analysis will be delayed for more than 8–10 h. Store all fractions long term at −20 °C or preferably at −80 °C.

3.4 Second Dimension Separation of Proteins

The second dimension module of the PF2D system includes a "High Performance Reverse Phase" (HPRP) column, where a hydrophobicity gradient is established and proteins are separated according to their hydrophobic properties. While the hydrophobicity gradient is obtained by water and acetonitrile flow through the column in changing proportions, TFA (trifluoracetic acid) is used as ion-suppressor. The solutions used in this stage are given in Subheading 2.3. The steps of second dimension separation are as follows:

1. Buffer A and Buffer B are prepared fresh. They are rid of air bubbles by degassing in ultrasonic bath for 5 min.
2. Buffer A and Buffer B are put into their places on the device, and the lines that will provide flow to the system are placed inside the buffers.
3. 100% Buffer A is passed through the system at a flow rate of 0.75 mL/0.5 min, and it is ensured that all lines are open, fluid flow is achieved comfortably, and the waste line is open and operational.
4. The second dimension column (HPRP) is mounted onto the system and the heater set to 50 °C is placed inside the block.
5. 100% Buffer A is passed through the column for 10 min at a flow rate of 0.75 mL/0.5 min.
6. The UV-2 detector is calibrated to 214 nm wavelength.
7. For balancing the HPRP column, Buffer B is passed through for 5 min, followed by 5 min of Buffer A flow.
8. Among the protein mixtures separated in the first dimension and fractioned in the 96-deep-well plate, 200 μL of each fraction is automatically transferred to the HPRP module for second dimension separation.
9. The protein mixture in each fraction is separated in the HPRP column based on their hydrophobicity properties.
10. For this, while separation of hydrophilic proteins is achieved by 100% Buffer A flow at a rate of 0.75 mL/0.5 min through the column at first, then a gradient flow toward 100% acetonitrile is achieved in the column by increasing proportion of Buffer B flow. Thus, an environment changing from polar to nonpolar is established in the column and starting with the proteins with the lowest hydrophobicity, separation is achieved in proteins with increasing hydrophobicity along the gradient (Fig. 2).

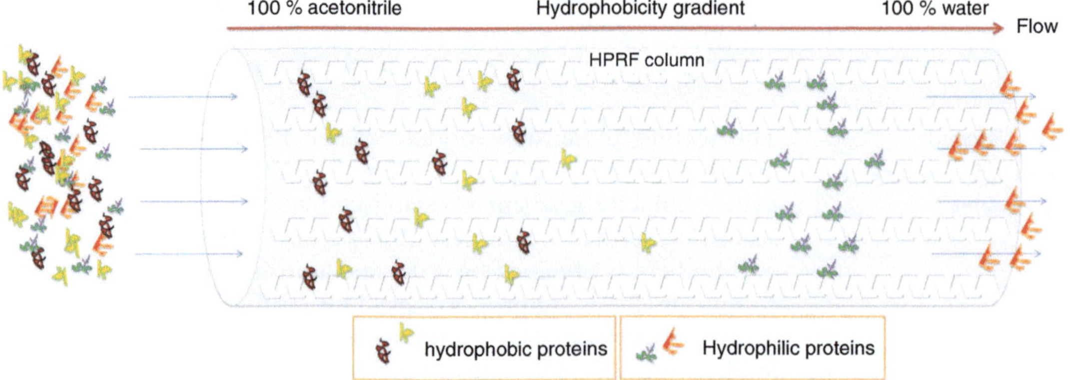

Fig. 2 Separation of proteins in the second dimension column based on their hydrophobicity

11. The proteins leaving the column, based on time (0.75 mL/0.5 min) are fractioned into 96-well microplates in the second dimension collector. Finally, the separation of the proteins with the highest hydrophobicity is achieved by 100% acetonitrile flow and the second dimension separation is completed.
12. The UV-2 detector in this module achieves more sensitive protein detection by measuring the absorbance of peptide bonds of proteins at 214 nm wavelength.
13. Each of the obtained UV peaks represents a single protein and "UV-2 chromatogram" is established for each fraction from these peaks.
14. **Steps 7–13** are repeated for each fraction separated in the first dimension and transferred to the second dimension.

3.5 Selection of Differentially Expressed Proteins

After the completion of two-dimensional separation, the obtained data (pH intervals, UV1 and UV-2 chromatograms) are processed by the ProteoVue software developed specially for the PF2D system and these data is converted to virtual gel maps. For this, steps for usage of the software are followed. Thus, for a single sample, pH interval values of fractions collected in the first dimension separation and UV-2 chromatograms of these fractions obtained during the second dimension separation are matched one to one. This way, protein profiles of all fractions belonging to a single sample are organized in virtual gel maps and obtained in the form of two-dimensional gel image (Fig. 3). This step should be completed first, in order to comparison of control and infected samples peak to peak.

DeltaVue software, again, developed for the PF2D system is used in comparing the protein profiles of control and infected samples. For this, virtual gel maps prepared with ProteoVue belonging to two samples to be compared are opened on the

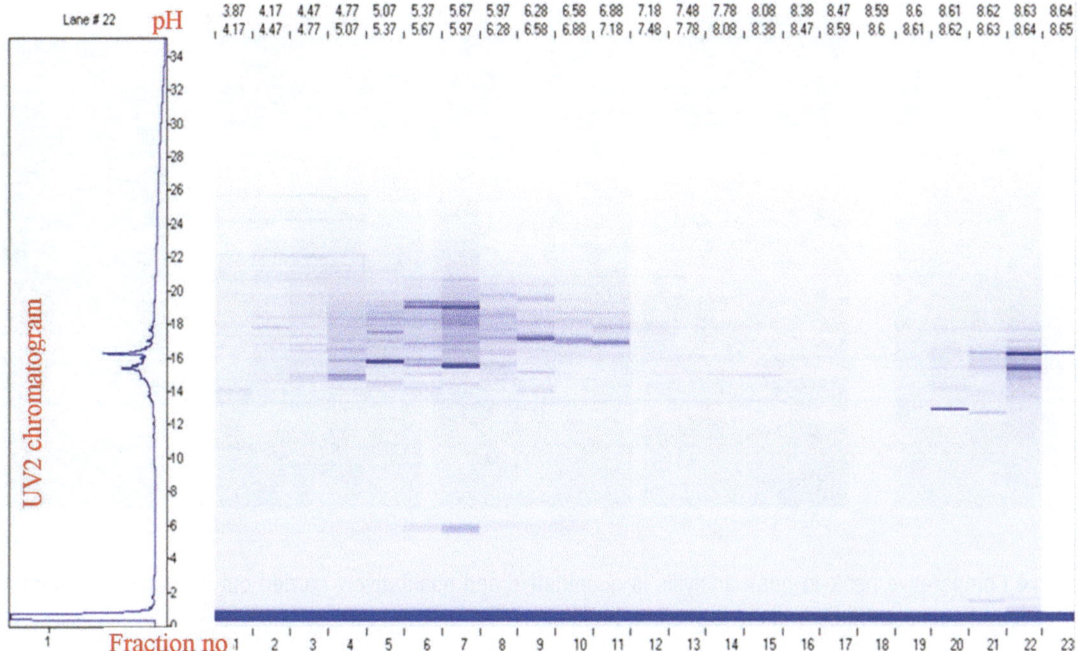

Fig. 3 Two-dimensional virtual gel map

DeltaVue software so that one sample profile would be placed on the right, and the other would be on the left side. Both chromatograms and virtual gel images are followed together in the own map of each sample. In the center, two profiles are analyzed by overlapping. For a more detailed comparison, protein bands of each couple of samples in the same pH interval are compared on raw chromatogram profiles peak to peak. Each protein peak in the UV-2 chromatograms are matched one to one and numbered. Peak area calculations are made for all numbered peaks, and expression amounts are determined. Detailed information such as pH interval, amount, ratio by the matched protein is obtained for the selected and numbered peaks. Consequently, all fractions of the two compared samples and all single proteins in all fractions are comparatively analyzed one by one. Hence, protein peaks with different expression levels are easily observed (Fig. 4). UV-2 chromatograms between two samples can be matched and analyzed on Offline mode of the PF2D operation program [4] (Fig. 5). Later, proteins with expression difference ratios ≥ 2 are selected and statistically confirmed before mass spectrometry (MS) annalysis. Almost 700–900 protein peaks can be observed and compared for each sample in this system.

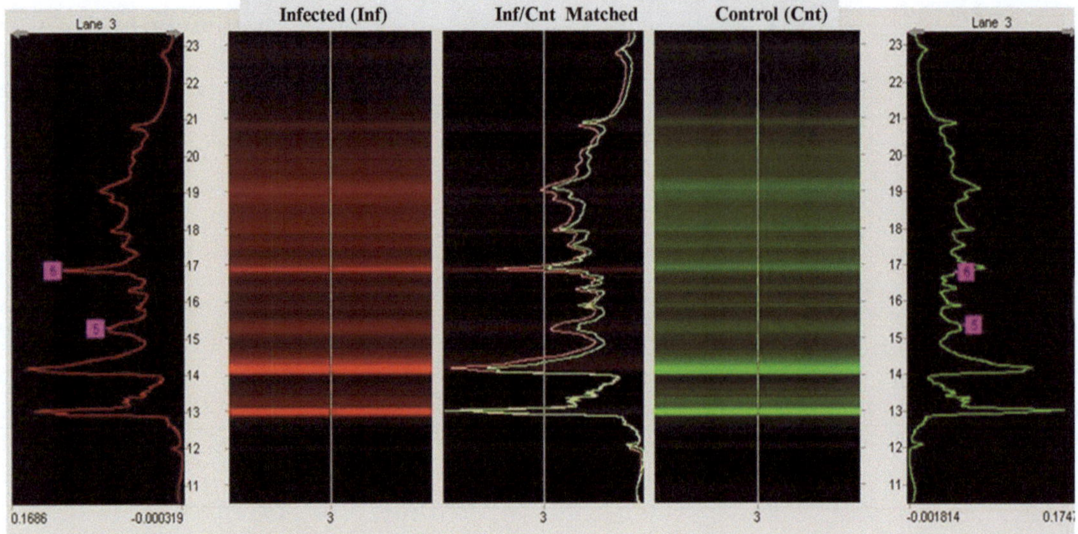

Fig. 4 Comparative peak-to-peak analysis is quantitative and qualitatively carried out by DeltaVue program (numbered peaks are differentially expressed proteins)

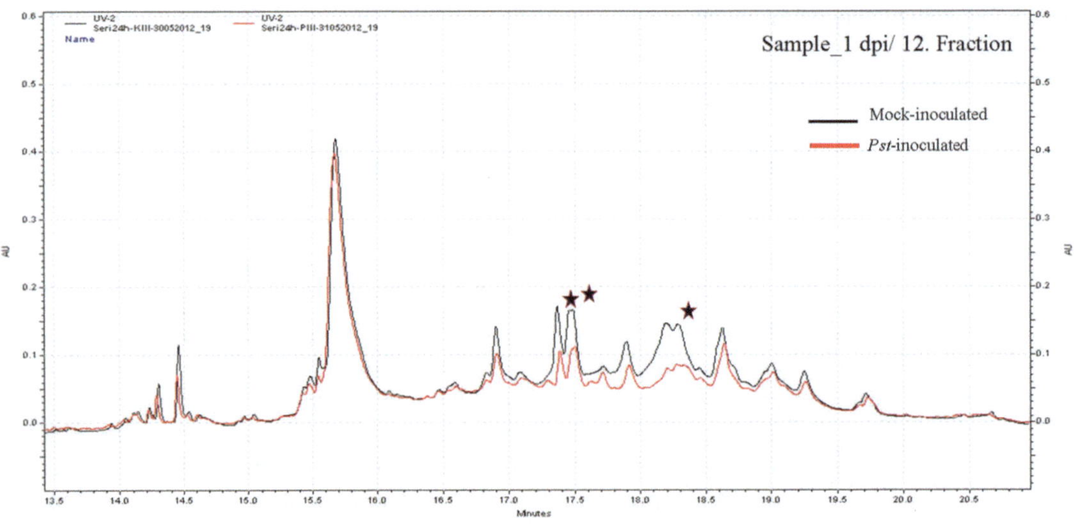

Fig. 5 Comparison of UV-2 chromatograms between *Pst*- and mock-inoculated samples (each *peak* represents a protein in the profiles of one fraction, and *stars* represent differentially expressed proteins)

3.6 Tryptic Digestion

The proteins with observed expression differences selected for MS/MS analysis are taken into 1.5 mL LoBind tubes and the following steps are carried out.

1. The samples are dried in a vacuum concentrator.
2. Ten microliters of 50 mM NH_4HCO_3 is added onto the dried protein samples, and dissolved by mixing in thermoshaker at 25 °C.

3. In order to weaken the disulfide bonds of the protein, 5.5 μL DTT (100 mM) is added and incubated at 60 °C for 15 min.
4. Then, in order to modify the weakened cysteine side chains, 6.1 μL iodoacetamide (200 mM) is added and incubated for 30 min at room temperature in dark.
5. Trypsin is then added to the samples (enzyme–protein ratio would be 1:50) and incubated overnight at 37 °C.
6. The digested tryptic peptides are concentrated by using ZipTip (Millipore, ZTC18M096) which have MicroC18 column at the tip, and cleaned of salts and reaction compounds.
7. For this, samples are centrifuged at low speeds for 10–15 sc.
8. The C18 column at the ZipTip pipette tip is cleaned by pipetting up and down in 70% acetonitrile containing 0.1% (v/v) TFA three times, and then five times in HPLC grade water containing 0.1% (v/v) TFA, and is made ready to bind the peptides.
9. Peptides are bond to the column by slowly pipetting in the sample for 30 times.
10. Then it is moved up and down three times in HPLC grade water containing 0.1% TFA, removing the salts.
11. Finally, the elution of peptides is achieved by pipetting 30 times in 70% (v/v) acetonitrile containing 0.1% (v/v) TFA in a separate tube.
12. Eluted peptides are dried in a vacuum concentrator and dissolved in 5 μL HPLC grade water containing 0.1% (v/v) formic acid.
13. 0.5 μL internal calibrant (50 fmol ADH1_YEAST-Waters MassPrep Enolase Digestion Standard) is added onto the obtained tryptic mixture, vortexed for 3–5 s, and transferred to the MS/MS device using special vials so that no air bubbles remained.
14. They are sent to mass spectrometry analysis.

3.7 Bioinformatic Analysis

Protein identification with mass spectrometry is based on partial sequence analysis, that is, it does not require knowledge of the entire amino acid series of the protein. However, the higher the number of definitions of peptides of a protein is, the wider its scope of sequence, which increases the reliability of identifying a protein. This method significantly widened data processing volume by achieving identification of thousands of proteins simultaneously. Nevertheless, the fact that large amounts of data have been gathered with mass spectrometry does not mean that the data will have meaningful correspondence, because protein identification with mass spectrometry involves at least three different stages independent from each other.

In the first two stages of experimental processes, proteins are broken by enzyme and from the end products: MS and/or MS/MS spectra are obtained, where signal magnitude in on the vertical axis and mass/load ratio is on the horizontal axis. Scanning these spectra in databases via different matching software is the subject of bioinformatics, which is concerned with computer systems.

Actually, precision is not present in protein identification with mass spectrometry, because all software that does database scanning provides the results with a certain probability value. The ones in the confidence interval among these statistical calculations are assumed to have made correct matching. While scanning peptide fractionation spectra in appropriate databased via scanning engines has become a standard practice today, databases unfortunately have not yet been established for all living species. Thus, researchers conducting the study should select a database suitable for mass/load data they obtained about tryptic peptides, amino acid series information and their own research material. In the study on wheat–stripe rust interaction, Uniprot wheat and Broad Institute Puccinia databases are used. IdentityE is calibrated by assuming trypsin as the breakdown enzyme so that fragment ion mass tolerance is 0.028 Da; parent ion tolerance is 0.011 Da. Apex3D data preparation parameters are set as 0.2 min chromatographic peak width, 10.000 MS TOF resolution, 150 counts of low energy threshold, 50 counts of high energy threshold and density threshold of 1200 counts. The search query for the database is set as at least three fragment ions for each peptide, at least seven fragment ions for each protein, at least one peptide match for each protein and one missed section. Carbamidomethyl-cysteine modification, acetyl N-TERM, asparagine and glutamine deamidation, and methionine oxidation are defined as variable modifications.

4 Notes

1. In general, three biological replicates must be made while preparing plant material.
2. Time point selection following inoculation is a very important point. Particularly first 12 h should be important because most of resistance responses occur at an early stage following plant–-pathogen interaction.
3. Air bubble is one of the most serious problems, therefore, degassing of buffers must not be neglected.
4. Loading onto the PF2D is performed by application of directions by the manufacturer. As differences in expression are directly related of protein amount, it is crucial that equal amounts of protein must be loaded for samples to be compared.

References

1. Suberbielle E, Gonzalez-Dunia D, Pont F (2008) High reproducibility of two-dimensional liquid chromatography using pH-driven fractionation with a pressure-resistant electrode. J Chromatogr B 871:125–129
2. McDonald T, Sheng S, Stanley B, Chen D, Ko Y, Cole RN, Pedersen P, Van Eyk JE (2006) Expanding the subproteome of the inner mitochondria using protein separation technologies: one- and two-dimensional liquid chromatography and two-dimensional gel electrophoresis. Mol Cell Proteomics 5:2392–2411
3. Wu WW, Wang G, Yu M, Knepper MA, Shen RF (2007) Identification and quantification of basic and acidic proteins using solution-based two dimensional protein fractionation and label-free or 18O-labeling mass spectrometry. J Proteome Res 6:2447–2459
4. Barre O, Solioz M (2006) Improved protocol for chromatofocusing on the ProteomeLab PF2D. Proteomics 6:5096–5098

Chapter 10

Investigating Gene Function in Cereal Rust Fungi by Plant-Mediated Virus-Induced Gene Silencing

Vinay Panwar and Guus Bakkeren

Abstract

Cereal rust fungi are destructive pathogens, threatening grain production worldwide. Targeted breeding for resistance utilizing host resistance genes has been effective. However, breakdown of resistance occurs frequently and continued efforts are needed to understand how these fungi overcome resistance and to expand the range of available resistance genes. Whole genome sequencing, transcriptomic and proteomic studies followed by genome-wide computational and comparative analyses have identified large repertoire of genes in rust fungi among which are candidates predicted to code for pathogenicity and virulence factors. Some of these genes represent defence triggering avirulence effectors. However, functions of most genes still needs to be assessed to understand the biology of these obligate biotrophic pathogens. Since genetic manipulations such as gene deletion and genetic transformation are not yet feasible in rust fungi, performing functional gene studies is challenging. Recently, Host-induced gene silencing (HIGS) has emerged as a useful tool to characterize gene function in rust fungi while infecting and growing in host plants. We utilized *Barley stripe mosaic virus*-mediated virus induced gene silencing (BSMV-VIGS) to induce HIGS of candidate rust fungal genes in the wheat host to determine their role in plant–fungal interactions. Here, we describe the methods for using BSMV-VIGS in wheat for functional genomics study in cereal rust fungi.

Key words Virus-induced gene silencing, VIGS, Host-induced gene silencing, HIGS, Functional genomics, Wheat rust fungi, Barley stripe mosaic virus, *Puccinia* gene silencing

1 Introduction

Wheat production is severely affected by rust fungi, belonging to the genus *Puccinia*, despite continued efforts in understanding rust fungus epidemiology, in breeding for resistance, and chemical control [1]. Recent advances in structural genomics of *Puccinia* species have shed some light on various aspects of their complex lifestyle with their cereal hosts. Genome sequence data for the three *Puccinia* species that attack wheat, namely leaf or brown rust (*P. triticina*), stem or black rust (*P. graminis*) and stripe or yellow rust (*P. striiformis*), are now available [2–4]. The generation of these genomic resources, and their computational and comparative analyses

Sambasivam Periyannan (ed.), *Wheat Rust Diseases: Methods and Protocols*, Methods in Molecular Biology, vol. 1659, DOI 10.1007/978-1-4939-7249-4_10, © Springer Science+Business Media LLC 2017

have laid the groundwork allowing the prediction of a wide array of genes [2, 5–9]. However, the strict obligate biotrophic nature of rust fungi and their recalcitrance to genetic transformation precludes the application of most commonly available genetic methods to study the biological function of these genes. Recently, an RNA interference (RNAi)-based concept called host-induced gene silencing (HIGS) has emerged as an effective tool to characterize gene function in biotrophic fungi [10–13]. The concept behind this method is the downregulation of the target gene transcript in the colonizing fungus by the uptake of siRNAs/dsRNA produced by the host plant expressing hairpin RNA (hpRNA) specific to the targeted fungal gene sequence [10]. The silencing of genes that are vital for the pathogen can ultimately have a major effect on phenotypic outcomes, such as altered growth morphology or disease suppression. We have demonstrated that HIGS induced by *Barley stripe mosaic virus*-mediated virus-induced gene silencing (BSMV-VIGS) is a robust approach for high-throughput functional genomics analysis of candidate genes in rust fungi [11].

The VIGS system is a powerful forward and reverse genetics tool for creating transient gene knockdown phenotypes from which gene function can be inferred and is particularly useful for species which are difficult to transform genetically [14, 15]. The mechanism of VIGS is based on the fact that plants defend themselves against invading viruses which act as a trigger to induce RNA-mediated gene silencing [16]. By inserting a fragment of a gene of interest into the viral RNA genome, transcripts of this gene fragment are also targeted for degradation during the defence response of plant, resulting in the downregulation of the corresponding gene by sequence-specific posttranscriptional gene silencing [17]. This leads to a reduction or in some cases the complete abolition of target gene function, which in turn can result in phenotypic changes. Compared with other reverse genetics approaches for associating genes with traits, VIGS provides a quick functional assessment or validation of candidate genes. VIGS is well established for studying plant–pathogen interactions in dicotyledonous plants, but the development of new viral vectors based on BSMV has expanded its utility to monocotyledonous plants such as wheat [18, 19]. BSMV is a single-stranded RNA virus of the genus *Hordeivirus* that infects many monocot species important to agriculture [20]. It has a tripartite positive sense genome, consisting of three RNAs termed α, β, and γ which are required for infection [21]. The BSMV-VIGS system has been successfully implemented for functional characterization of genes required for disease resistance in wheat and barley [18, 19, 22, 23]. Recently, we utilized BSMV as a vector to induce RNAi in wheat leaves for silencing wheat leaf rust fungus *P. triticina* genes involved in pathogenesis [11]. In this chapter we provide the protocol for performing HIGS in rust fungi using BSMV-VIGS.

2 Materials

2.1 Construction of Recombinant BSMV γRNA Vector

1. BSMV γ vector DNA.
2. Restriction enzymes: PacI, NotI.
3. *Escherichia coli* (*E. coli*) DH5α transformation competent cells.
4. Ampicillin.
5. Agarose.
6. Plasmid DNA extraction kit.
7. Luria-Bertani (LB) media (liquid and agar plates): To prepare 1 l LB: Add 10 g bacto tryptone, 5 g yeast extract, and 10 g NaCl to 800 ml of distilled H_2O. Dissolve and adjust pH to 7.0 with NaOH. Adjust volume to 1 l and sterilize by autoclaving. This can be stored at room temperature. For solid media, add 15 g of Bacto agar per liter and autoclave.
8. 1 kb DNA ladder.
9. 10× Tris–borate–EDTA (TBE) gel electrophoresis buffer: To prepare 1 l 10× TBE: Dissolve 121.1 g Tris base, 61.8 g boric acid, and 7.2 g EDTA in 800 ml of RNAse-free H_2O. Make up to 1 l and autoclave. This can be stored for 6 months at room temperature. Dilute with sterile distilled H_2O (dH_2O) to make 1× working solution.
10. Gel DNA extraction Kit.
11. TE buffer (10 mM Tris–HCl, 1 mM EDTA).
12. T4 DNA Ligase.

2.2 Preparation of BSMV In Vitro Transcription Reactions

1. BSMV α, β, and γ plasmids.
2. Restriction enzymes: MluI, SpeI, and BssHII.
3. mMESSAGE mMACHINE® High Yield Capped RNA Transcription Kit.
4. RNase inhibitor.

2.3 Plant Inoculation with Viral Transcripts

1. Seeds of wheat (*Triticum aestivum*).
2. Square Dura Pots (3.5″) and germination trays.
3. Standard germination soil (substrate no. 1) and potting soil (no. 3) for plant growth.
4. 10× Glycine Phosphate (GP) buffer: Dissolve 18.77 g Glycine and 23.13 g of K_2HPO_4 (dipotassium phosphate) in 500 ml dH_2O and autoclave.
5. FES inoculation buffer. To prepare 250 ml FES: Dissolve 2.5 g sodium pyrophosphate, 2.5 g Bentonite , 2.5 g Celite in 50 ml of 10× GP buffer. Bring volume to 250 ml with ddH_2O and autoclave.
6. BSMV α, β, and γ in vitro RNA transcripts.

3 Methods

3.1 Preparation of BSMV Plasmids and Construction of Recombinant γ RNA Vector

1. Streak *E. coli* glycerol stocks carrying BSMV plasmids α, β, and γ on LB agar plates supplemented with ampicillin (100 mg/l) and culture overnight at 37 °C.
2. Isolate a single colony from each α, β, and γ plasmid plate and inoculate a 20 ml overnight LB culture containing Ampicillin (100 mg/l) at 37 °C with constant shaking (200–250 rpm).
3. Carry out plasmid extraction using a plasmid miniprep kit as per product instructions (*see* **Note 1**). Check the quality of each plasmid by running 1 μl of the eluted product on a 1% w/v agarose–TAE gel. Determine the concentration of each plasmid using a spectrophotometer (e.g., NanoDrop).
4. Select a candidate fungal gene (*see* **Note 2**) and PCR-amplify a segment of the gene-of-interest (GOI) using gene-specific forward and reverse primers harboring an NotI and a PacI restriction site, respectively (*see* **Note 3**).
5. Digest 5 μg of PCR-amplified product of the GOI with NotI enzyme. After digestion, run a sample on a 1.5% (w/v) agarose–TAE gel (*see* **Note 4**) along with a DNA ladder of appropriate size markers to confirm the expected size. Excise the desired DNA fragment from the gel and purify the DNA using the gel DNA extraction kit (*see* **Note 5**); elute the NotI digested PCR segment using TE buffer. Measure the concentration using a spectrophotometer and check the integrity and purity of the eluted fragment by running (1–2 μl) on a 1% agarose–TAE gel. Now, digest this fragment (1–5 μg) with the second (PacI) enzyme. Perform gel electrophoresis to analyze the result of this restriction digest reaction. Elute the DNA from the desired, excised gel band using the gel DNA extraction kit. Measure the concentration of the eluted DNA using a spectrophotometer and check the integrity by running (1–2 μl) on 1% agarose–TAE gel.
6. Similarly, treat the BSMV γ vector with PacI and NotI restriction enzyme to create compatible ends at the multiple cloning sites for cloning of the PacI- and NotI-digested fragment of the GOI.
7. Set up the ligation reaction using a molar vector to insert ratio of 1:3 (*see* **Note 6**). Ligate the restriction enzyme digested BSMV γ vector and segment of the GOI with 1 μl T4 DNA ligase and 2 μl 10× T4 DNA Ligase buffer overnight at 16 °C in a total volume of 20 μl.

8. Transform the ligated mixture into transformation competent *E. coli* DH5α as per the manufacturer instructions. Incubate at 37 °C for 1 h by shaking at 200 rpm. Spread the transformation mixture (10–100 μl) onto a LB agar plate supplemented with ampicillin (100 mg/l), and incubate at 37 °C overnight.
9. Next day, pick 10–15 colonies (*see* **Note 7**) and start 5 ml LB + ampicillin (100 mg/l) cultures for 16 h at 37 °C. Use 4 ml of bacterial culture to extract plasmid DNA and store the remaining at 4 °C under sterile conditions.
10. Set up a diagnostic restriction enzyme digest with Notl and PacI to determine which plasmid contains the desired fragment of the GOI in the correct orientation (*see* **Note 8**). After identifying the correct construct, make a master plate using the remaining stored culture and carry out a large-scale preparation of plasmid DNA (*see* **Note 9**). Freeze your clone by adding 15% glycerol and storing at −80 °C.

3.2 Germinate Wheat Seeds

Germinate wheat seeds in 3.5″ Square Dura pots containing standard germination soil at 25 °C with 16 h light and 8 h dark period with 74 μmol/m^{2}s light intensity and 55–65% relative humidity. Label each pot with transcripts to be inoculated.

3.3 Preparation of In Vitro Transcripts

1. The α, β, and γ plasmid vectors are linearized by restriction enzyme digestion with MluI, SpeI, and BssHII, respectively, and used as templates for in vitro transcription using the mMessage and mMachine transcription kits, following the manufacturer instructions.
2. Run 1 μl of each of the digested plasmids on a 1% agarose–TAE gel to confirm that linearization is complete (*see* **Note 10**; Fig. 1). Having confirmed complete digestion, inactivate the reaction by heating at 65 °C for 20–30 min (*see* **Note 11**).
3. Treat each linearized plasmid reaction with RNAse inhibitor to prepare for in vitro transcription. Use 40 units RNAse inhibitor per 20 μl linearized plasmid reaction.
4. Set up the in vitro transcription reaction using the mMESSAGE mMACHINE® High Yield Capped RNA Transcription Kit following the manufacturer protocol and incubate at 37 °C for 2 h (*see* **Note 12**). Determine completion of transcription by running 1 μl of each reaction with 9 μl of RNase free H_2O and 10 μl of loading dye provided in the mMessage and mMachine transcription kit (*see* **Note 13**). A successful in vitro transcription reaction should yield intact bands and any smearing indicates degradation of the RNA transcripts (*see* **Note 14**).

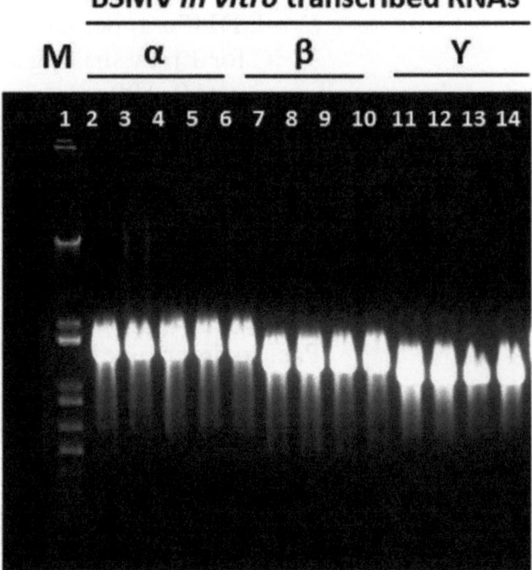

Fig. 1 Gel image of in vitro-synthesized RNA transcripts from linearized BSMV plasmid templates. *Lane 1*, 1 kb DNA ladder: *Lanes 2–6*, BSMVα in vitro-transcribed (IVT) RNAs; *Lanes 9–10*, BSMV β IVT RNAs; *Lanes 11–14*, BSMVγ IVT RNAs. 1 μl of IVT full-length RNA product loaded in each lane. Plasmids α, β, and γ linearized with restriction enzymes MluI, SpeI, and BssHII, respectively. Transcription reaction run on 1% agarose–TAE gel

3.4 Plant Inoculation with Viral Transcripts

1. For BSMV inoculation, combine the three transcripts (α, β, and γ) in equimolar ratio (1:1:1) using 1 μl of each in vitro-transcribed RNA in 22.5 μl of inoculation buffer (FES).
2. Apply the freshly prepared inoculum (transcript mix in FES buffer) on the first leaf of 10 day-old wheat seedlings using a pipette. Gently hold the base of the leaf with one hand and, while firmly holding it between the thumb and index finger of your other hand, rub the surface of the leaf with the inoculation mixture from the base to the tip in a single motion. Repeat the process one or two times as required (*see* **Note 15**).
3. Keep inoculated plants in the growth chamber at 25 °C with 16 h light–8 h dark cycle.

3.5 Symptom Observations and Fungal Inoculations

1. In wheat, BSMV symptoms can be seen as yellow mottling or small streaks on the leaves at 7–8 days post inoculation (dpi) (*see* **Note 16**; Fig. 2).
2. Once BSMV symptoms are observed, plants are challenged with rust urediniospores at 10 dpi and observed for disease or growth phenotype (*see* **Note 17**). Rust inoculations should be done away from the control plants. Thoroughly spray urediniospores suspended in Soltrol 170 on to the leaf surface using an airbrush.

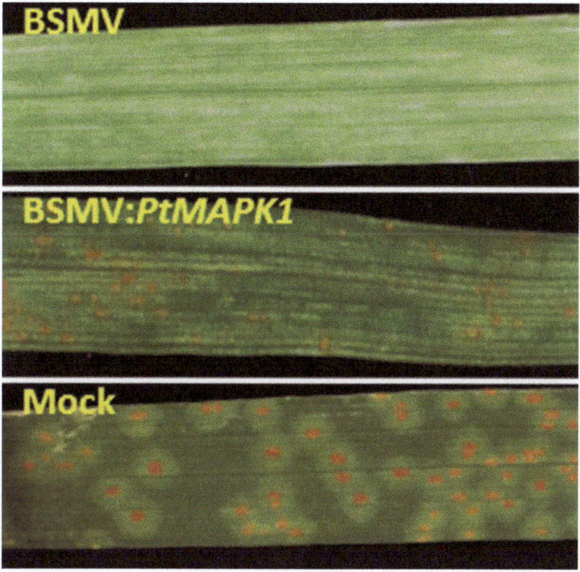

Fig. 2 Wheat plant inoculated with BSMV vector alone and derivative of recombinant γ vector carrying *P. triticina PtMAPK1* gene segment. Wheat plants inoculated with empty vectors showing typical BSMV symptoms of white mottling, spotting, and streaking in leaves (*upper panel*). Plants inoculated with BSMV vectors harboring *P. triticina* gene show disease suppression (*middle panel*) whereas FES treated (mock) controls are heavily infected with fungus (*lower panel*). Photographs were taken 10 days after fungal inoculation

3. Incubate plants in a dew chamber with near 100% relative humidity overnight. Next day, remove plants from the dew chamber and return to the growth chamber. For wheat leaf rust (*P. triticina*), inoculated plants will display discolored infected spots strating from 4 to 5 days post urediniospore inoculation, depending on the pathogen isolate used. For molecular analyses, rust fungus-challenged wheat tissues can be harvested at different time points as desired by experiments.

4 Notes

1. Avoid using RNase A during BSMV plasmid preparation as it may interfere with in vitro transcription. Any residual RNase will degrade the in vitro-transcribed RNAs produced from these plasmids.
2. HIGS relies on careful selection of fungal gene sequence as to avoid off-target RNAi effects. Different gene fragments can show variability in VIGS experiments. Fragments of 300–1500 bp are maintained in the γ-genome and have been

used successfully to induce VIGS. Silencing efficiency is reduced by shorter fragments, whereas longer inserted fragments run a strong risk of being lost from the recombinant virus.

3. It is recommended to use cDNA as template for the PCR amplification of the candidate fungal gene as it has no noncoding sequences. Alternatively, the fragment with the flanking restriction enzyme sites is synthesized.
4. Ethidium bromide used to stain nucleic acids in agarose gel is a carcinogen and should be handled carefully. Always wear disposable gloves when working with ethidium bromide. Anything coming in contact with it must be handled as a hazardous waste and disposed of accordingly.
5. PCR-amplified product can also be purified using any standard PCR Clean-UP System following the manufacturer instructions.
6. It is generally recommended to use 100 ng of total DNA in a standard ligation reaction. When setting up the ligation reaction, make sure to include a positive control (vector without insert) and a negative control (vector DNA without T4 DNA ligase) in parallel. This will provide information on how much background level of uncut or self-ligating recipient plasmid backbone is present. The plate with the ligated mixture should contain more colonies as compared with the control plates.
7. Pick colonies depending on the number of background colonies on the control plate. The higher the background, the more colonies need to be checked.
8. It is highly recommended that the selected positive clones are sequenced to confirm the presence of the correct insert.
9. All reagents and materials used should be nuclease-free to avoid degradation of RNAs during in vitro-transcript preparation.
10. Partial linearization will result in production of less viral RNA by T7 RNA polymerase. It is therefore important that complete restriction digestion is obtained for optimum results.
11. Alternatively, restriction enzyme digestion can be followed by DNA purification since any contamination in the digestion reaction may inhibit subsequent transcription. If using phenol–chloroform extraction, add one volume of phenol–chloroform–isoamyl alcohol (25:24:1) to the digested sample and vortex thoroughly by hand for 30 s. Centrifuge at room temperature for 5 min at 16,000 × *g* and carefully transfer the upper aqueous phase to a new tube. Precipitate the linearized plasmid by adding one tenth of a volume of ammonium acetate (5 M concentration) and two volumes of absolute ethanol and storing at −80 °C for 1 h. Collect the pellet by centrifuging at

16,000 × *g* for 15 min at 4 °C. Remove the supernatant and wash the DNA pellet with 70% ethanol, air-dry, and resuspend in TE to achieve a concentration of approximately 1 μg/μl.

12. Amplification of the transcribed RNA may require removal of any DNA by addition of 1 μl of DNase (supplied in the kit) and further incubation at 37 °C for 15 min.
13. The in vitro-transcribed product can be checked by running on standard 1% agarose–TAE gel. However, as you are dealing with RNA, make sure that the gel running buffer and electrophoresis unit is free of RNase contamination. Use gloves and filter-pipette tips when working with RNA.
14. Any smearing of the bands indicates degradation of the RNA transcripts. RNAs can be stored short term at −20 °C and for longer at −80 °C.
15. Do not damage the leaf by squeezing it too hard or applying too much force. Label each plant after inoculation to separate it from non-inoculated plants. When using two or more different constructs, make sure to change gloves after each application to prevent cross-contamination.
16. For BSMV-based VIGS in wheat, the apparent virus phenotype can usually be observed 10 days post-infection The timing of onset and region of initial silencing can vary between different genetic backgrounds. Since VIGS induces viral symptoms, an empty virus vector-infected plant has to be included as a negative control in each experiment.
17. If the target fungal gene is essential for the fungus, then HIGS will result in altered fungal growth in the host plant and/or a changed disease phenotype (*see* Fig. 2). The silencing phenotype obtained in experiments might indicate a possible function of the target gene. However, repeats to obtain reproducible phenotypes are often desirable. Silencing efficiency can differ from plant to plant even if all conditions are adjusted and standardized.

References

1. McCallum BD, Hiebert CW, Cloutier S et al (2016) A review of wheat leaf rust research and the development of resistant cultivars in Canada. Can J Plant Pathol 38:1–18
2. Cantu D, Govindarajulu M, Kozik A et al (2011) Next generation sequencing provides rapid access to the genome of *Puccinia striiformis* f. sp. *tritici*, the causal agent of wheat stripe rust. PLoS One 6:e24230
3. Duplessis S, Como C, Lin YC et al (2011) Obligate biotrophy features unraveled by the genomic analysis of rust fungi. Proc Nat Acad Sci U S A 108:9166–9171
4. Zheng W, Huang L, Huang J et al (2013) High genome heterozygosity and endemic genetic recombination in the wheat stripe rust fungus. Nat Commun 4:2673
5. Xu J, Linning J, Feller M et al (2011) Gene discovery in EST sequences from the wheat leaf rust fungus *Puccinia triticina* sexual spores, asexual spores and haustoria, compared to other rust and corn smut fungi. BMC Genomics 12:161

6. Fellers JP, Soltani BM, Bruce M et al (2013) Conserved loci of leaf and stem rust fungi of wheat share synteny interrupted by lineage-specific influx of repeat elements. BMC Genomics 14:60
7. Cuomo C, Bakkeren G, Khalil HB et al (2017) Comparative analysis highlights variable genome content of wheat rusts and divergence of the mating loci. Genes Genomes Genet 7:361–376
8. Cantu D, Segovia V, MacLean D et al (2013) Genome analyses of the wheat yellow (stripe) rust pathogen *Puccinia striiformis* f. sp. *tritici* reveal polymorphic and haustorial expressed secreted proteins as candidate effectors. BMC Genomics 14:270
9. Upadhyaya NM, Garnica DP, Karaoglu H et al (2014) Comparative genomics of Australian isolates of the wheat stem rust pathogen *Puccinia graminis* f. sp. *tritici* reveals extensive polymorphism in candidate effector genes. Front Plant Sci 5:759
10. Panwar V, McCallum B, Bakkeren G (2013) Endogenous silencing of *Puccinia triticina* pathogenicity genes through in planta-expressed sequences leads to the suppression of rust diseases on wheat. Plant J 73:521–532
11. Panwar V, McCallum B, Bakkeren G (2013) Host-induced gene silencing of wheat leaf rust fungus *Puccinia triticina* pathogenicity genes mediated by the Barley stripe mosaic virus. Plant Mol Biol 81:595–608
12. Panwar V, McCallum B, Bakkeren G (2015) A functional genomics method for assaying gene function in phytopathogenic fungi through host-induced gene silencing mediated by Agroinfiltration. In: Mysore KS, Senthil-Kumar M (eds) Plant gene silencing, vol 1287. Springer, New York, NY, pp 179–189
13. Koch A, Kumar N, Weber L et al (2013) Host-induced gene silencing of cytochrome P450 lanosterol C14α-demethylase-encoding genes confers strong resistance to *Fusarium* species. Proc Natl Acad Sci U S A 110:19324–19329
14. Dalmay T (2007) Virus-induced gene silencing. In: Meyer P (ed) Annual plant reviews. Plant epigenetics, vol 19. Blackwell Publishing Ltd, Oxford, pp 223–243
15. Baulcombe DC (2015) VIGS, HIGS and FIGS: small RNA silencing in the interactions of viruses or filamentous organisms with their plant hosts. Curr Opin Plant Biol 26:141–146
16. Voinnet O (2001) RNA silencing as a plant immune system against viruses. Trends Genet 17(8):449–459
17. Baulcombe DC (1999) Fast forward genetics based on virus-induced gene silencing. Current Opin Plant Biol 2(2):109–113
18. Holzberg S, Brosio P, Gross C et al (2002) Barley stripe mosaic virus-induced gene silencing in a monocot plant. Plant J 30:315–327
19. Lee WS, Hammond-Kosack KE, Kanyuka K (2012) Barley stripe mosaic virus-mediated tools for investigating gene function in cereal plants and their pathogens: virus-induced gene silencing, host-mediated gene silencing, and virus-mediated overexpression of heterologous protein. Plant Physiol 160:582–590
20. Jackson AO, Lim HS, Bragg J et al (2009) Hordeivirus replication, movement, and pathogenesis. Annu Rev Phytopathol 47:385–422
21. Smith O, Clapham A, Rose P et al (2014) A complete ancient RNA genome: identification, reconstruction and evolutionary history of archaeological barley stripe mosaic virus. Sci Rep 4:4003
22. Hein I, Pacak-Braciszewska M, Hrubikova K et al (2005) Virus-induced gene silencing-based functional characterization of genes associated with powdery mildew resistance in barley. Plant Physiol 138:2155–2164
23. Scofield SR, Haung L, Brandt AS et al (2005) Development of a virus-induced gene-silencing system for hexaploid wheat and its use in functional analysis of the *Lr21*-mediated leaf rust resistance pathway. Plant Physiol 138:2165–2173

Chapter 11

Apoplastic Sugar Extraction and Quantification from Wheat Leaves Infected with Biotrophic Fungi

Veronica Roman-Reyna and John P. Rathjen

Abstract

Biotrophic fungi such as rusts modify the nutrient status of their hosts by extracting sugars. Hemibiotrophic and biotrophic fungi obtain nutrients from the cytoplasm of host cells and/or the apoplastic spaces. Uptake of nutrients from the cytoplasm is via intracellular hyphae or more complex structures such as haustoria. Apoplastic nutrients are taken up by intercellular hyphae. Overall the infection creates a sink causing remobilization of nutrients from local and distal tissues. The main mobile sugar in plants is sucrose which is absorbed via plant or fungal transporters once unloaded into the cytoplasm or the apoplast. Infection by fungal pathogens alters the apoplastic sugar contents and stimulates the influx of nutrients towards the site of infection as the host tissue transitions to sink. Quantification of solutes in the apoplast can help to understand the allocation of nutrients during infection. However, separation of apoplastic fluids from whole tissue is not straightforward and leakage from damaged cells can alter the results of the extraction. Here, we describe how variation in cytoplasmic contamination and infiltrated leaf volumes must be controlled when extracting apoplastic fluids from healthy and rust-infected wheat leaves. We show the importance of correcting the data for these parameters to measure sugar concentrations accurately.

Key words Rust fungi, *Puccinia striiformis* f. sp. *tritici*, Apoplastic fluids, Hexoses, Cytoplasmic contamination

1 Introduction

The obligate biotrophic fungus *Puccinia striiformis* f. sp. *tritici* (*Pst*) is present on all continents where wheat is grown and causes widespread and severe stripe rust epidemics [1]. The fungus invades the wheat leaf to obtain energy from the living plant. The major sites of nutrient uptake are mesophyll cells and the leaf apoplast [2, 3]. The apoplast along with the conductive tissues is one of two major dissemination routes for water and nutrients. Changes in apoplastic nutrient concentrations, among other parameters, determine whether tissue behaves as source exporting nutrients to distal tissues, or as sink that imports nutrients. For example, high apoplastic hexoses to sucrose ratios trigger nutrient import, whereas low ratios or high levels of sucrose stimulate export to

Sambasivam Periyannan (ed.), *Wheat Rust Diseases: Methods and Protocols*, Methods in Molecular Biology, vol. 1659, DOI 10.1007/978-1-4939-7249-4_11, © Springer Science+Business Media LLC 2017

other tissues [4]. Consequently, changes in apoplastic nutrients provide information about their allocation dynamics during infection.

Apoplastic nutrients are commonly extracted by first infiltrating leaves with water or a buffered solution, before centrifuging the excised leaves to recover the apoplastic fluids. There are several variables that affect measurement of apoplastic sugars that must be carefully accounted for during experimentation. These include the temperature at which leaves are treated, the osmolarity of the infiltration solution, and careful leaf handling and solution infiltration procedures [5]. These variables can cause leakage from cells due to changes in osmolarity or wounding and by induction of plant responses to abiotic stress. Beside these technical aspects, the leaf cell wall thickness, the number of stomata and the extent of the fungal infection affect the infiltrated volume and as a consequence the final concentration of apoplastic solutes. Here, we present an improved extraction method for apoplastic solutes and show how quantified sugar values should be corrected for accurate interpretation of the data.

2 Materials

2.1 Extraction of Apoplastic Sugars

1. Infected leaves (*see* **Note 1**).
2. 70% ethanol (v/v).
3. Vacuum desiccator and Vacuubrand diaphragm vacuum pump.
4. Infiltration solution: 0.2 M sorbitol in sterile pure water (*see* **Note 2**).
5. Refrigerated centrifuge containing a swinging bucket rotor for 50 mL Falcon tubes.
6. 10 mL disposable syringes.
7. Sterile disposable 50 mL Falcon tubes and 1.5 mL polypropylene Eppendorf tubes.

2.2 Estimation of Cytoplasmic Contamination

1. Malate dehydrogenase buffer solution (MBS): 100 mM Tris–HCl, pH 7.5, 0.1 mM β-nicotinamide adenine dinucleotide, reduced disodium salt hydrate.
2. Malate dehydrogenase substrate: 0.17 mM oxaloacetic acid dissolved in sterile water. Can be stored at −20 °C.
3. 96-well plastic plates, clear, flat-bottom wells.
4. Plate reader for measuring absorbance at 340 nm.

2.3 Quantification of Hexoses and Sucrose

1. Digestion of sugars: D-hexose-6-phosphotransferase, glucose-6-phosphate dehydrogenase, phosphoglucose isomerase, β-D-fructofuranosidase.

2. Assay buffer: 100 mM HEPES, pH 7.5, 5 mM $MgCl_2$, 1 mM DTT, 0.02% (w/v) BSA, 8 mM NAD^+, 4 mM ATP. These components are prepared for a final volume of 200 μL but are made up initially to a volume of 180 μL. Prepare fresh.
3. Glucose, fructose, and sucrose standards: Prepare standards fresh from stock solutions as sugars oxidize at low concentrations. Mix fructose, glucose, and sucrose solutions together for final concentrations of 0.05 mM, 0.1 mM, 0.2 mM, 0.4 mM, 0.6 mM, and 1 mM.

3 Methods

3.1 Extraction of Apoplastic Fluids

1. Wipe leaves with 70% ethanol to remove contaminating microbes or fungal spores.
2. Use at least seven leaves for each treatment evaluated. Harvest leaves 1 h before the commencement of the night cycle to ensure maximal accumulation of sugars in the leaves. Extract the apoplastic fluids immediately. Always collect tissue at the same time relative to the night cycle.
3. Peel the wheat leaf from the blade. Do not wound the leaf by cutting it.
4. Weigh leaves prior to infiltration.
5. Submerge leaves in infiltration solution within a vacuum desiccator of 14 mm diameter at room temperature (*see* **Note 3**). Leaves normally float; to ensure that they are fully immersed in the infiltration solution, cover the leaves with a light object such as a plastic petri dish. Apply a vacuum for 30 s and then release the pressure slowly (*see* **Notes 4** and **5**). Repeat this two to three times until the leaves become saturated (they will appear darker in color). Wheat flag leaves often require a further cycle of infiltration.
6. Blot the leaves quickly on absorbent paper to dry them and reweigh (*see* **Note 6**).
7. Carefully bend the leaves so as not to damage the tissue and place them within a 10 mL disposable syringe. Place the syringe into a 50 mL polypropylene Falcon tube containing a 1.5 mL Eppendorf tube such that the tip of the syringe sits in the 1.5 mL tube (Fig. 1).
8. Apoplastic fluids are then extracted from the leaves by centrifugation. To reduce cytoplasmic contamination, flag leaves are centrifuged at 400 × *g* for 5 min at 4 °C, and other leaves at 200 × *g* for 15 min (*see* **Note 7**). The liquid collected in the 1.5 mL tube is the apoplastic fluid. You should expect about 10–15 μL from 170–200 mg of leaf tissue.

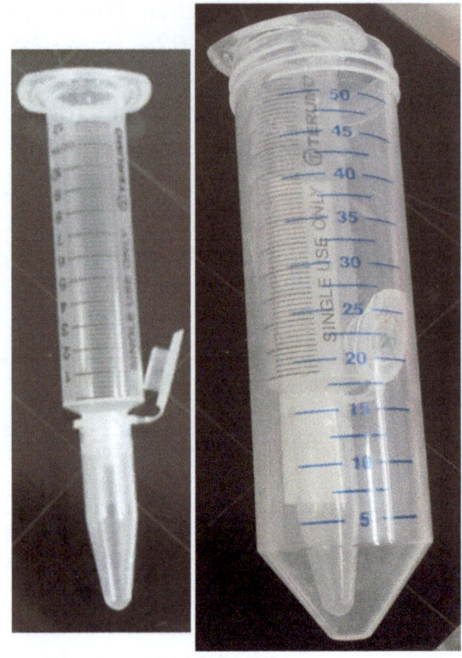

Fig. 1 Apparatus for extracting apoplastic fluids from infiltrated leaves

9. Aliquot apoplastic fluid for assays: 2 μL to assay cytoplasmic contamination, and 5 μL for quantification of sugars. Snap freeze the fluids and keep them at −20 °C until use.

3.2 Quantification of Cytoplasmic Contamination

The activity of the cytoplasmic enzyme malate dehydrogenase (MDH, 1.1.1.37) can be quantified as an indicator of contamination of apoplastic fluids by cytoplasmic contents. The method was modified from Dani et al. [6]. It measures the decrease in absorbance due to NADH oxidation by MDH as it reduces OAA.

1. Grind 100 mg of leaf (healthy or infected) in 1 mL Infiltration solution at room temperature in a mortar and pestle. This is considered as the 100% contamination reference. Sample reference leaves at each experimental time point.
2. Mix 2 μL of each apoplast extract with 196 μL of MBS in a single well within a 96-well plate. Measure the absorbance at 340 nm every min for 5 min using the plate reader. Record the final most stable value as A_1.
3. Add 2 μL of OAA to each well and mix. Measure absorbance again for 4 min. If absorbance values decrease, record the final stable value as A_2. If there is no change in absorbance, consider the sample as not contaminated.
4. Use the following equation to calculate the percentage cytoplasmic contamination (*see* **Note 8**; Fig. 2):

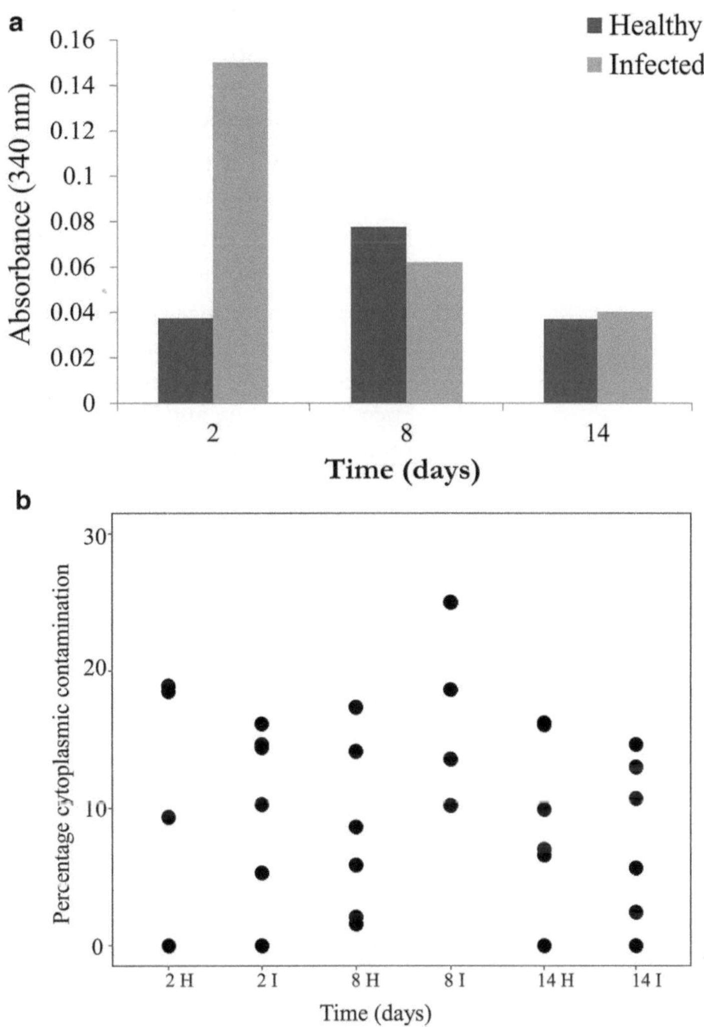

Fig. 2 Variation of cytoplasmic contamination levels in wheat leaves. (**a**) MDH activity (NADH absorbance) in reference samples. (**b**) Percentage cytoplasmic contamination based on MDH activity in samples as a proportion of the references. The X-axis represents healthy (H) and infected (I) leaves 2, 8, and 14 days after infection or mock infection, $n = 6$

$$\text{Contamination percentage} = \frac{A1_{\text{sample}} - A2_{\text{sample}}}{A1_{\text{reference}} - A2_{\text{reference}}}$$

3.3 Quantification of Hexoses and Sucrose

The method is a modification of Scholes et al. [7].

1. Mix 5 μL of standard or sample with 185 μL of freshly prepared assay buffer in 96-well plates.
2. Measure absorbance at 340 nm for 5 min. Record the last stable value as A_0.

3. Add 5 μL of HKX + G6PDH solution (9 units of HXK and 3.2 units of G6PDH in water, prepare fresh) to each well, mix and measure absorbance for 10–16 min. Record the last stable value as A_G (*see* **Note 9**).
4. Repeat **step 3** with 5 μL PGI fresh solution (2 units of PGI in water). Record the last stable value as A_F.
5. Repeat **step 3** with 5 μL INV fresh solution (85 units of INV in water). Record the last stable value as A_S.
6. The glucose concentration is derived from the difference between A_G and A_0, fructose concentration from the difference between A_F and A_G, and sucrose concentration from the difference between A_S and A_F.
7. Plot the data against a linear regression based on the absorbance values of the standards to calculate each molarity.

3.4 Cytoplasmic Contamination Correction for Apoplast Sugar Concentrations

Sugar values should be corrected based on all sources of variation, including the percentage of cytoplasmic contamination and the infiltrated volume (leaf weight difference before and after infiltration).

1. Use the following equation to calculate the sugar concentration:

$$\text{Apoplastic sugar} = \frac{(\text{Sugar concentration} - (\text{Sugar concentration} \times \text{Cytoplasmic contamination}))}{\text{Infiltrated volume}}$$

2. Measurement of sugar concentrations will be incorrect if corrections are not taken into account (Fig. 3).

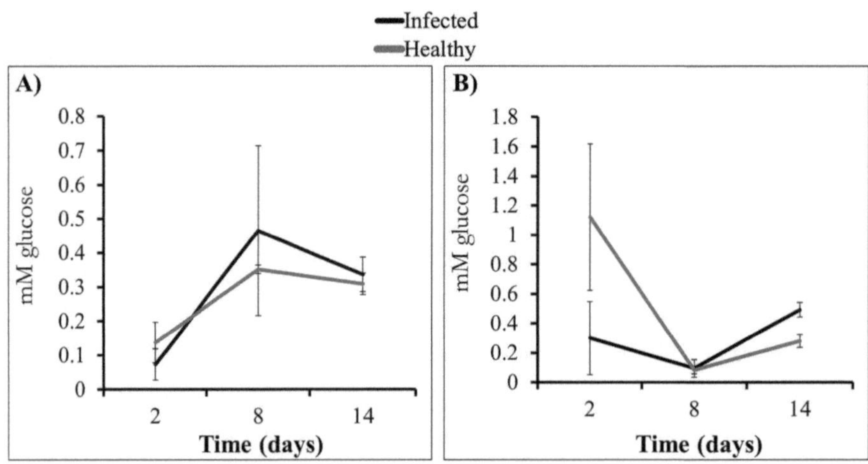

Fig. 3 Apoplastic glucose quantification in *Pst*-infected and healthy wheat leaves without (**a**) and with (**b**) corrections. Error bars represent standard error, $n = 5$

4 Notes

1. To ensure homogeneous infection across the leaf, use a paintbrush to spread the spores which are first mixed 1:5 with talc powder. Here, talc alone was used for mock inoculation. Cover the infected leaf with aluminum foil and keep in the dark at 8 °C for 1 day. Transfer the plants to a growth chamber at 21 °C, 150 μE of light, 65% humidity, 400 ppm CO_2, with a 16 h light–8 h dark cycle. Remove the foil and let them acclimate for at least 24 h before harvest. Wheat flag leaves are technically more challenging than older leaves to extract apoplastic fluids from, because the flag leaf epidermis is thicker and a smaller volume of extraction buffer penetrates the leaves. For these leaves use 7–10 biological replicates per treatment to increase the extracted volumes.
2. Do not use water as infiltration solution because the consequent changes in osmotic potential can cause the release of cytoplasmic sugars [8].
3. In respond to cold temperatures, plants accumulate more hexoses as osmolytes to protect their membranes [9].
4. To reduce the damage caused by bubbles that form under negative pressure, be sure to shake the desiccator regularly.
5. A syringe vacuum method was also tested for the extraction of apoplastic fluids. Leaves were placed inside 50 mL syringes with 20 mL of infiltration solution. Negative pressure was applied (by moving the plunger outwards) to infiltrate the solution. However, this method is more time consuming and gives higher cytoplasmic contamination levels than the vacuum pump method (Fig. 4).
6. The amount of infiltrated solution varies among different leaves under different conditions, as shown in Fig. 5. Intercellular hyphae occupy more of the apoplastic space as they multiply during infection. Therefore, ensure that leaves are weighed before and after infiltration to estimate the apoplastic volume of each leaf. Another approach is to normalize volumes with a dilution factor calculated by changes in indigo carmine absorbance [8]. Indigo carmine dye solution does not cross membranes therefore can be used to calculate the apoplastic volume. The dye is infiltrated into the leaves and after centrifugation; the recovered dye will be diluted due to mixing with apoplastic fluids. The extent of dilution can be converted to a factor and used to modify the sugar concentration values. This method was tested for *Pst* infected leaves at three time points: 2, 8, and 14 dpi. However, at 14 dpi, the absorbance increased and the

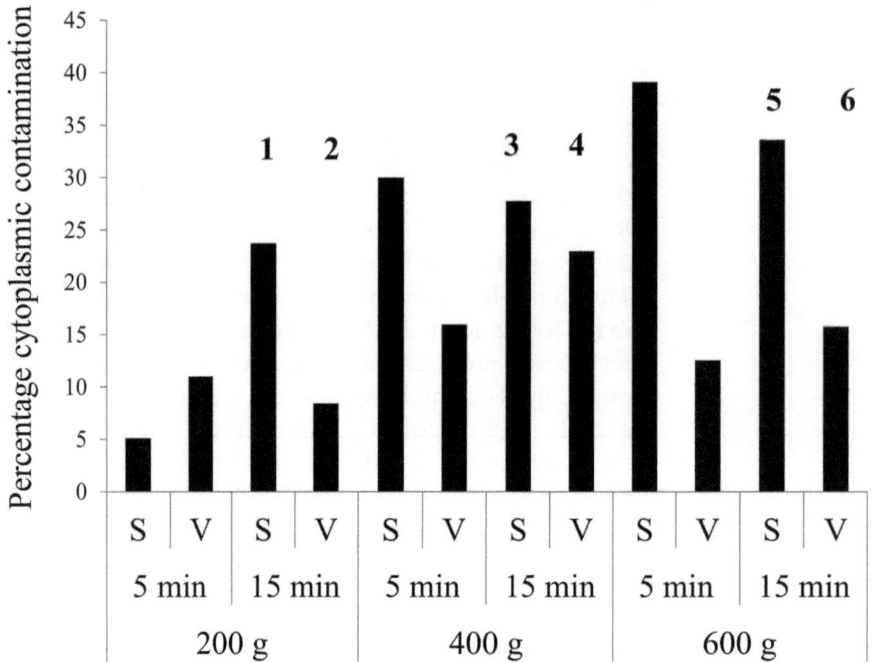

Fig. 4 The influence of extraction conditions on cytoplasmic leakage in wheat leaves. Leaves were infiltrated using vacuum (V) or syringe (S) infiltration methods described in Subheading 3 and in **Note 5**, respectively. Apoplastic fluids were recovered using different centrifugation speeds and times as indicated. The extracts labeled 1–6 were evaluated for protein content in Fig. 6

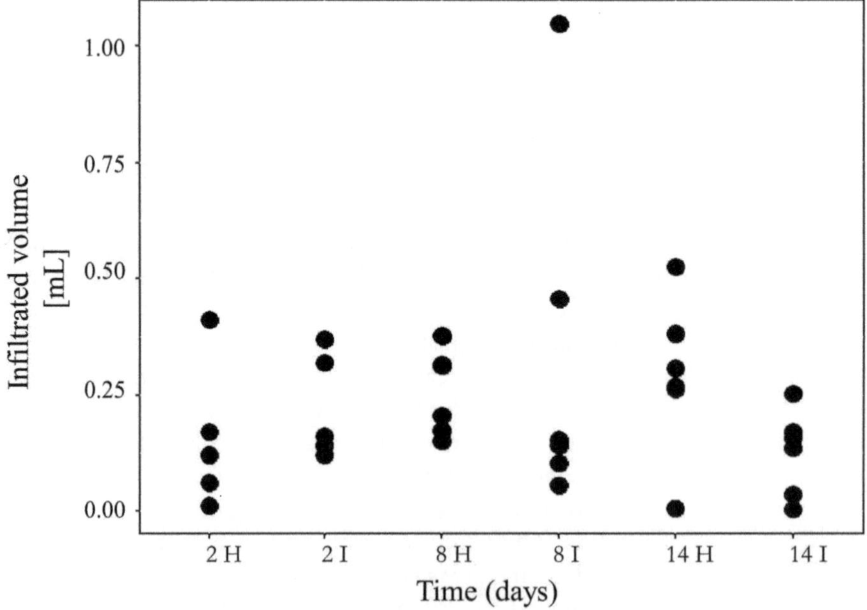

Fig. 5 Leaf volume differences before and after infiltration in healthy (H) and infected (I) wheat leaves 2, 8, and 14 days after infection with *Pst* or with mock (healthy leaves). The Y-axis represents the change in leaf volume after infiltration, where the infiltration solution has a density of 1 mg/mL

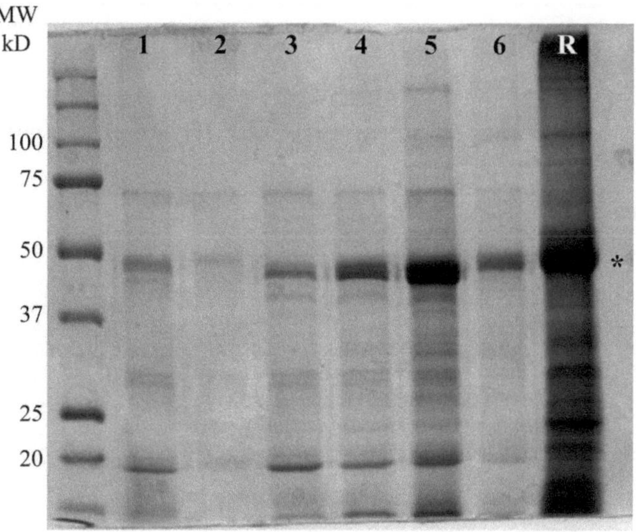

Fig. 6 Protein content of apoplastic fluids from extracts 1–6 in Fig. 4. Samples were run on a denaturing 12% SDS-PAGE gel. *MW* molecular weight standards, *R* Reference sample (whole tissue extract). *Star* represents RuBisCO

dye changed color. This suggests that the apoplast was alkaline or a plant compound is reacting with the dye changing he color; therefore, a dilution factor cannot be calculated late in the infection period.

7. Different centrifugation speeds (equivalent to 80 × *g*, 200 × *g*, 400 × *g*, and 600 × *g*) and times (5 and 15 min) were tested. 80 × *g* was discarded because less than 5 μL of fluid was collected which is insufficient for the sugar assays. For some of these treatments, the presence of chloroplastic RuBisCO was checked by SDS-PAGE gel electrophoresis (Fig. 6) as an indicator of contamination. However, the MDH test was used for routine assays because it is quantitative so can be used to normalize the data.
8. Calculate the percentage with respect to fully homogenized tissue (apoplastic + cytoplasmic sugars) at each time point. It is recommended not to use samples with contamination values higher than 25%. Centrifugation at low temperatures seems to reduce the sugar export to the apoplast [10].
9. In contrast to MDH assays, beware of absorbance increases in this experiment. Wait until the absorbance stabilizes which indicates that all of the sugar substrate has been consumed.

Acknowledgments

We thank Prof. Peter Solomon (The Australian National University) for a critical review of the manuscript.

References

1. Ali S, Gladieux P, Rahman H, Saqib MS, Fiaz M, Ahmad H, Leconte M, Gautier A, Justesen AF, Hovmøller MS, Enjalbert J, de Vallavieille-Pope C (2014) Inferring the contribution of sexual reproduction, migration and off-season survival to the temporal maintenance of microbial populations: a case study on the wheat fungal pathogen *Puccinia striiformis* f.sp. *tritici*. Mol Ecol 23:603–617
2. Voegele RT, Struck C, Hahn M, Mendgen K (2001) The role of haustoria in sugar supply during infection of broad bean by the rust fungus *Uromyces fabae*. Proc Natl Acad Sci U S A 98:8133–8138
3. Kneale J, Farrar JF (1985) The localization and frequency of Haustoria in colonies of Brown rust on barley leaves. New Phytol 101:495–505
4. Chiou T-J, Bush DR (1998) Sucrose is a signal molecule in assimilate partitioning. Proc Natl Acad Sci U S A 95:4784–4788
5. Minchin PEH, Thorpe MR, Farrar JF, Koroleva OA (2002) Source–sink coupling in young barley plants and control of phloem loading. J Exp Bot 53:1671–1676
6. Dani V, Simon WJ, Duranti M, Croy RRD (2005) Changes in the tobacco leaf apoplast proteome in response to salt stress. Proteomics 5:737–745
7. Scholes JD, Lee PJ, Horton P, Lewis DH (1994) Invertase: understanding changes in the photosynthetic and carbohydrate metabolism of barley leaves infected with powdery mildew. New Phytol 126:213–222
8. Solomon P, Oliver R (2001) The nitrogen content of the tomato leaf apoplast increases during infection by *Cladosporium fulvum*. Planta 213:241–249
9. Olien CR, Clark JL (1995) Freeze-induced changes in carbohydrates associated with hardiness of barley and rye. Crop Sci 35:496
10. Tetlow IJ, Farrar JF (1993) Apoplastic sugar concentration and pH in barley leaves infected with Brown rust. J Exp Bot 44:929–936

Part III

Genetics of Rust Resistance

Chapter 12

Genetic Analysis of Resistance to Wheat Rusts

Caixia Lan, Mandeep Singh Randhawa, Julio Huerta-Espino, and Ravi P. Singh

Abstract

Leaf rust, stripe rust, and stem rust pose a significant threat to global wheat production. Growing rust resistant cultivars is the most efficient and environment friendly method to reduce yield losses. Genetic analysis is undertaken to identify genes and study their roles in conferring rust resistance in a given wheat background. This chapter summarizes the protocol for genetic analysis of rust resistance at both seedling and adult plant stages. Additionally, it examines statistical analysis and related software to characterize quantitative trait loci (QTL) linked with rust resistance.

Key words Wheat, Leaf rust, Stem rust, Stripe Rust, Resistance, Identification, Mapping

1 Introduction

Rusts are the most important diseases of wheat which cause significant losses in the absence of proper chemical or genetic control measures. Leaf (brown) rust, stripe (yellow) rust, and stem (black) rust, which are caused by *Puccinia triticina* (*Pt*), *P. striiformis* f. sp. *tritici* (*Pst*), and *P. graminis* f. sp. *tritici* (*Pgt*), respectively, are the three important rust diseases of wheat. If any of these rusts reach an epidemic level, devastating yield losses can occur and wipe out as much as 100% of the crop in an individual field with susceptible varieties. Leaf rust has caused serious epidemics throughout wheat growing regions in Europe, North America, Oceania, Southern Africa, South America, and South Asia [1]. Recently, its incidence increased in China due to warmer temperatures favoring disease development and the widespread cultivation of susceptible varieties. Stripe rust is generally found in northern latitudes, highlands, and wheat-growing regions with cooler temperatures during early growth stages. However, recent large-scale epidemics are in warmer wheat-growing areas with the emergence of two closely related *Pst* strains with increased aggressiveness and tolerance to warm temperatures [2]. Highly virulent stem rust is also moving into new

Sambasivam Periyannan (ed.), *Wheat Rust Diseases: Methods and Protocols*, Methods in Molecular Biology, vol. 1659,
DOI 10.1007/978-1-4939-7249-4_12, © Springer Science+Business Media LLC 2017

areas with the emergency of a new *Pgt* race, Ug99 (designated as TTKSK using the North American differentials set), which was detected in Uganda in 1998 and has virulence to most of the widely deployed race-specific resistance genes. It was recognized as a significant threat to global food security [3]. There are many ways to manage these diseases; however, development and cultivation of resistant varieties are the most efficient control methods particularly for low-income smallholder farmers of developing countries.

Two types of rust resistance genes are often defined in wheat. Race-specific, or major, resistance genes which usually confer protection throughout the growth cycle are also referred to as "all-stage resistance" [4]. These resistance genes cause hypersensitive reactions in the host when infected with rust isolates carrying corresponding avirulence alleles [5]. In contrast, race nonspecific or minor genes confer adult plant resistance (APR) and are normally present together with other similar effect genes and are therefore associated with quantitative inheritance [6, 7]. Most cultivars with multiple genes for APR are susceptible at the seedling stage but later display resistance to a number of races, as its name (APR) indicates [8].

2 Materials

2.1 Phenotyping at Seedling Stage

1. Greenhouse: Light and temperature controlled according to the rust type.
2. Trays: Plastic trays filled with soil (a standard size which will allow us to evaluate 24 entries for yellow and stem rust or 48 entries in the case of leaf rust).
3. Soil: Steam a mixture of raw soil and peat moss (3:1) at 95 °C for 8 h, add fertilizer (for example, for 100 kg soil, add 22 g of Urea, 11 g of Scat, 5 g of KCl, 10 g of Magnesium sulfate, and 0.3 g of Ultra mix) after 24 h to the cooled down soil. We mix the prepared soil with peat moss (1:1) again before planting.
4. Inoculum: Dried urediniospores are usually kept at −70 °C. Before inoculation, urediniospores of leaf rust and stem rust have to be placed in a water bath (heat shock) at 40 °C for 4 min, then placed in a humid chamber (40 °C, at least 4 h) to gain moisture, whereas stripe rust urediniospores can be put in the humid chamber (40 °C, at least 4 h) directly.
5. Mineral oil: Soltrol 170. Specialized inoculators have been designed to use compressed air to spray spore suspensions in mineral oil with fine droplet sizes [9].
6. Chamber: High humidity is generated by humidity chambers, which can be either permanent or temporary in design; the temperature can be controlled.

7. Several gelatin capsules or small glass bottles used to keep urediniospores of purified rust pathotypes in ultrafreezer.
8. Test material: Seeds of wheat varieties, advanced breeding lines, recombinant inbred lines, and a set of differentials carrying known resistance genes.

2.2 Phenotyping at Adult Plant Stage

1. Seeds: Wheat varieties, advanced breeding lines, recombinant inbred lines, and a set of differentials carrying known resistance genes.
2. Field: The experiment area should be in the "hot spot" of disease.
3. Spreaders: Mixture of susceptible varieties or lines with specific known resistance genes for specific disease.
4. Inoculum: Urediniospores can be mixed with carrier talcum powder or paraffinic mineral oil such as Soltrol. Urediniospores are very hydrophobic, so they do not mix readily with water; however, water-based urediniospore suspensions can be injected into elongating wheat stems to infect plants in the field without the need for exogenous moisture. Suspensions of urediniospores in mineral oil can be efficiently applied to plants using sprayers of various kinds. Typically, handheld sprayers are used [10, 11].

2.3 Genotyping

1. Genomic DNA of each line.
2. SSR markers and KASP assays.
3. SNP and GBS genotype platforms.

2.4 Softwares for Analysis

1. MapManager QTXb20 [12] from http://mapmgr.roswellpark.org/mmQTX.html.
2. Kyazma B.V. software from Wageningen University [13].
3. MapChart [14] from http://www.wur.nl/en/Expertise-Services/Research-Institutes/plant-research/Download-MapChart.htm.
4. Inclusive composite interval mapping (ICIM) [15] from http://www.isbreeding.net/software/?type=detail&id=18.
5. QTL Cartographer software [16] from http://statgen.ncsu.edu/qtlcart/WQTLCart.htm.

3 Methods

3.1 Phenotyping at Seedling Stage

3.1.1 Sowing

For phenotyping of wheat varieties, about 8–10 seeds of advanced breeding lines or recombinant inbred lines are sown as hills in a tray or pot. Spacing of hills varies depending on the expected growth of the seedlings and number of days required before notes to be taken.

If the objective is to screen segregating populations, then the number of plants per population or line is higher and depends on the number of expected resistance genes. Simultaneously, a set of differentials carrying known resistance genes is also sown to determine the race use in the study and the expression of resistance genes. Differential sets are available as near-isogenic lines, single resistance genes carrying lines, or varieties with known or temporarily designated genes (*see* **Note 1**). The trays are placed in a greenhouse room or growth chamber with good light and optimum temperatures (15–25 °C), allowing good seedling growth and vigor. Seedlings are watered once every 2 days using a handheld water sprinkler. A single dose (4 g/L) of Nitrogen fertilizer (Ultrasol) is applied to 7-day-old seedlings. About 9- to 12-day-old seedlings with their second leaf one-third to halfway expanded is the most common stage for carrying out inoculation.

3.1.2 Inoculation

Nine- to twelve-day-old seedlings (two-leaf stage) are inoculated by using an atomizer to spray urediniospores of purified rust pathotypes suspended in light-weight mineral oil, e.g., Soltrol 170, at concentrations of 2–5 mg spores/mL or by observing the light brown or yellow color of the suspension. For stripe rust, the trays/pots carrying inoculated seedlings are subsequently placed in a mist chamber in trays filled with water, covered with plastic hoods and incubated at 7–9 °C for 24 h. However, for leaf and stem rust, trays/pots are placed in chambers at 18–22 °C and subjected to continuous mist produced by ultrasonic or other devices (*see* **Note 2**). The next day, seedlings are moved into a greenhouse maintained at 18–25 °C for leaf rust, 10–18 °C for stripe rust, and 18–28 °C for stem rust phenotyping. Disease scoring is conducted 10–14 days after inoculation; leaf rust about 10 days after inoculation, whereas for the other two rusts it is conducted for about 14 days.

3.1.3 Scoring

Leaf rust and stem rust infection types on wheat seedlings are recorded after 10 and 14 days post-inoculation, respectively, and are usually based on a 0–4 scale as described in Roelfs et al. [11], where "0" = no visible symptoms; ";" = only necrotic/chlorotic flecks without any uredinia; "1" = small uredinia surrounded by necrosis; "2" = small to medium uredinia surrounded by chlorosis or necrosis; "3" = medium-sized uredinia without chlorosis or necrosis; "4" = large-sized uredinia without chlorosis or necrosis; "X" = random distribution of variable-sized uredinia; and "+" and "−" were used when uredinia were somewhat larger or smaller than normal for the infection types (ITs). Seedling ITs of 0, ;, 1, 2, and X are generally considered as resistant, whereas 3 and 4 are susceptible.

For stripe rust the infection types are recorded about 14 days post-inoculation using a 0–9 scale [17], where "0" = no visible infection; "1" = necrotic/chlorotic flecks without sporulation;

"2" = necrotic/chlorotic stripes without sporulation; "3" = necrotic/chlorotic stripes with trace sporulation; "4" = necrotic/chlorotic stripes with light sporulation; "5" = necrotic/chlorotic stripes with intermediate sporulation; "6" = chlorotic stripes with moderate sporulation; "7" = stripes without chlorosis or necrosis and with moderate sporulation; "8" = stripes without chlorosis or necrosis and with sufficient sporulation; and "9" = stripes without chlorosis or necrosis and abundant sporulation. ITs of 0–5, 6, and 7–9 are categorized as resistant, intermediate, and susceptible, respectively.

3.2 Phenotyping at Adult Plant Stage

Although phenotyping of plants at post-seedling growth stages can be conducted in the greenhouse using the methods described for seedlings, it is more common to conduct adult plant studies in field trials where plants are exposed to multicycling disease progression under natural conditions. These tests are important to determine the effectiveness of race-specific as well as adult-pant resistance genes in reducing the disease severity and crop losses.

3.2.1 Field Trials Layout

For phenotyping varieties, advanced breeding lines and recombinant inbred lines, the size of field plots can vary from hills to short rows depending on the resources and objective. At CIMMYT we use paired rows plots, i.e., two short rows of 0.7 m length, 30 cm apart sown on top of 80 cm wide raised beds. This allows 50 cm spacing between the rows of two different lines. If planting is done on flat land, furrows are opened 30 cm apart; two furrows are used for planting, and a furrow is left unplanted between each plot. We also leave a 0.3 m alleyway and sow a hill plot using 8–10 seeds of susceptible spreader on one side of the plot in the alleyway. This allows each plot to have a spreader adjacent to it for a uniform disease buildup and spread. About 60–80 seeds (about 3–5 g) of each line are planted. If hill plots are sown, then using 8–10 seeds of each line is recommended. The experimental field should also be surrounded with two rows of spreader plots. The spreader plots are often a mixture of susceptible varieties depending on the objective, spreaders with differential susceptibility can also be used, e.g., varieties susceptible to leaf rust, but resistant to stripe rust and vice versa, or varieties susceptible to a specific pathotype but resistant to the other. For the screening of breeding materials, a mixture of susceptible varieties allowing the establishment and multiplication of the most important and relevant pathotypes is recommended for the field nurseries.

3.2.2 Field Inoculation

The most efficient method of inoculation is spraying 4–8-week-old spreaders with rust urediniospores suspended in Soltrol 170, similar to inoculations in the greenhouse. Stripe rust inoculations are carried out earlier in the season and should coincide with cool night temperatures with good dew formation or wet conditions.

Leaf rust and stem rust inoculations can be done at later growth stages but should be completed by heading stage. Dew formation, or availability of free moisture, on leaves and stems following the inoculation is critical for establishing infections, however, temperature can also play role, especially for stripe rust. If Soltrol or another similar lightweight mineral oil is not available, urediniospores can be suspended in water with a few drops of Tween 20 to break the surface tension, and can be used for spraying or injecting in plants through a syringe as described in Roelfs et al. [11]. Syringe inoculation can be done throughout the day, and is the safest method to establish disease. However, new suspensions are recommended every 3–4 h. The spore–water–Tween 20 suspension should be made fresh and sprayed late in the evening so that water does not evaporate and the process coincides with dew formation. Another way to inoculate the nursery is to mix urediniospores with talcum powder and dust the spreaders using a handheld duster or a cloth bag. Finally, pots of spreader seedling inoculated in greenhouse can also be placed at regular intervals in the field to establish rust diseases on field spreaders. Depending on the objective, urediniospores of different pathotypes of the same rust fungi or different rust fungi can be mixed (*see* **Note 3**).

3.2.3 Disease Evaluation

Disease evaluation in the field involves two components namely, disease severity visually estimated by percentage using the modified Cobb's Scale [18], and host response to infection as described in Roelfs et al. [11]. Host responses commonly used are: R (resistant)—small uredinia (or stripes) with chlorosis or necrosis and little sporulation; MR (moderately resistant)—medium sized uredinia (or stripes) with chlorosis or necrosis and some sporulation; MS (moderately susceptible)—medium to large sized uredinia (or stripes) with slight or no chlorosis and without necrosis and with moderate sporulation; and S (susceptible)—large sized uredinia (or stripes) without chlorosis or necrosis and with profuse sporulation. Disease severity and reaction can be recorded at the first appearance of the disease, but it is common to wait until the susceptible checks, parents or some lines in mapping population have close to a 60–80% disease severity. Repeated disease data can be recorded at weekly to 10-day intervals until plants reach close to physiological maturity. For breeding materials it is common to record one disease data at an appropriate stage of disease development when data is more relevant for the selection of resistant materials.

3.3 Methods for Genotyping

With the advancement of genotyping technologies over the past two decades, molecular markers have been widely used in the mapping and identification of resistance genes to wheat rusts. Several molecular technologies have been used, including restriction fragment length polymorphisms (RFLP), random amplified polymorphic DNA (RAPD), amplified fragment length polymorphisms (AFLP), simple sequence repeats (SSR), diversity arrays technology (DArT), single nucleotide polymorphisms (SNP), and genotyping-by-sequencing (GBS). RFLP markers were mostly co-dominant and restricted to regions with low-copy sequences [19]. RAPD markers are usually dominant, in combinaiton with low levels of polymorphism, and have reproducibility problems in wheat [20]. Although AFLP markers offer higher reproducibility and resolution at the whole genome level, the procedure of AFLP analysis is complex and costly [21]. SSR marker was the preferred system due to co-dominance, accuracy, high repeatability, high levels of polymorphism, chromosome specificity, and ease of manipulation at the turn of the century [22]. In the last 5 years, high-throughput technologies, DArT, SNP, and GBS, have become the main genotyping platforms. DArT have two genotyping platforms, including DArT-array and DArT-GBS. The former system is dominant and needs SSR markers to confirm their exact chromosomal locations. Some DArT markers have been located on the physical map (http://www.cerealsdb.uk.net/CerealsDB/Documents/FORM_ DArT_1A.php). DArT-GBS can be read by two methods, viz., SNP assays and Slico (presence–absence variance, PAV). SNP assays directly interrogate sequence variation, reducing genotyping errors compared to assays based on size discrimination. SNPs are ideally suited for construction of high-resolution genetic maps, investigations of population evolutionary history, and the discovery of marker–trait associations [23, 24, *see* **Note 4**]. With the development of sequencing technologies, the various methods applied to genetic variation analysis will shift from SNP-based genotyping to direct sequencing of all individuals in populations [25]. The GBS approach was recently applied to construct genetic maps of crops with large genomes [26, 27], allowing direct analysis of genetic variation and reducing the effect of ascertainment bias caused by the SNP discovery process.

3.4 Single Gene Analysis

3.4.1 Mendelian Analysis of Segregation Ratios in Different Generations

Many genetic analyses have been conducted historically using the segregation ratios in different segregating generations. Using Mendelian genetics, the observed segregation ratio is tested with the expected segregation ratio to determine the number of segregating genes and establish independence or linkage if more than one gene involved. Monosomic analysis was then done to identify chromosome location, followed by telocentric mapping to establish chromosome arm location. Reliable phenotyping in the seedling or adult plant stage played a crucial role in these studies.

When inheritance of resistance is complex, i.e., based on minor genes with additive effects, also known as quantitative trait loci (QTL), inheritance studies are more reliable when conducted using mapping populations of recombinant inbred lines in F_5 or F_6 generations, or double-haploid populations. Phenotyping is more accurate in these populations due to high homozygosity. As a result, most of the remaining genetic effects are additive. The number of rust resistance genes in the RIL population can be estimated using Mendelian segregation analysis [28, 29], where

Table 1
Expected segregation ratios of resistance genes in F_5 and F_6 generation

Number of genes	Generation	Lines (%)		
		HTPR[a]	HTPS[b]	Other[c]
1	F_5	43.75	43.75	12.5
1	F_6	46.875	46.875	6.25
2	F_5	19.1	19.1	61.8
2	F_6	22.0	22.0	56.0
3	F_5	8.4	8.4	83.2
3	F_6	10.3	10.3	79.4
4	F_5	3.7	3.7	92.6
4	F_6	4.8	4.8	90.4
5	F_5	1.6	1.6	97.4
5	F_6	2.3	2.3	95.4

[a] HTPR = Homozygous parental type resistant
[b] HTPS = Homozygous parental type susceptible
[c] Other = Lines with intermediate levels of disease severity

the observed frequencies for each phenotypic category is tested against the expected frequencies (Table 1) for different numbers of additive genes using Chi-squared (χ^2) analysis.

Moreover, the minimum number of resistance genes can also be estimated using the quantitative approach described by Wright [30] as $n = (\text{GR})^2/4.57(\sigma^2 \text{g})$, where GR (genotypic range) = phenotype range $\times$ h^2 (narrow-sense heritability), σ^2g = genetic variance of the F_5 RILs in the present population, $h^2 = \sigma^2 \text{g}/\sigma^2 \text{g} + \sigma^2 \text{e}$. The analysis of variance (ANOVA) is carried out using SAS 9.2 (SAS Institute, Cary, NC) with the final disease severity (FDS) in each environment.

3.4.2 Single Gene Mapping Using Molecular Methods

The prerequisite for any molecular mapping is consideration of both phenotypic and genotypic data. Any relevant genotyping platforms can be used and the cost and accessibility often dictates the selection. It has become common to outsource genotyping as it is less expensive than any alternative. Chi-square (χ^2) tests are performed to evaluate the goodness of the fit of observed segregation with expected genetic ratios and to detect marker–trait linkages. Recombination fractions are calculated using the software MapManager QTXb20 [12] or Joinmap 4.0 [13] and converted to centiMorgens using the Kosambi mapping function [31]. Logarithm of odds (LOD) scores of 3–4 is usually used to determine significance of genetic linkages. Linkage maps were graphically visualized with MapChart [14]. The workflows of MapManager QTXb20, Joinmap 4.0, and MapChart are listed in Figs. 1, 2, and 3, respectively.

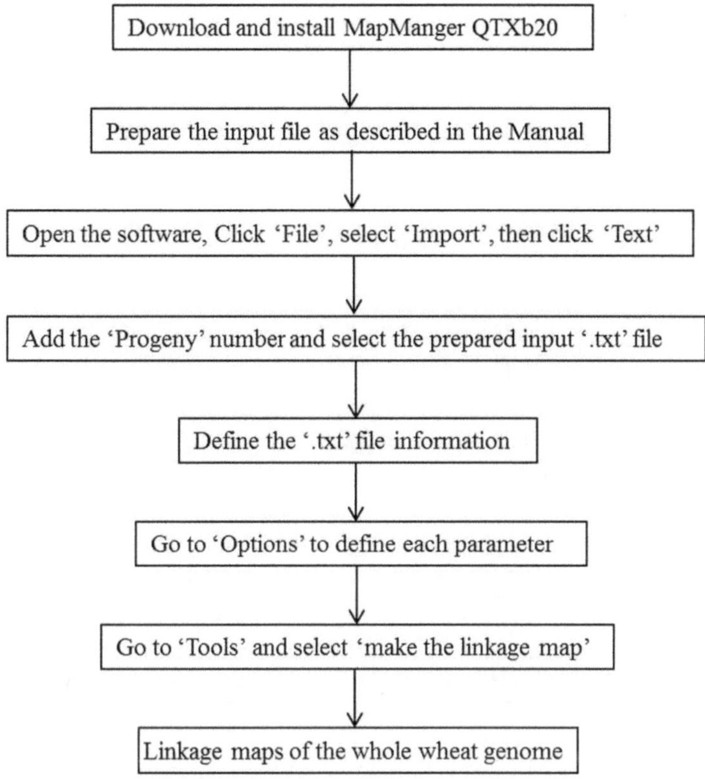

Fig. 1 Workflow of Mapmanger QTXb20 to construct linkage maps

3.5 Methods for QTL Analysis

Quantitative trait locus (QTL) mapping of biparental population is conducted on the basis of available linkage maps with whole wheat genome molecular markers and phenotypic data from multiple environments/locations. The linkage maps can be constructed by JoinMap software as mentioned above. QTL mapping using disease severity from each experiment can be carried out using inclusive composite interval mapping (ICIM) [15] and QTL Cartographer software [16]. The workflows of IciMapping 4.1 and QTL Cartographer are listed in Figs. 4 and 5, respectively.

4 Notes

1. The number of lines in differential sets varies between labs and rust diseases but often have about 50 lines for leaf rust, 30 lines for stripe rust, and 50 lines for stem rust.
2. For stem rust, it is also important to add light about 12 h after misting while seedlings are still wet and to let them dry slowly.

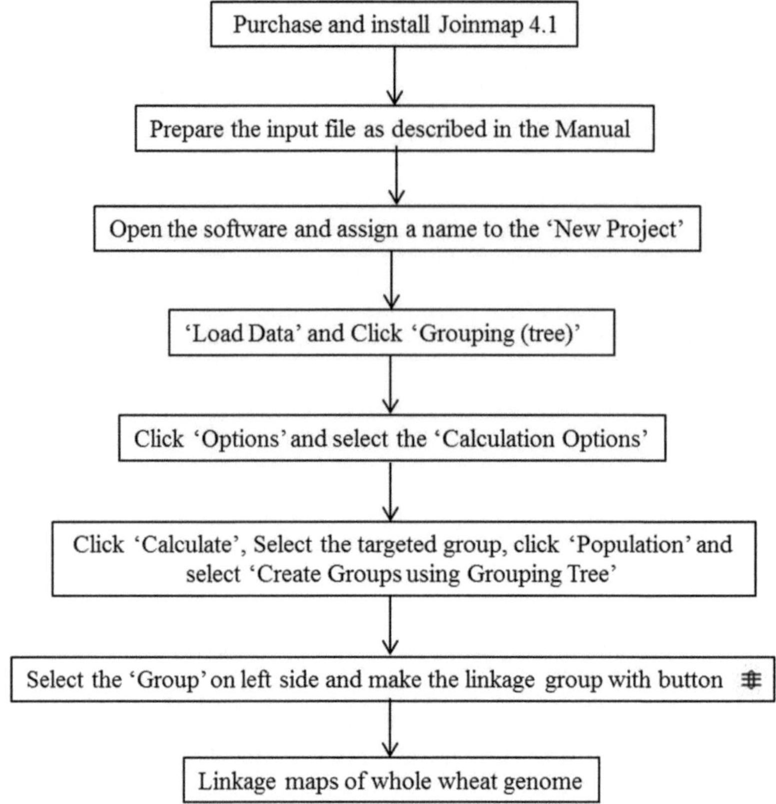

Fig. 2 Workflow of Joinmap 4.1 to construct linkage maps

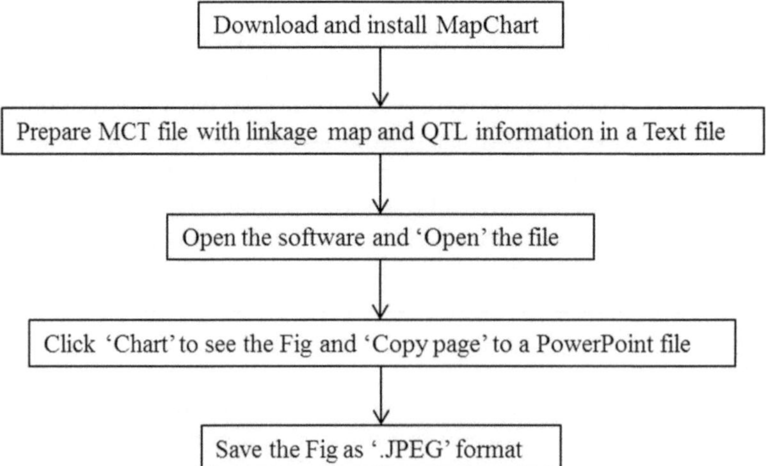

Fig. 3 Workflow of Mapchart to draw maps

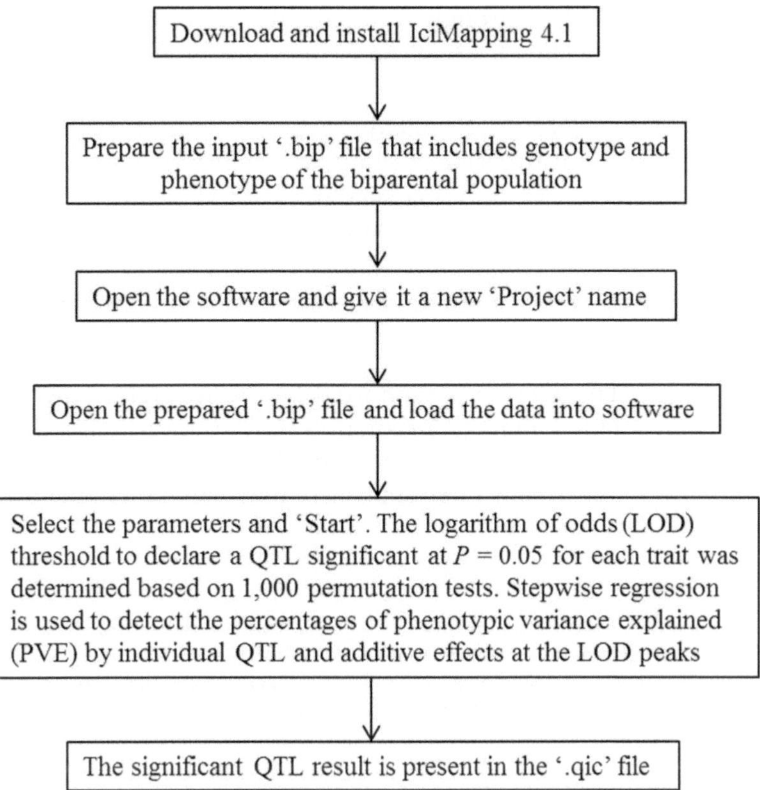

Fig. 4 Workflow of Icimapping 4.1 to do QTL analysis

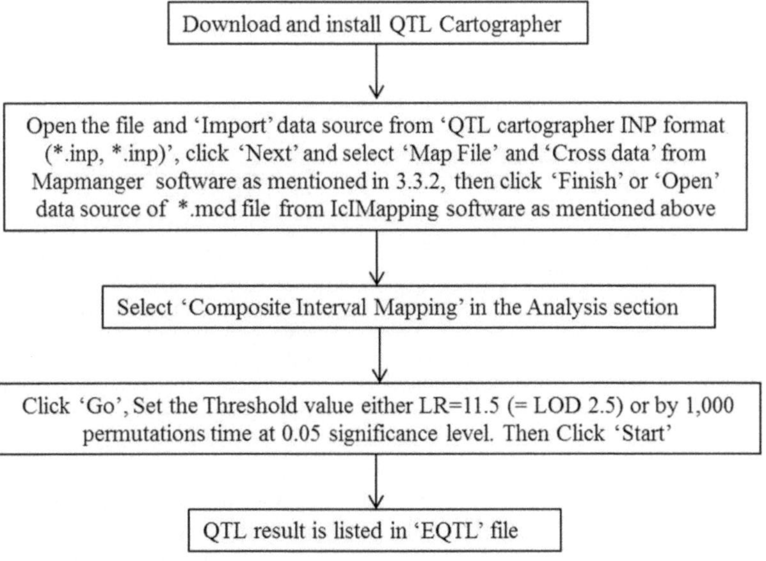

Fig. 5 Workflow of QTL Cartographer to do QTL analysis

3. It is strongly recommended that single or a few pathotypes of known virulences of the same fungi should be used for genetic studies.
4. However, limited D genome markers, too many repeat markers at the same position, and confusing chromosome information from SNP markers in 90K SNP assays may affect widespread use in genetic analysis (Per. Comm. Jizeng Jia).

Acknowledgments

This work was supported by the Australian Grains Research and Development Corporation (GRDC), Australian Cereal Rust Control Program, and National Natural Science Foundation of China (31301309). We appreciate the technical editing by Julie Mollins.

References

1. Dubin HJ, Brennan JP (2009) Fighting a 'Shifty Enemy', the international collaboration to contain wheat rusts. In: Spielman DJ, Pandya-Lorch R (eds) Millions fed: proven successes in agricultural development. International Food Policy Research Institute, Washington. http://ebrary.ifpri.org/cdm/ref/collection/p15738coll2/id/130812
2. Hovmøller MS, Sørensen CK, Walter S, Justesen AF (2011) Diversity of Puccinia striiformis on cereals and grasses. Annu Rev Phytopathol 49:197–217
3. Singh RP, Hodson DP, Jin Y, Huerta-Espino J, Kinyua M, Wanyera R, Njau P, Ward RW (2006) Current status, likely migration and strategies to mitigate the threat to wheat production from race Ug99 (TTKS) of stem rust pathogen. CAB Rev Perspect Agric Vet Sci Nutr Nat Resour 1:54
4. Chen XM (2013) Review article: high-temperature adult-plant resistance, key for sustainable control of stripe rust. Am J Plant Sci 4:608–627
5. Flor HH (1942) Inheritance of pathogenicity in *Melampsora lini.* Phytopathology 32:653–669
6. Das MK, Rajaram S, Mundt CC, Kronstad WE, Singh RP (1992) Inheritance of slow rusting resistance in wheat. Crop Sci 32:1452–1456
7. Johnson R, Law CN (1973) Cytogenetic studies on the resistance of the wheat variety Bersée to *Puccinia striiformis.* Cereal Rusts Bull 1:38–43
8. Bjarko ME, Line RF (1988) Heritability and number of genes controlling leaf rust resistance in four cultivars of wheat. Phytopathology 78:457–461
9. Browder LE (1971) Pathogenic specialization in cereal rust fungi, especially *Puccinia recondite* f. sp. *tritici*: Concepts, methods of study and application. US Dep Agric Tech Bull 1432: pp51
10. McIntosh RA, Wellings CR, Park RF (1995) Wheat rusts: an atlas of resistance genes. CSIRO Publications, East Melbourne, Australia
11. Roelfs AP, Singh RP, Saari EE (1992) Rust diseases of wheat: concepts and methods of disease management. CIMMYT, Mexico, D.F.
12. Manly KF, Cudmore RH Jr, Meer JM (2011) Map manager QTX, cross-platform software for genetic mapping. Mamm Genome 12 (12):930–932
13. Van Ooijen JW (2006) Join map 4, software for the calculation of genetic linkage maps in experimental population. Plant Research International, Wageningen, Netherlands. http://www.joinmap.nl
14. Voorrips RE (2002) MapChart: software for the graphical presentation of linkage maps and QTLs. Heredity 93:77–78
15. Meng L, Li HH, Zhang LY, Wang JK (2015) QTL IciMapping: integrated software for genetic linkage map construction and quantitative trait locus mapping in biparental populations. Crop J 3:269–283
16. Wang S, Basten CJ, Zeng ZB (2012) Windows QTL Cartographer 2.5. Department of Statistics, North Carolina State University, Raleigh,

NC. (http://statgen.ncsu.edu/qtlcart/WQTLCart.htm)

17. McNeal FH, Konzak CF, Smith EP, Tate WS, Russell TS (1971) A uniform system for recording and processing cereal research data. USDA-ARS Bull, Washington, pp 34–121
18. Peterson RF, Campbell AB, Hannah AE (1948) A dia-grammatic scale for estimating rust intensity on leaves and stems of cereals. Can J Res 26:496–500
19. Saiki RK, Scharf S, Faloona F, Mullis KB, Erlich HA, Arnheim N (1985) Enzymatic amplification of beta-globin genomic sequences and restriction site analysis for diagnosis of sickle cell anemia. Science 230:1350–1354
20. Devos KM, Gale MD (1992) The use of random amplified polymorphic DNA markers in wheat. Theor Appl Genet 84:567–572
21. Mueller UG, Wolfenbarger LL (1999) AFLP genotyping and fingerprinting. Trends Ecol Evol 14:389–394
22. Röder MS, Korzun V, Wendehake K, Plaschke J, Tixier MH, Leroy P, Ganal MW (1998) A microsatellite map of wheat. Genetics 149:2007–2023
23. Aranzana MJ, Kim S, Zhao K, Bakker E, Horton M, Jakob K, Lister C, Molitor J, Shindo C, Tang C, Toomajian C, Traw B, Zheng H, Bergelson J, Dean C, Marjoram P, Nordborg M (2005) Genomewide association mapping in *Arabidopsis* identifies previously known flowering time and pathogen resistance genes. PLoS Genet 1:e60
24. Zhao K, Aranzana MJ, Kim S, Lister C, Shindo C, Tang C, Toomajian C, Zheng H, Dean C, Marjoram P, Nordborg M (2007) An Arabidopsis example of association mapping in structured samples. PLoS Genet 3:e4
25. Hamilton JP, Buell CR (2012) Advances in plant genome sequencing. Plant J 70:177–190
26. Elshire RJ, Glaubitz JC, SunQ PJA, Kawamoto K, Buckler ES, Mitchel SE (2011) A robust, simple genotyping-by-sequencing (GBS) approach for high diversity species. PLoS One 6:e19379
27. Poland JA, Brown PJ, Sorrells ME, Jannink JL (2012) Development of high-density genetic maps for barley and wheat using a novel two-enzyme genotyping-by-sequencing approach. PLoS One 7:e32253
28. Knott DR, Padidam M (1988) Inheritance of resistance to stem rust in six wheat lines having adult plant resistance. Genome 30:283–288
29. Singh RP, Rajaram S (1992) Genetics of adult-plant resistance to leaf rust in 'Frontana' and three CIMMYT wheats. Genome 35:24–31
30. Wright S (1968) Evolution and the genetics of populations. In: Genetic and biometric foundations, vol 1. University of Chicago Press, Chicago, IL
31. Kosambi DD (1944) The estimation of map distances from recombination values. Ann Eugenics 12:172–175

Chapter 13

Advances in Identification and Mapping of Rust Resistance Genes in Wheat

Urmil Bansal and Harbans Bariana

Abstract

Genetic characterisation of new rust resistance loci in wheat using cytogenetic/low-throughput genotyping systems required at least 5 years. Development of next-generation sequencing (NGS) based molecular marker genotyping platforms in the last decade has provided scientists with the genomic resources to expedite precise mapping of target loci. Here, we describe methodologies for genetic analysis and application of NGS-based resources to determine the precise genomic locations of rust resistance loci in wheat and development of closely linked markers for marker assisted selection.

Key words KASP assay, SNP genotyping, DArTseq marker, Mapping populations, Molecular markers

1 Introduction

Wheat is grown on 220 million hectares worldwide. More than 50% increase in global wheat production is necessary to achieve food security. Wheat is attacked by many diseases and pests, and based on comparative production losses and geographic distribution, rust diseases, caused by *Puccinia* species, are considered very important. Rust diseases have the potential to cause complete crop failures. Breeding for rust resistance has been performed to sustain/increase global wheat production [1]. Resistance to rust diseases can be defeated by evolution of virulence in pathogen populations and therefore characterization of new sources of resistance is necessary. Genetically diverse sources of resistance against economically important plant pathogens have been discovered and formally named in many crops; however, a precise and systematic catalog of genes is available only for wheat [2–4].

The modern wheat rust resistance gene discovery pipeline involves identification of stocks carrying putatively new loci through phenotypic assays, development of segregating populations and molecular mapping to determine genomic locations of the target loci. It is then followed by detailed mapping of the target

Sambasivam Periyannan (ed.), *Wheat Rust Diseases: Methods and Protocols*, Methods in Molecular Biology, vol. 1659,
DOI 10.1007/978-1-4939-7249-4_13, © Springer Science+Business Media LLC 2017

region to develop user-friendly markers for the deployment of identified gene in future wheat cultivars through marker-assisted selection (MAS). This chapter summarizes methodologies that can be used for fast characterization of putatively new rust resistance loci in wheat. The methodology can however be applied to disease resistance/trait discovery work in other crops.

2 Materials

1. Germplasm (national or international nurseries, landraces).
2. Set of differentials (genotypes carrying known genes/historic sets).
3. Well characterized pathotypes or isolates or races (*Puccinia striiformis* f. sp. *tritici*, *P triticina*, and *P. graminis* f. sp. *tritici*).
4. Good greenhouse facility for seedling response tests.
5. Irrigated experimental area (field) for evaluation of stocks carrying the target resistance gene at the adult plant stage.
6. KASP master mix containing the two universal FRET cassettes (FAM and HEX), TAQ polymerase, and buffer.

3 Methods

3.1 Identification of Genotype(s) Carrying Putatively New Rust Resistance Locus (Multi-Pathotype Tests)

1. Grow test entries and a set of known gene differentials described in McIntosh et al. [2].
2. Inoculate 10 to 12-day-old seedlings with the appropriate rust pathotype at the seedling stage [2].
3. Incubate inoculated plants for 24 h for leaf rust and stripe rust, and 24 to 48 h for stem rust [2].
4. Move inoculated plants to microclimate rooms running at specific temperatures for development of the respective disease (Leaf rust 23 °C; Stripe rust 17 °C; Stem rust 25 °C).
5. Assess rust responses 12–16 days after inoculation using the scales (for respective pathogen) given in McIntosh et al. [2].

   Assessment of leaf and stem rust seedling responses:

   (a) Infectiontype (IT) "0" is given when there is no visible sign of infection, representing a immune response.

   (b) IT ";" - strong hypersensitive response and no uredia formation.

   (c) IT "1" - small uredia accompanied with necrosis and/or chlorosis are present.

   (d) IT "2" - small to medium uredia with green islands.

(e) IT "3" - medium to large sized sporulating uredia with or without chlorosis.

(f) IT "4" - large sporulating uredia.

(g) IT "X" - presence of more than two infection types on the same leaf.

(h) IT 0,1,2,3 and X are regarded as resistant and IT 4 as susceptible.

Assessment of seedling stripe rust responses
Two scales are being used and details are provided below:

1. Description of 0 to 4 scale [*see* 2]:
   (a) IT "0" - no visible signs of infection.
   (b) IT ";" - necrotic flecks.
   (c) ";N" necrotic areas without sporulation.
   (d) "1" - necrotic and chlorotic areas with restricted sporulation.
   (e) "2" - moderate sporulation with necrosis and chlorosis.
   (f) "3" - sporulation with chlorosis.
   (g) "4" - abundant sporulation without chlorosis and necrosis.
   (h) ITs from 0 to 3 are considered resistant and 4 is susceptible.
2. Description of 0–9 scale [5].
3. Compare infection types produced by test cultivars with the known genes represented in the differential set as shown in Table 1 [6].
4. For example in Table 1, Cultivar 9 was resistant against all pathotypes indicating the presence of putatively new resistance gene(s).

3.2 Characterization of the Target Resistance Locus

1. Cross the resistant genotype 9 with the susceptible genotype 10 and harvest F_1 seed.
2. Grow F_1 plants and harvest each F_1 plant individually (F_2 seed).
3. Grow 200 plants and harvest individually to produce F_3 population.
4. Advance generations by growing single seed from each family (F_3 to F_4) and harvest single spike.
5. Sow single seed from each F_5 line and harvest whole plant (F_6) to name the resultant generation as recombinant inbred line (RIL) population (Fig. 1; *see* **Note 1**).
6. Rust tests can be carried out on the F_3 or RIL population.

Table 1
Identification of a genotype carrying putatively new resistance gene(s) through multi-pathotype tests

Genotype/Pathotype	1	2	3	4	5	6	7	8	Gene postulation
Test entries									
1	R	S	S	S	S	S	S	S	Gene 1
2	S	R	S	S	S	S	S	S	Gene 2
3	S	S	R	S	S	S	S	S	Gene 3
4	S	S	S	R	S	S	S	S	Gene 4
5	S	S	S	S	R	S	S	S	Gene 5
6	S	S	S	S	S	R	S	S	Gene 6
7	S	S	S	S	S	S	R	S	Gene 7
8	S	S	S	S	S	S	S	R	Gene 8
9	R	R	R	R	R	R	R	R	New gene
10	S	S	S	S	S	S	S	S	NIL
Controls									
Differential 1	R	S	S	S	S	S	S	S	Gene 1
Differential 2	S	R	S	S	S	S	S	S	Gene 2
Differential 3	S	S	R	S	S	S	S	S	Gene 3
Differential 4	S	S	S	R	S	S	S	S	Gene 4
Differential 5	S	S	S	S	R	S	S	S	Gene 5
Differential 6	S	S	S	S	S	R	S	S	Gene 6
Differential 7	S	S	S	S	S	S	R	S	Gene 7
Differential 8	S	S	S	S	S	S	S	R	Gene 8

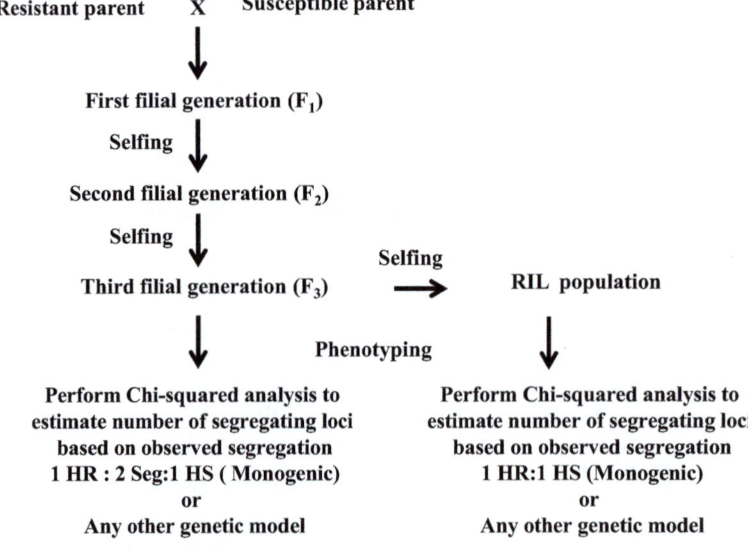

Fig. 1 Schematic presentation of population development (F_3 and RIL) and genetic analysis

Table 2
Stripe rust response variation among F_3 and RIL populations depicting monogenic and digenic segregation models

Population/ frequency	F_3			RIL		
	Observed	Expected	χ^2	Observed	Expected	χ^2
Monogenic			1:2:1			1:1
HR	27	24.5	0.26	54	49	0.51
Seg	48	49	0.02	–	–	–
HS	23	24.5	0.09	44	49	0.51
Total	98	98	0.37	98	98	1.02
Digenic			7:8:1			3:1
HR	40	42.9	0.19	74	73.5	0.003
Seg	50	49	0.02	–	–	–
HS	8	6.1	0.57	24	24.5	0.01
Total	98	98	0.78	98	98	0.013

Table values of χ^2 at $P = 0.05$ (1 d.f.) is 3.84 and (2 d.f.) is 5.99

7. Classify the population into different categories based on the phenotypic results (Table 2; *see* **Note 2**).
   (a) *Homozygous resistant (HR)*: All plants among F_3 or an RIL population produce resistant responses.
   (b) *Homozygous susceptible (HS)*: All plants among F_3 or RIL population produce susceptible responses.
   (c) *Segregating (Seg)*: F_3 families or RILs include both resistant and susceptible plants.
8. Genetic analyses results based on F_3 and RIL populations are given in Table 2. Chi-squared analysis of segregation was used to predict the number of observed segregating loci. If the calculated value of χ^2 for a tested genetic ratio was less than the table value at a given degree of freedom, then the segregation data were considered to conform to the genetic model based on that ratio. In the case of segregation at two or more loci development of single locus segregating population (SLSP) for detailed molecular mapping should be performed [7].

3.3 Molecular Mapping

3.3.1 DNA Extraction

1. Extract DNA from the eight or more plants from each F_3 family or RIL using a standard DNA isolation method such as modified CTAB method [8].
2. Quantify DNA using Nano-Drop (http://tools.thermofisher.com/content/sfs/manuals/nd-1000-v3.8).

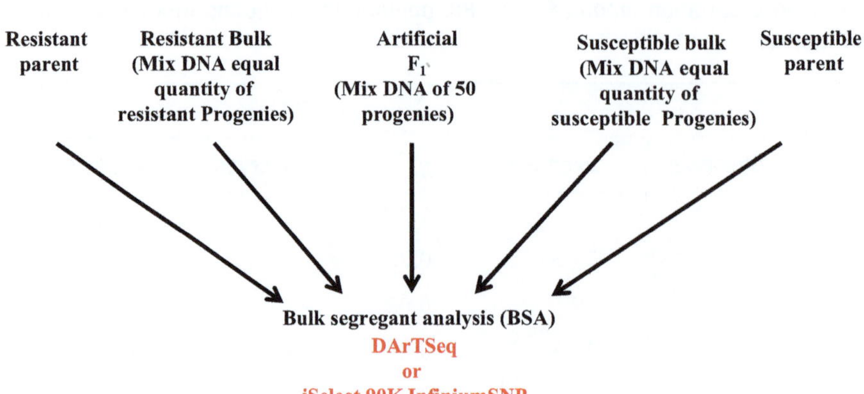

Fig. 2 Diagram depicting the bulked segregant analysis

3. Check the quality by running DNA on the agarose gel (*see* **Note 3**).
4. Make the working dilution at the concentration of 30 ng/μL for setting PCRs.

3.3.2 Bulked-Segregant Analysis (BSA)

1. Pool equal quantity of DNA from 8 or more HR and HS lines to constitute the resistant bulk and the susceptible bulk, respectively (*see* **Note 4**).
2. Pool equal quantity of DNA from 50 random lines to constitute an artificial F_1 (Fig. 2)
3. DNA from the F_1 plant can also be used instead of artificial F_1.
4. Set final concentration of DNA at 1 μg per sample.
5. Send DNA of bulks, parents and artificial F_1 of a mapping population for high throughput analysis to DArTseq (http://www.diversityarrays.com/dart-application-dartseq) or iSelect 90K SNP array for genotyping [9].
6. BSA will infer the putative chromosomal location(s) of the target resistance locus/loci based on marker–trait association (s). Confirm chromosomal location(s) by genotyping linked markers on the entire mapping population. Comparison of these results with the genomic location(s) of known gene(s) will be required to confirm the uniqueness of resistance gene(s) carried by cultivar 9.

3.3.3 Marker Development from 90K SNP Results (See Notes 5 and 6)

1. The results of 90K SNP based BSA are given in Table 3. Marker-trait linkages are calculated using Genome Studio software developed by Illumina.
2. Each linked SNP has ~101 bp sequence and SNP is marked in parenthesis; for example IWB22020.

Table 3
List of markers linked with resistance gene in the 90K SNP BSA

SNP ID	Chromosome	Position (cM)	SNP	Infinium sequence	Diff in Norm Theta B1-B2	Diff in Norm Theta P1-P2	(B1-B2) /(P1-P2)	(B1-F1) /(B1-B2)	(B2-F1) /(B1-B2)	Linkage Evidence	Parent1	BulkP1	ArtF1	BulkP2	Parent2
IWB22020	2B	88.439297	[A/G]	AAGTATGAGAACGCC	0.2	0.2	1.1	0.4	0.6	Strong	AA	AA	AA	BB	BB
IWB6607	2B	88.863392	[A/C]	AAGAAATTTACCACT(	0.2	0.2	0.9	0.3	0.7	Strong	AA	AA	NC	BB	BB
IWB29853	2B	90.241699	[T/C]	GCTTCATGTTCTTGTG	0.9	1.0	0.9	0.4	0.6	Strong	AA	AA	NC	BB	BB
IWB6822	2B	90.971391	[T/C]	TCGCCGTCGACCAGC	0.4	0.4	1.0	0.4	0.6	Strong	AA	AA	NC	BB	BB
IWB1602	2B	91.102362	[T/C]	AACTTCGAAGGATAG	0.4	0.3	1.1	0.5	0.6	Strong	AA	AA	NC	BB	BB
IWB7331	2B	92.281095	[T/C]	GACGTGATGGATCGC	-0.3	-0.3	1.1	0.5	0.5	Strong	AA	AA	NC	BB	BB
IWB6330	2B	95.823531	[A/G]	CGTCGCGCAGCACCC	0.9	0.9	1.0	0.5	0.5	Strong	AA	AA	NC	BB	BB
IWB4614	2B	96.135366	[A/G]	GGCGTTGATGTTTATG	0.9	0.9	1.0	0.3	0.7	Strong	AA	AA	NC	BB	BB
IWB28408	2B	99.160157	[T/C]	AAACCACAAAAGCA(	0.5	0.5	1.0	0.6	0.4	Strong	AA	AA	NC	BB	BB

IWB22020: AAGTATGAGAACGCCTCATGGAGCGCTCCTCTTGCTCTTCATCCTGGCGT[A/G]CC TCCGCGGTGCGTACTCCGGTGGCGGCCGGAGCCGCTTCACCTCCATAA
IWB22020-A1: 5′ GAAGGTGACCAAGTTCATGCTTGCTCTTCATCCTGGCGTA 3′
IWB22020-A2: 5′ GAAGGTCGGAGTCAACGGATTTGCTCTTCATCCTGGCGTG 3′
IWB22020-C: 5′ TTATGGAGGTGAAGCGGCTC 3′

3. Design three primers for each SNP using batch primer3 software (http://batchprimer3.bioinformatics.ucdavis.edu/cgi-bin/batchprimer3/batchprimer3.cgi).
   (a) Allele 1
   (b) Allele 2 and
   (c) Common or reverse primer
4. Development of kompetitive allele specific PCR (KASP) markers for the linked SNPs is shown at the end of Table 3.
5. Allele specific primers were prefixed with FAM and HEX sequences (21 bp each highlighted in red font). This corresponds to one of the two universal FRET (fluorescent resonance energy transfer) cassettes present in the KASP master mix.
6. Similarly KASP markers can be developed from other linked SNPs

KASP assay composition

DNA	3.0 μL (30 ng/μL)
KASP mix	4.0 μL
Primer*	0.11 μL
Water	0.89 μL
Total	8.0 μL

*12 μM of each allele specific primer and 30 μM of the common or reverse primer

PCR profile: This is a two-step PCR and profile is given below (*see* **Note 7**):

94 °C 15 min	
94 °C 20 s 65 °C 60 s	10× (dropping down 0.8 °C/cycle)
94 °C 20 s 57 °C 60 s	30–40× depending upon SNP
40 °C 30 s	Fluorescence detection (End point reading; *see* **Note 8**)

3.3.4 Genotyping of F_3 or RIL Population

1. Test all KASP markers developed from the linked SNPs on parents.
2. Select KASP markers showing clear clusters between parents.
3. Run polymorphic KASP markers on the entire F_3 or RIL population and observe clusters for two alleles representing parental types and mixture of both alleles representing heterozygous families (Fig. 3).

3.3.5 Genotyping by Sequencing Results (See Note 9)

1. Results of BSA with DArTseq markers are given in Table 4.
2. BLAST DArTseq marker(s) (~30–40 bp) sequence (column 2) closely linked with the resistance gene(s) against the flow-sorted chromosome survey sequence (CSS) contigs of Chinese Spring (CS) using BLASTN (https://urgi.versailles.inra.fr/blast/?dbgroup=wheat_all&program=blastn) or identify SNP between DARTseq seq and CSS contig. For example, DArTseq marker sequence for 1257291 showed 100% identity CSS contig CSS_1AS_3277420 ($4e^{-28}$).
3. Use CSS_1AS_3277420 sequence to design SSR markers using batch primer3 software (http://batchprimer3.bioinformatics.ucdavis.edu/cgi-in/batchprimer3/batchprimer3.cgi). This software will identify the repeat motif during the development of primers (Table 5).
4. Add M13 sequence tail to each forward primer at 5′ end. This will help in resolving small differences using the 4300 DNA Analysis system (LI-COR Biosciences).
5. Run polymorphic markers on the entire population, tabulate data, and generate map as described in the previous section.

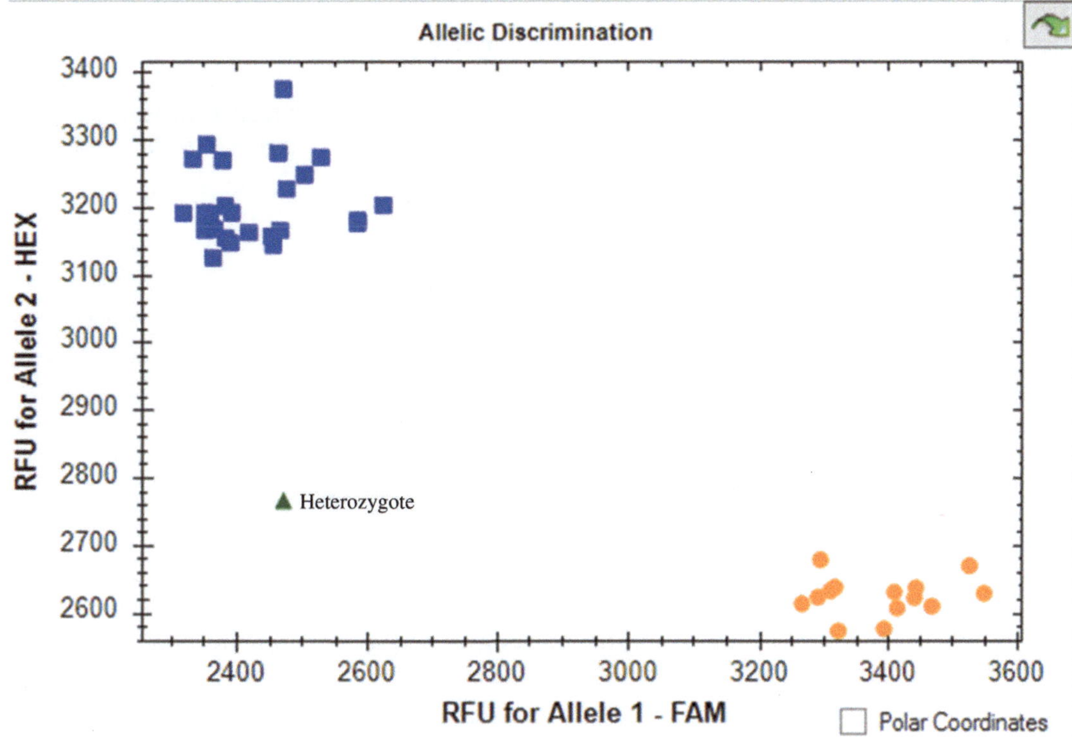

Fig. 3 A snapshot of KASP assay using real time PCR (Allele 1, Allele 2, and heterozygote)

Table 4
List of DArTseq markers linked with the resistance gene

Clone ID	Sequence	Chromosome	Position (cM)
1104400	TGCAGTTCCATGGCACTCAAGTGCTGAAA TTACCCAGATCTGATGATAATGGCACCCG AGATCGGAAGA	1A	33
1257291	TGCAGCGAGCTTGCCAAGATCCGCGAGGAT GCAGCGTCCGAGCCATGCCGTGCCCT CATCACCACCTCG	1A	34
4404828	TGCAGCTCGCTGGAGTATTAGTCGGAATCG AGCTAAATTGGGGATCGCCGAGATCG GAAGAGCGGTTCA	1A	36
3222161	TGCAGAATTTGAAAATACGTCACCAATCCAT TGTCGTCCAGCCCGAGATCGGAAGAGCG GTTCAGCAGG	1A	36
2292464	TGCAGCTCCTCGACGGAGACGCCTTGCG CCAGTTCAACATCAGCCTCAACGACGAAA GCCGAGATCGGA	1A	37

Table 5
List of primers designed from the CSS_1AS_3277420 contig

No.	Orientation (5′–3′)	Start	Length	Temperature	GC%	Seq	Product size	Repeat motif	Repeats
1	FORWARD	3	22	54.34	31.82	GACAATTATCAT GAACAAGGAA	148	ATA	4
	REVERSE	150	21	54.61	33.33	ACAAAACTTGT TGATTTGCTC			
2	FORWARD	7673	20	55.12	50.00	CTCCTGACAAG GAAGCATAG	131	CTC	4
	REVERSE	7803	21	55.02	47.62	CTATAGCGATC GAGACGAGTA			
3	FORWARD	4861	21	55.08	47.62	GTAAGAGCTGC ACATTACCAC	144	AGCT	4
	REVERSE	5004	20	56.09	45.00	AGCTCATTGCT TGTGGAGTA			

PCR profile for SSR markers: This is a three-step PCR and profile is given below:

95°C-10 min	
94 °C30 s 65 °C-30 s 72 °C-30 s	5×
94 °C-30 s 72 °C-30 s 60 °C-30 s	35×
72 °C-5 min	

3.3.6 Construction of Linkage Group

1. Export the allele discrimination data for each marker or score marker data from gel and tabulate in an Excel sheet for mapping using an appropriate statistical software.
2. Use any of the following software programs for preparing genetic map: MapManager [10], MapDisto [11], ICIM mapping (https://www.integratedbreeding.net/supplementary-toolbox/genetic-mapping-and-qtl/icimapping), etc., using Kosambi mapping function [12].
3. Prepare the genetic linkage map using software MapChart [13] as shown in Fig. 4.

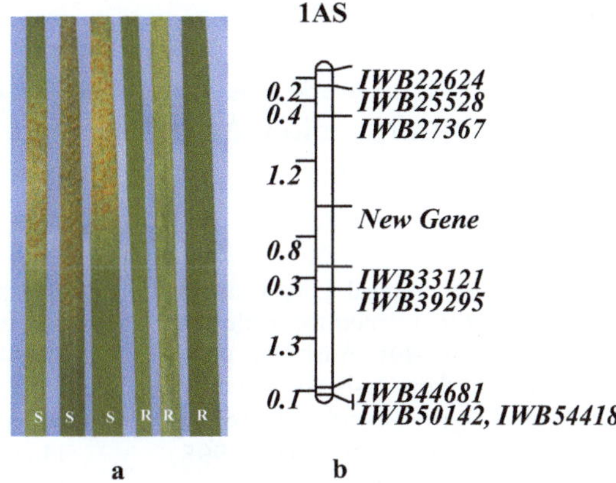

Fig. 4 (**a**) Infection types produced by susceptible (S) and resistant (R) parents. (**b**) genetic linkage map of a population showing location of a new gene

4 Notes

1. Population size of 100–150 individuals is enough if it is a single seedling resistance gene.
2. Some genes may be difficult to phenotype due to environmental sensitivity.
3. Make sure that DNA does not get degraded.
4. Bulks based on higher number of lines with contrasting phenotypes improve detection of marker–trait associations.
5. The advantage of the 90K SNP assay is that linked SNPs can be converted into KASP markers without any further search.
6. This platform will test SNPs which were present in the array and will miss the SNPs specific to the parents in a given bi-parental population.
7. PCR for KASP markers can be carried out in 96-well thermocycler and are also amenable to be used in robotic systems.
8. Allelic discrimination can be carried out using either a real-time PCR machine or a plate reader.
9. The DArTseq method deploys sequencing of the samples using the Next-Generation Sequencing (NGS) platforms and polymorphism is relevant to parental genotypes, therefore it results in more precise marker development and less false positives; however, their conversion to PCR-based markers involves an extra step compared to designing of KASP markers from trait-linked SNPs. If there is any SNP present in the contig, then KASP markers can be developed as explained earlier.

Acknowledgments

This work was supported by Grains Research and Development Corporation, Australia.

References

1. Bariana HS, Brown GN, Bansal UK, Miah H, Standen GE, Lu M (2007) Breeding triple rust resistant wheat cultivars for Australia using conventional and marker assisted selection technologies. Aust J Agric Res 58:576–587
2. McIntosh RA, Welling CR, Park RF (1995) Wheat rusts: an atlas of resistance genes. CSIRO, Melbourne, Australia, p 200
3. McIntosh RA, Dubcovsky J, Rogers WJ, Morris C, Appels R, Xia XC (2011) Catalogue of gene symbols for wheat: 2011 supplement. http://shigen.nig.ac.jp/wheat/komugi/genes/symbolClassList.jsp
4. McIntosh RA, Dubcovsky J, Rogers WJ, Morris C, Appels R, Xia XC (2015–2016) Catalogue of gene symbols for wheat: http://shigen.nig.ac.jp/wheat/komugi/genes/symbolClassList.jsp
5. McNeal FH, Konzak CF, Smith EP, Tate WS, Russell TS (1971) In: A uniform system for recording and processing cereal research data. US Dep Agric Agric Res Serv Bull, Washington DC. pp 34–412
6. Pathan AK, Park RF (2006) Evaluation of seedling and adult plant resistance to leaf rust in European wheat cultivars. Euphytica 149:327–342
7. Bariana HS, Bansal UK (2017) Breeding for rust resistance. In: Encyclopedia of applied plant sciences, 2nd edn. Elsevier, Amsterdam. http://www.elsevier.com/locate/permissionusematerial
8. Bansal UK, Kazi AG, Singh B, Hare RA, Bariana HS (2014) Mapping of durable stripe rust resistance in a durum wheat cultivar Wollaroi. Mol Breed 33:51–59
9. Wang S, Wong D, Forrest K, Allen A, Chao S, Huang BE, Maccaferri M, Salvi S, Milner SG, Cattivelli L, Mastrangelo AM, Whan A, Stephen S, Barker G, Wieseke R, Plieske J, International Wheat Genome Sequencing C, Lillemo M, Mather D, Appels R, Dolferus R, Brown-Guedira G, Korol A, Akhunova AR, Feuillet C, Salse J, Morgante M, Pozniak C, Luo MC, Dvorak J, Morell M, Dubcovsky J, Ganal M, Tuberosa R, Lawley C, Mikoulitch I, Cavanagh C, Edwards KJ, Hayden M, Akhunov E (2014) Characterization of polyploid wheat genomic diversity using a high-density 90,000 single nucleotide polymorphism array. Plant Biotechnol J 12:787–796
10. Manly KF, Cudmore Jr RH, Meer JM (2001) Map manager QTX, cross-platform software for genetic mapping. Mamm Genome 12:930–932
11. Lorieux M (2012) MapDisto: fast and efficient computation of genetic linkage maps. Mol Breed 30:1231–1235
12. Kosambi DD (1943) The estimation of map distances from recombination values. Ann Eugen 12:172–175
13. Voorrips R (2002) MapChart: software for the graphical presentation of linkage maps and QTLs. J Hered 93:77–78

Chapter 14

Chromosome Engineering Techniques for Targeted Introgression of Rust Resistance from Wild Wheat Relatives

Peng Zhang, Ian S. Dundas, Steven S. Xu, Bernd Friebe, Robert A. McIntosh, and W. John Raupp

Abstract

Hexaploid wheat has relatively narrow genetic diversity due to its evolution and domestication history compared to its wild relatives that often carry agronomically important traits including resistance to biotic and abiotic stresses. Many genes have been introgressed into wheat from wild relatives using various strategies and protocols. One of the important issues with these introgressions is linkage drag, i.e., in addition to beneficial genes, undesirable or deleterious genes that negatively influence end-use quality and grain yield are also introgressed. Linkage drag is responsible for limiting the use of alien genes in breeding programs. Therefore, a lot of effort has been devoted to reduce linkage drag. If a gene of interest is in the primary gene pool or on a homologous chromosome from species in the secondary gene pool, it can be introgressed into common wheat by direct crosses and homologous recombination. However, if a gene of interest is on a homoeologous chromosome of a species belonging to the secondary or tertiary gene pools, chromosome engineering is required to make the transfer and to break any linkage drag. Four general approaches are used to transfer genes from homoeologous chromosomes of wild relatives to wheat chromosomes, namely, spontaneous translocations, radiation, tissue culture, and induced homoeologous recombination. The last is the method of choice provided the target gene(s) is not located near the centromere where recombination is lacking or is suppressed, and synteny between the alien chromosome carrying the gene and the recipient wheat chromosome is conserved. In this chapter, we focus on the homoeologous recombination-based chromosome engineering approach and use rust resistance genes in wild relatives of wheat as examples. The methodology will be applicable to other alien genes and other crops.

Key words Wheat, *Triticum aestivum*, Wild relatives, Chromosome engineering, Homoeologous recombination, Interspecies gene transfer, Linkage drag, Rust resistance

1 Introduction

Hexaploid, common or bread wheat (*Triticum aestivum* L., 2n = 6x = 42, AABBDD), belongs to the *Triticum* genus of the tribe Triticeae in family Poaceae or Gramineae. Up to 36 genera have been described by various taxonomists in the tribe Triticeae [1]. The wild relatives of wheat represent a vast resource for

Sambasivam Periyannan (ed.), *Wheat Rust Diseases: Methods and Protocols*, Methods in Molecular Biology, vol. 1659, DOI 10.1007/978-1-4939-7249-4_14, © Springer Science+Business Media LLC 2017

agriculturally useful genes. For the rusts alone, which are the most damaging foliar diseases of wheat, there are 38 cataloged leaf rust, 36 stem rust, and 17 stripe rust resistance genes that have been transferred to wheat from the primary, secondary, and tertiary gene pools [2]. These genes are from *Triticum*, *Aegilops*, *Secale*, *Thinopyrum*, *Dasypyrum*, and *Elymus* species, and some have been developed into widely cultivated, high yielding varieties [3]. Hybrids (amphiploids) between genera in the Triticeae rarely result in agriculturally acceptable crops, the notable exceptions being tetraploid wheat (*T. turgidum* ssp. *dicoccum* and *durum*, 2n = 4x = 28, AABB), common wheat, and triticale (×*Triticosecale* Witt., 2n = 6x = 42 or 2n = 8x = 56, AABBRR or AABBDDRR). In the vast majority of cases, hybrids between wheat and wild relatives contain too many deleterious or undesirable characters from the wild species to be suitable for direct use in agriculture.

Chromosome engineering is directed at inducing transfer of an alien chromosome segment carrying a targeted resistance gene to a related wheat chromosome so that minimal alien chromatin lacking deleterious or undesirable genes is transferred without loss of indispensable wheat genes. This chapter describes the procedures of utilizing modern chromosome engineering techniques for targeted introgression from wild relatives into wheat using rust resistance genes as examples.

2 Materials

1. Murashige and Skoog (MS) basal medium: 50 g/L sucrose, 8% agar. In 750 mL of ddH_2O, add 50 mL of 10× MS Basal Salt Macronutrient Solution (Sigma, M0654), 100 mL of 10× MS Basal Salt Micronutrient Solution (Sigma, M0529), 1.5 mL of 1000× MS Vitamin Solution (Sigma, M3900), 1 mL of 1 mg/mL naphthalene acetic acid (Sigma, N1641), and 50 g of sucrose; adjust pH to 5.8 using 0.5 M of KOH or HCl (*see* **Note 1**). Add ddH_2O to adjust the volume to 1 L. Add 8 g of agar, autoclave for 20 min at 120 °C. Pour the medium into petri dishes after cooling (500 mL should be enough for 20 plates).
2. 0.03% colchicine solution: Dissolve 300 mg of colchicine powder and 100 mg of gibberellic acid (GA_3) in 500 mL of ddH_2O water, add 20 mL of DMSO (dimethyl sulfoxide) and 300 μL of Tween 20; adjust pH to 5.5 and volume to 1 L by adding ddH_2O.

3 Methods

Steps involved in successful targeted transfer of a rust resistance gene from a wild relative to wheat are presented in Fig. 1. Details on the critical steps are as follows.

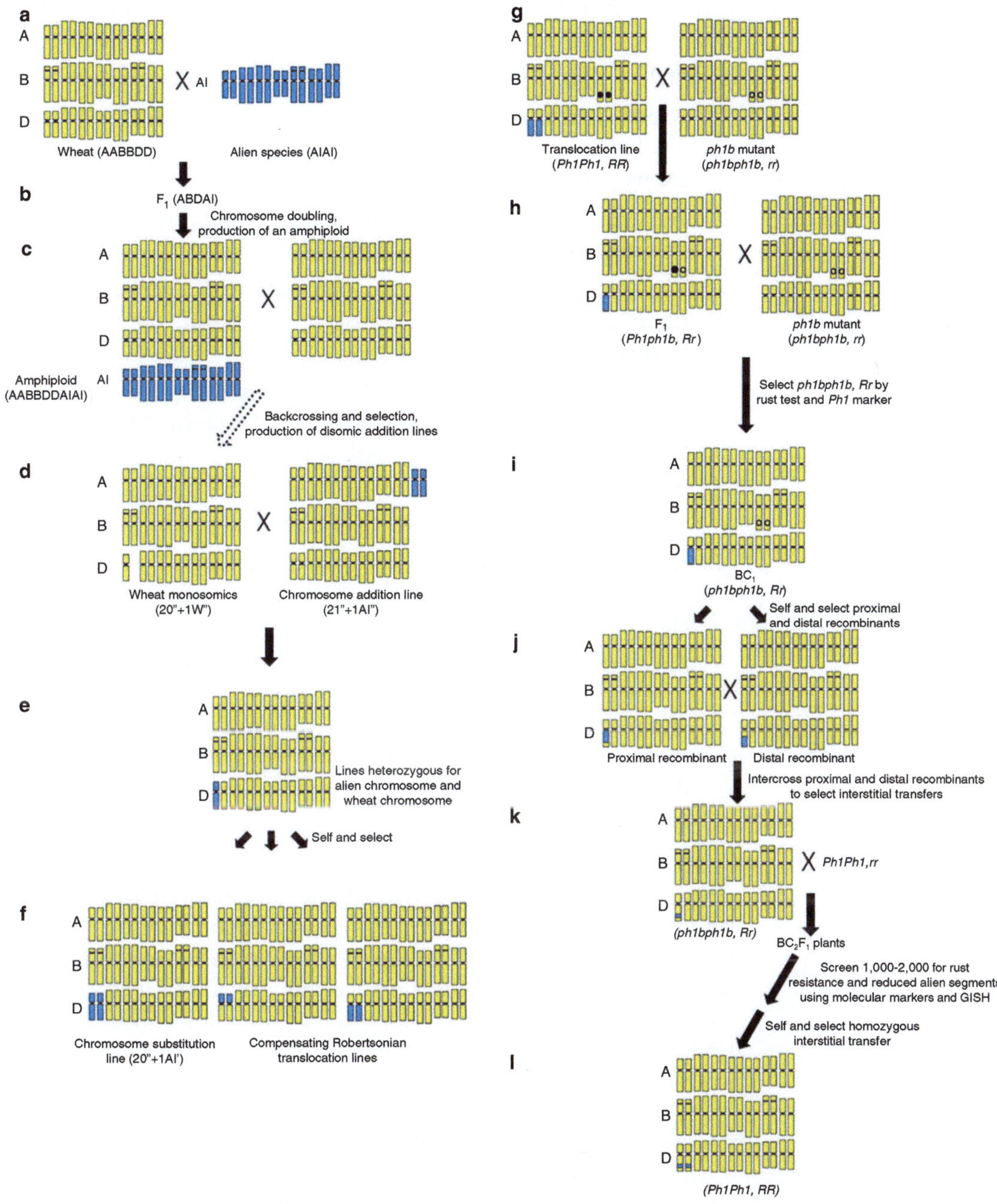

Fig. 1 Scheme for targeted transfer of a rust resistance gene from a diploid wild relative to wheat. Wheat and alien chromosomes are illustrated by *yellow* and *blue* colors, respectively. *Solid* and *open* circles on chromosome 5BL represent the *Ph1* and *ph1b* alleles, respectively (**g–i**)

3.1 Production of an Amphiploid

3.1.1 Hybridization Procedure

The first step in engineering alien chromosomes is to combine the genomes of the wild and cultivated species to produce a hybrid (Fig. 1a). Depending on the specific wild relative carrying the resistance gene, it may be necessary to grow the donor species and recipient wheat line under quarantine conditions. In this case, rust resistance needs to be tested in a Quarantine Containment Level 2 (QC2) facility. There are a few criteria in choosing the recipient wheat variety (*see* **Note 2**).

1. Select a wheat spike just emerging from the flag leaf. Remove a few lower and upper spikelets and retain middle spikelets that are well developed.
2. With fine forceps, remove the central florets of each spikelet, retaining the primary and secondary florets.
3. With fine forceps, carefully open each floret and remove all three anthers prior to dehiscence without damaging the stigma.
4. With fine scissors, cut off the top one-third of the florets. Cover the entire emasculated spike with a light-weight glassine bag to exclude extraneous pollen/pollinators.
5. When stigmas become feathery (1–3 days after emasculation), collect nondehisced anthers from the male parent (*see* **Note 3**).
6. Apply pollen to each female floret using forceps. Re-cover the pollinated spike with the glassine bag and secure with a paper clip.
7. Successful fertilization should be obvious 2 days after pollination when the stigma appears withered. The glumes, lemmas and paleae of unfertilized florets will open wider. After 5 days, the young developing grain will become obvious. At this stage, it is unconfirmed whether it is a genuine hybrid or a product of self-pollination or haploidy.

Sometimes direct hybridization between wheat and a wild species is unsuccessful irrespective of the direction of the cross. A bridging cross may be used. For instance, hybridization between tetraploid *T. dicoccoides* and diploid *Ae. umbellulata* followed by crosses with *T. aestivum* was a successful means of incorporating rust resistance from *Ae. umbellulata* into hexaploid wheat [4]. The bridging host genotype must permit tracking of the target gene from the donor parent.

3.1.2 Embryo Rescue

1. About 2 weeks after fertilization, the developing hybrid grain is harvested.
2. Sterilize the seeds with 70% ethanol for 1 min and then 20% Clorox® (commercial solution) for 15 min, and rinse three times with sterilized distilled H_2O.
3. Aseptically excise the embryos, and culture the large embryos directly on MS basal medium in petri dishes. Culture the small embryos on the MS media with transplanted nursing endosperm. Keep the petri dishes with embryos at room temperature (20–24 °C) in darkness for 1–2 weeks.

4. After the embryos germinate, transfer the small seedlings to test tubes or jars containing half strength MS media with 30 g/L sucrose and 8 g/L agar. Keep them at 20–24 °C, 16 h light/8 h darkness for about 2 weeks.
5. When the plants grow to 5–6 cm, transfer them to small pots (10 cm diameter) containing sterile potting mix and fertilize with a slow-release fertilizer. Keep them in a high humidity growth chamber at 18–20 °C and 16 h day/8 h night photoperiod.

3.1.3 Chromosome Doubling and Amphiploid Production

As the seedlings grow, it is possible to verify whether the plantlets are hybrids or products of accidental self-pollination (*see* **Note 4**). An amphiploid is produced by using colchicine, which acts as a spindle fiber formation inhibitor and disturbs normal polar chromosomal migration in mitotic cells, thereby doubling the chromosome number [5]. The procedure is as follows (Fig. 1b) [6]:

1. When seedlings are at the 2–3 tiller stage (approximately 4 weeks after planting), remove them from the pot, wash the roots thoroughly, and place in a beaker containing an aerated solution of 0.03% colchicine that completely covers the roots and stem base and place at 20–22 °C in darkness for 8 h (*see* **Note 5**).
2. Remove the seedlings from the colchicine solution, thoroughly rinse the plants with running water overnight, replant in sterile potting mix in a small pot (10 cm diameter), and fertilize with a slow-release fertilizer.
3. Keep the plants in a growth chamber (16 h day/8 h night photoperiod at 14–16 °C) for about 2 weeks until recovery (new tillers grow out) and then repot and move the plants to the greenhouse (16 h day/8 h night photoperiod) at 20–24 °C until maturity (*see* **Note 6**).

3.2 Backcrossing, Resistance Testing, and Production of Alien Addition Lines

Attempts to produce amphiploids through colchicine treatment may not be successful and the interspecific hybrids may be partially or completely sterile. Maintenance of the hybrid line is essential and can be best assisted by backcrossing to the hybrid using copious amounts of pollen from a rust susceptible wheat variety (Fig. 1c). Several generations of backcrossing may be required before the lines become self-fertile.

During the backcrossing procedure, the chromosomes derived from the wild species will be in a hemizygous state and will randomly assort at meiosis to produce gametes with variable chromosome numbers. After several backcrosses or selfed generations, a set of lines will eventually be produced each containing alien-derived chromosomes ranging in number from the full genome complement to zero (= euploid wheat). These alien chromosome addition lines are not a "weed risk" and are able to be grown outside of quarantine glasshouse facilities.

Rust resistance testing should take place during backcrossing or at this stage to (a) confirm that the resistance gene detected in the wild species effectively functions in the wheat background, and (b) to identify rust resistant addition lines carrying single alien chromosomes. Markers specific to the seven homoeologous groups in Triticeae species should be used to identify the alien chromosome [7, 8]. The barley chromosome map [9] is a reliable starting point as a guide for suitable molecular markers to identify chromosomes from *Aegilops* and *Thinopyrum* spp. [8, 10]. When the alien chromosome carrying the rust resistance has been identified, a range of markers spread across both arms must then be selected. At this stage, the exact location of the resistance gene is unknown.

3.3 Introgression of Alien Chromosome Segments into Wheat Chromosomes

Transfer of a resistance gene from an alien chromosome to a wheat chromosome can be achieved by (a) tissue culture or spontaneous translocation, (b) radiation-induced translocation, or (c) homoeologous chromosome pairing. Spontaneous (centric or Robertsonian, and noncentric) translocations occur after breakage of the alien chromosome and any wheat chromosome followed by random fusion of chromosome arms. The most widely grown spontaneous translocations involve the 1RS chromosome arm from rye. Despite the huge success of wheat–rye T1BL•1RS and T1AL•1RS translocations, such translocations occur at low frequency and are mostly noncompensating. Radiation-induced translocations occur following chromosome breakage caused by radiation [4, 12]. Breakage of the alien and wheat chromosomes is completely random and may occur in interstitial regions of the chromosomes. Fusion of broken chromosome ends may occur randomly. The resulting translocations are usually genetically unbalanced and show low fertility. However, if the targeted gene is in the proximal region of a chromosome or the synteny between the alien chromosome carrying the target gene and wheat chromosome is not conserved, these two methods will provide opportunities for introgression. This procedure needs to be accompanied by strong selection for compensating translocations [13].

The highest chance of producing an alien introgression line with a balanced genetic content involves homoeologous recombination where genetically related alien and wheat chromosomes pair. This is achieved by removing the *Ph1* gene on wheat chromosome 5BL through the use of nulli-5B wheat lines or use of the *ph1b* mutant [14], which has a deleted interstitial segment of chromosome 5BL that carries *Ph1*. Recombination rates are beyond the control of researchers and the pairing frequency of the alien and wheat chromosomes cannot be accurately predicted, because it varies among the chromosomes and genotypes. However, it can be estimated by analyzing meiotic metaphase I pairing in plants homozygous for *ph1b* and heterozygous for the homoeologous wheat and alien chromosomes or chromosome segments. The homoeologous recombination procedure involving *ph1b* is:

1. Cross the alien chromosome addition line, alien chromosome substitution line or translocation line with a *ph1bph1b* wheat genotype (Fig. 1g uses a chromosome translocation line as an example) (*see* **Note 7**).
2. F_1 plants should be screened and selected for the presence of the alien chromosome using an appropriate rust testing procedure.
3. Selected F_1 plants are backcrossed to a *ph1bph1b* wheat genotype (Fig. 1h). BC_1F_1 seedlings from this cross should be screened for the alien chromosome and *ph1bph1b* genotype using rust testing and molecular markers, respectively. The *ph1bph1b* plants can be identified using the PCR marker ABC302.3 (F: 5′-ATAAAGGAGAAGATTGAGTC-3′, R: 5′-ATAAGGAACAGGAACAGAGT-3′) with annealing at 51 °C for 60 s and extension at 72 °C for 70 s [15].
4. Self BC_1F_1 plants (Fig. 1i). Screening for dissociation of markers linked to the alien chromosome is performed on BC_1F_2 seedlings carrying the alien chromosome and *ph1bph1b* genotype. Proximal and distal primary recombinants are recovered (*see* **Note 8**).
5. Intercross proximal and distal recombinants to select interstitial secondary recombinants (Fig. 1j) (*see* **Note 9**).
6. Plants showing dissociation of widely separated alien chromosome-specific markers must be (a) progeny tested to confirm the marker results, and (b) crossed with normal wheat *Ph1Ph1* genotypes to restore regular chromosome pairing (Fig. 1k).
7. Screen 1000–2000 BC_2F_1 plants for rust reaction and reduced alien segments using molecular markers and genomic in situ hybridization (GISH) (*see* **Note 10**).
8. Self and select homozygotes with distal and interstitial transfers (Fig. 1l).
9. Putative recombinants should be rust tested to determine if the resistance gene is present.

3.4 Verification of the Recombinant Structure of Dissociation Lines

Researchers should be cautious in assuming that dissociation of alien markers represents the production of true recombinants. Deletion mutants (including spontaneous production of telocentric chromosomes from non-paired chromosomes) will also show loss of some alien markers. Verification that wheat–alien chromosome interchanges have occurred involves proof by genetic and cytological evidence.

1. In lines segregating for the alien chromosome segment, markers associated with the wheat chromosome involved in recombination/translocation events will cosegregate. By using a codominant molecular marker specific to the alien and wheat homoeologues, the researcher should look for segregation of both the alien and wheat genome-related markers. For example, self-fertilization of a plant monosomic for a recombinant chromosome and its normal wheat homoeologue might show 25% of progeny with only the alien-specific markers, 50% showing both the alien and wheat markers and 25% showing only the wheat marker. This is the "expected" scenario only; in many cases, the chromosomes with alien segments will show biased transmission rates.
2. In lines homozygous for shortened alien chromosome segments, the researcher should determine which wheat-specific markers are simultaneously absent. In this way, the wheat chromosome involved in the recombination/translocation event can be identified.
3. Testcrossing the putative recombinant with the corresponding wheat line (or Chinese Spring) that is nullisomic for the homoeologue (or the corresponding arm) will produce a plant monosomic for the wheat–alien chromosome recombinant. Selfed progeny of that plant will show linkage of alien chromosome markers and wheat genome-specific markers [10].
4. GISH using labeled alien DNA in excess of unlabeled wheat DNA can provide visual proof of the existence of a wheat–alien recombinant chromosome and also will be the simplest and quickest way to identify the wheat chromosome involved. GISH can be used to determine the size of the transferred alien segment and the size of the wheat segment that it replaced.

4 Notes

1. Add only one drop at a time. The pH will change dramatically.
2. Cultivated wheat is usually used as the female parent for several reasons, (a) the florets are larger than those of most wild species and hence are more easily emasculated, (b) wild relatives are usually prolific pollen producers with large anthers making them ideal male parents, and (c) retention of wheat cytoplasm will reduce the chance of loss of important but perhaps undocumented maternal influences or the introduction of deleterious genes. Although not commercial cultivars, Chinese Spring and many Chinese landrace genotypes carry one or more crossability genes (*kr1* to *kr4*) [2], thereby increasing the success of producing interspecific hybrid embryos. In addition, it is

important to choose a recipient wheat variety that is rust susceptible so that the presence of the resistance gene(s) from the wild species can be verified easily in subsequent generations.

3. Anthers ready for pollination will be evenly yellow (not green) and will dehisce within seconds of contact with a warm hand.
4. It is possible to verify whether the plantlet is a true hybrid by the presence of (a) morphological characteristics from the male parent and/or (b) molecular markers derived from the wild species.
5. During the treatment, provide a gentle air flow into the colchicine solution to supply the roots with oxygen. A small air pump suitable for a domestic aquarium is ideal for aeration.
6. When the plants are growing vigorously, transplant them to 25 cm diameter pots to allow tillering, which is very important because late tillers may be derived from cells with doubled chromosome numbers and may show improved fertility over early tillers.
7. Although the original *ph1b* mutant is in the Chinese Spring [14], the researcher is encouraged to transfer the *ph1b* mutant allele to a more agriculturally acceptable cultivar by backcrossing as was done for Pavon 76, Angas and Overley. This is to improve the chance of producing wheat–alien chromosome recombinants in agriculturally adapted genetic backgrounds.
8. Identification of putative wheat–alien chromosome recombinants/translocations is dependent on the alien chromosome being in a monosomic state. If BC_1F_2 seedlings do not show segregation of alien markers, recross those plants with the *ph1bph1b* wheat genotype and rescreen for populations segregating for the alien chromosome. If this is unsuccessful, the researcher may be dealing with a gametocidal (*Gc*) chromosome [16].
9. Alternatively, **steps 4** and **5** can be skipped. Instead, the resistant BC_1F_1 plants with *ph1bph1b* are crossed with normal susceptible wheat with genotype *Ph1Ph1* to produce a large BC_2F_1 population, a procedure that can be more efficient in some cases [6, 17].
10. Integrated use of molecular markers and cytogenetic techniques, such as GISH and C-banding, improve the efficiency and precision of homoeologous-based chromosome engineering.

Acknowledgments

The authors thank the support by the Grains Research and Development Corporation, Australia (PZ, ISD, RAM), USDA-ARS CRIS Project No. 3060-520-037-00D (SSX), and the WGRC I/

UCRC NSF contract 1338897 (BF, WJR). This is contribution number 17-040-B from the Kansas Agricultural Experiment Station, Kansas State University, Manhattan, KS 66506-5502, USA.

References

1. http://www.k-state.edu/wgrc/Taxonomy/Xtra tables/trit1930+.html. Accessed 9 June 2016
2. McIntosh RA, Yamazaki Y, Dubcovsky J, Rogers J, Morris C, Appels R, Xia XC (2013) Catalogue of gene symbols for wheat. In: Proceedings of the 12th international wheat genetics symposium, Yokohama, Japan, 8–13 September 2013. http://www.shigen.nig.ac.jp/wheat/komugi/genes/download.jsp
3. McIntosh RA, Wellings CR, Park RF (1995) Wheat rusts – an atlas of resistance genes. CSIRO Publications, Australia
4. Sears ER (1956) The transfer of leaf-rust resistance from *Aegilops umbellulata* to wheat. Brookhaven Symp Biol 9:1–22
5. Jensen CJ (1974) Chromosome doubling techniques in haploids. In: Kasha KL (ed) Haploids in higher plants – advances and potential. University of Guelph, Guelph, Canada, pp 153–190
6. Niu Z, Jiang A, Hammad WA, Oladzadabbasabadi A, SS X, Mergoum M, Elias EM (2014) Review of doubled haploid production in durum and common wheat through wheat × maize hybridization. Plant Breed 133:313–320
7. Sharp PJ, Desai S, Chao S, Gale MD (1988) Isolation, characterization and applications of a set of 14 RFLP probes identifying each homoeologous chromosome arm in the Triticeae. In: Miller TE, Koebner RMD (eds) Proceedings of the seventh international wheat genetics symposium, Cambridge, UK, 13–19 July 1988. Institute of Plant Sciences Research, Cambridge, UK, pp 639–646
8. Dundas IS, Verlin DC, Park RF, Bariana HS, Butt M, Capio E, Islam AKMR, Manisterski J (2008) Rust resistance in *Aegilops speltoides* var. *ligustica*. Proceedings of the 11th international wheat genetics symposium, vol 1, Brisbane, Australia, 24–29 August 2008, pp 200–202. http://ses.library.usyd.edu.au/handle/2123/3372
9. Langridge P, Karakousis A, Collins N, Kretschmer J, Manning S (1995) A consensus linkage map of barley. Mol Breed 1:389–395
10. Dundas I, Zhang P, Verlin D, Graner A, Shepherd K (2015) Chromosome engineering and physical mapping of the *Thinopyrum ponticum* translocation in wheat carrying the rust resistance gene *Sr26*. Crop Sci 55:648–657
11. Schlegel R, Korzun V (1997) About the origin of 1RS 1BL chromosome translocations from Germany. Plant Breed 116:537–540
12. Knott DR (1961) The inheritance of rust resistance. VI. The transfer of stem rust resistance from *Agropyron elongatum* to common wheat. Can J Plant Sci 41:109–123
13. Sears ER (1993) Use of radiation to transfer alien segments to wheat. Crop Sci 33:897–901
14. Sears ER (1977) An induced mutant with homoeologous pairing in common wheat. Can J Genet Cytol 19:585–593
15. Wang X, Lai J, Liu G, Chen F (2002) Development of a SCAR marker for the *Ph1* locus in common wheat and its application. Crop Sci 42:1365–1368
16. Endo TR (1978) On the *Aegilops* chromosomes having gametocidal action on common wheat. In: Ramanujam S (ed) Proceedings of the 5th international wheat genetics symposium, vol 1, New Delhi, India, pp 306–314
17. Niu Z, Klindworth DL, Friesen TL, Chao S, Jin Y, Cai X, SS X (2011) Targeted introgression of a wheat stem rust resistance gene by DNA marker-assisted chromosome engineering. Genetics 187:1011–1021

Chapter 15

Applications of Genomic Selection in Breeding Wheat for Rust Resistance

Leonardo Ornella, Juan Manuel González-Camacho, Susanne Dreisigacker, and Jose Crossa

Abstract

There are a lot of methods developed to predict untested phenotypes in schemes commonly used in genomic selection (GS) breeding. The use of GS for predicting disease resistance has its own particularities: (a) most populations shows additivity in quantitative adult plant resistance (APR); (b) resistance needs effective combinations of major and minor genes; and (c) phenotype is commonly expressed in ordinal categorical traits, whereas most parametric applications assume that the response variable is continuous and normally distributed. Machine learning methods (MLM) can take advantage of examples (data) that capture characteristics of interest from an unknown underlying probability distribution (i.e., data-driven). We introduce some state-of-the-art MLM capable to predict rust resistance in wheat. We also present two parametric R packages for the reader to be able to compare.

Key words Rust resistance, Genomic selection, Machine learning

1 Introduction

The development of low-cost, high-throughput, genotyping strategies has made it possible for genomic selection (GS) to offer new opportunities for increasing the efficiency of plant breeding programs [1].

Using GS for predict quantitative adult plant resistance or APR has its own particularities:

1. Although there is general consensus about the additive nature of APR to rust, some populations present epistasis. Parametric (linear) GS approaches are restricted to modeling additive effects, whereas most MLM also allow modeling epistasis.

Electronic supplementary material: The online version of this chapter (doi:10.1007/978-1-4939-7249-4_15) contains supplementary material, which is available to authorized users.

Sambasivam Periyannan (ed.), *Wheat Rust Diseases: Methods and Protocols*, Methods in Molecular Biology, vol. 1659, DOI 10.1007/978-1-4939-7249-4_15, © Springer Science+Business Media LLC 2017

2. Most traits are affected by large numbers of small-effect genes, whereas APR is based on effective combinations of major and minor genes. MLM provide a general suite of flexible methods for this purpose.
3. Disease resistance is commonly expressed in ordinal categorical traits (e.g., 1–9, or 1–5. Even if the data are transformed, the aforementioned problems may remain in the model. In MLM there is no imposition regarding the distribution of both predictor and response variables, avoiding the problems that may arise in parametric GS.

The objective of this work is to present several state-of-the-art machine learning solutions developed for GS, able to deal with the particularities of the architecture of wheat diseases, and plant diseases in general. We also present the protocols of BGLR and rrBLUP packages, which are also suitable for predicting disease resistance.

2 Materials

All scripts included herein are to be executed in R [2]. Commands are shown in the boxes. Before starting, we recommend all required packages to be installed by typing install.packages() in the command line. Scripts in the boxes can be edited using the R-studio software (https://www.rstudio.com/products/rstudio/download/). In supplementary materials we attach two complete rust datasets for practice, Supplementary Tables 1 and 2. Additional data sets and R scripts can be found by exploring the CIMMYT repository (http://repository.cimmyt.org/) (*see* **Note 1**).

3 Methods

3.1 General Protocol for GS

To perform GS, a population that has been both genotyped and phenotyped, i.e., the training population (TP), is used to train or calibrate a GS model, which is in turn used to predict breeding or genotypic values of non-phenotyped selection candidates (Box 1). This second set of individuals can be referred to as the breeding population (BP) [1].

Before the GS procedure, it is necessary a protocol for load and process the data (Box 1). Usually, the genotype and phenotype data is presented in a numeric format (Table 1); however, raw marker data, e.g., alleles coded as pair of observed alleles “A/T,” “G/C,” or by genotypes “AA,” “BB,” “AB,” may be present, and it is necessary to recode it into the numeric format [3]. In our protocol, missing data is imputed by random forest [4]. If time is demanding and missing data is less than 20%, imputation by the mean or the mode of the nonmissing values at that marker is enough

Box 1
R script load and preprocess data from training and a breeding population:

```
  require(synbreed)
  require(missForest)
  setwd('.../workingDirectory/)
## loading a numeric matrix (e.g. Table 1)  a training and a breeding population
      datasT <- read.csv("trainingpopulation.csv",header=FALSE)
      datasB <- read.csv("breedingpopulation.csv",header=FALSE)
      MarkersTrain <- datasT [,-ncol(datasT)] ## subtract the column of
      phenotype
      MarkersBred <- datasB[,-ncol(datasB)]  ## there's not phenotypic data
      y <- datasT [,ncol(datasT)] ## extract the column of phenotypic data
      u <- datasB [,ncol(datasB)] ## extract the column of phenotypic data
## standardized in the format (-1, 0, 1)
      MarkersTrain  <-  MarkersTrain - 1  #    SNPs
      MarkersTrain <- 2* MarkersTrain - 1   #   DArT
      # DO THE SAME WITH THE MarkersBred DATA
      # MarkersBred   <-  MarkersTrain - 1  #    SNPs
      MarkersBred  <- 2* MarkersBred  - 1   #   DArT
## delete features with more than 50% missing data and imputing
## case A: when training and breeding populations are related
      miss <- c()
      data = rbind(MarkersTrain, MarkersBred)
      for(i in 1:ncol(data)) {
      if(length(which(is.na(data [,i]))) > 0.5*nrow(data)) miss <- append
      (miss,i)
      }
      if (!is.null(miss)) data <- data [,-miss]        newMarker  <-
      missForest(as.data.frame(data))
      impMarker  <-  newMarker$ximp
      trn <- 1:nrow(MarkersTrain)
      impMarkerTrain <- impMarker[trn,]
      impMarkerBreed <- impMarker[-trn,]
## delete features with more than 50% missing data and imputing
## case B: when training and breeding populations are unrelated
      miss <- c()
      for(i in 1:ncol(MarkersTrain)) {
      if(length(which(is.na(MarkersTrain [,i]))) > 0.5*nrow(MarkersTrain))
      miss <- append(miss,i)
      }
      if (!is.null(miss)) MarkersTrain <- MarkersTrain [,-miss]  miss <- c()
      for(i in 1:ncol(MarkersBred)) {
      if(length(which(is.na(MarkersBred[,i]))) > 0.5*nrow(MarkersBred)) miss
      <- append(miss,i)
      }
```

(continued)

Box 1 (continued)

```
        if (!is.null(miss)) MarkersBred <- MarkersBred[,-miss]  newMarkerT <-
         missForest(as.data.frame(MarkersTrain))
         impMarkerTrain  <-  newMarkerT $ximp
         newMarkerB <-    missForest(as.data.frame((MarkersBred))
         impMarkerBreed  <-  newMarkerB$ximp
## minor allele frequency e.g. MAF <= 5% case A and B
         S <- which(colMeans(impMarkerTrain) > 0.9)
         impMarkerTrain <- impMarkerTrain[,-S]
         impMarkerBreed <- impMarkerBreed [,-S]
         S <- which(colMeans(impMarkerTrain) < -0.9)
         impMarkerTrain <- impMarkerTrain[,-S]
         impMarkerBreed <- impMarkerBreed [,-S]
```

Table 1
Example of a dataset with genotype and phenotype, in numeric format, as used in this work

M_1	M_2	M_3	M_i	M_p	Y
1	−1	0	...	1	−14.67
0	1	1	...	NA	−3.14
...	...	...	...	...	...
−1	0	NA	...	0	49.55
−1	1	NA	...	1	29.87

(*see* **Note 2**). In package Synbreed there are other alternatives for imputation [3] (*see* **Note 3**). To fulfill the requirements of rrBLUP, in our protocol the numeric format is standardized in the format centered in zero (−1, 0, 1) to carry out analysis (Box 1).

Finally, in Box 2, we present a general MLM protocol of GS. Methods selected are random forest [5] and support vector regression with linear kernel [6] (*see* **Note 4**). Later, we also present some parametric alternatives (Subheading 3.3).

3.2 Cross-Validation to Assess the Performance of GS Methods

There is a plethora of methods developed to GS, each one adapted for a particular situation, the analyst does not know "a priori" which the best to predict the non-phenotyped BP. To have an idea of which method would perform better, we recommend to perform a cross-validation scheme on the TP (Box 3). Briefly, k random partitions are generated using a predefined random binary matrix of order n × k (n is the number of phenotyped individuals, and k is the number of folds, generally 10–50); each column of the matrix randomly assigns 80–90% individuals of the TP to a training

Box 2
R script for predict non-phenotyped individuals using MLM and a phenotyped training population:

```
  require(mlr)
  require(randomForest)
  require(kernlab)
  p <- ncol(impMarkerTrain)
  dataGG <- as.data.frame(cbind(impMarkerTrain,y))
  rdesc <- makeResampleDesc("CV", iters = 10L)
  taski <- makeRegrTask (data = dataGG, target = "y")
  ctrl <- makeTuneControlGrid()
###Random Forest Optimization
### Optimization is time demanding, default values may present good enough
predictions
  lrner <- makeLearner("regr.randomForest")
  ranMtry <- round (p/c(2,2.5,3,3.5,4))
  psRF  <- makeParamSet(
    makeDiscreteParam ("mtry", values = ranMtry),
    makeDiscreteParam("ntree", values = seq(400, 600, by = 50))
  )
  resRF = tuneParams( lrner, task = taski, resampling = rdesc, par.set = psRF,
  control = ctrl)
  myrf <- randomForest(impMarkerTrain, y, ntree= resRF[3] $x$ntree, mtry= resRF
  [3] $x$mtry)  #Random Forest Optimized
  #myrf <- randomForest(impMarkerTrain, y)  #  default mtry = p/3 ntree = 500
  RFpred <- as.numeric(predict(myrf, impMarkerBreed))
  print (RFpred)
  cor(u,RFpred) ## plot(u,RFpred)
###SVR Optimization
  lrner = makeLearner("regr.ksvm")
  psSVR = makeParamSet(
    makeDiscreteParam("kernel", values = "vanilladot"),
    makeDiscreteParam("C", values = 2^(-8:2))
  )
  res = tuneParams( lrner, task = taski, resampling = rdesc,par.set = psSVR,
  control = ctrl)
  mySVR <- ksvm(y~.,data=dataGG,kernel="vanilladot",C=res[3]$x$C)
  SVRpred = as.numeric(predict(mySVR, impMarkerBreed))
  print(SVRpred)
  cor(u, SVRpred) ## plot(u, SVRpred)
```

set and the 20–10% remaining to a testing set (Table 2) (in Box 6 we present a protocol to generate this random matrix). Each GS model is fitted to the training set; predictions for individuals in the test set are compared with the observed values using Pearson's correlation coefficient or mean squared error (*see* **Note 5**). Finally, the performances of different methods are contrasted by an appropriate statistical test, e.g., paired samples Wilcoxon test.

Box 3
Cross validation script to evaluate the performance of a gs method in a phenotyped population using RF and a parametric model of BGLR package:

```
  require(randomForest)
  require(BGLR)
  setwd('.../workingDirectory/)
## required perform imputing and MAF protocol of Box 1
  set <- read.csv("index.csv", header=FALSE)
  for (fold in 1:50){
        train <-  which(set[,fold]==1)
        pred  <-  which(set[,fold]==2)
## Bayes B
        yNa <- y
        yNa[pred] <- NA
        ETA<-list(list(X= impMarkerTrain, model='BayesB'))
        fit_BGLR<- BGLR(y=yNa, ETA=ETA, nIter=7000,
   burnIn=2500,thin=3,saveAt='',df0=5, S0 = NULL, weights=NULL,R2=0.5)
        resultsBGLR <- fit_BGLR $yHat[pred]
## Random Forest  ## optional: optimization protocol of box 1
        myrf <- randomForest(round(impMarker [train,]), y[train])
        Ypred <- as.numeric(predict(myrf,round(impMarker [pred,])))
        CorBGLR  <- cor(resultsBGLR,y[pred])
        CorRF <- cor(Ypred,y[pred])
        ouputCor <- cbind(fold, CorBGLR, CorRF)
        mseBGLR <- ((resultsBGLR -y[pred]) %*%(resultsBGLR -y[pred]))/length
        (y[pred])
        mseRF <- ((Ypred-y[pred]) %*%(Ypred-y[pred]))/length(y[pred])
        ouputMSE <- cbind(fold, mseBGLR, mseRF)
        write.table(ouputCor,"resultsCorrelation.csv", sep =",",append = TRUE)
        write.table(ouputMSE,"resultsMSE.csv",   sep =",",append = TRUE)
    }
```

Table 2
Example of a (n × k) binary matrix used to generate the k random partitions

$Fold_1$	$Fold_2$	...	$Fold_2$	$Fold_k$
1	2	1	...	1
2	1	1	...	2
...	...	...	...	...
1	2	2	...	1
2	1	2	...	1

n is the number of individuals in the training populations
If the ith element in the j-th column is 1, the i-th individual is assigned to the train set, and if 2, the i-th individual is assigned to the test set

Box 4
R script to perform GS using rrBLUP and BGLR:

```
## RUNNING rrBLUP
  require(BGLR)
  require(rrBLUP)
  yNa <- c(y,rep(NA,nrow(impMarkerBreed)) )
  impMarker <- rbind(impMarkerTrain, impMarkerBreed)
  A1 <- A.mat(impMarker,impute.method="EM",max.missing=0.5,shrink=FALSE)
  rownames(A1) = 1:nrow(A1)
  set.seed(456)
  data2 <- data.frame(y= yNa,gid= 1:nrow(A1))
  ans1 <- kin.blup(data2,K=A1,geno="gid",pheno="y")
  trn <- 1: length(y)
  results = ans1$g[-trn ]
  cor(u, results) ## plot(u, results)
## RUNNING BGLR options
  #ETA<-list(list(X= impMarker, model='BayesA'))
  #ETA<-list(list(X= impMarker, model='BayesB'))
  #ETA<-list(list(X= impMarker, model='BayesC'))
  #ETA<-list(list(X= impMarker, model=' BL '))
## RUNNING BGLR gaussian
  ETA<-list(list(X= impMarker, model='BayesB'))
  fit_BGLR=BGLR(y=yNa,ETA=ETA,nIter=7000,burnIn=2500,thin=3,saveAt='',df0=5,
  S0=NULL,weights=NULL,R2=0.5)
  results <- fit_BGLR$yHat[-trn]
  cor(u, results) ## plot(u, results)
```

3.3 Alternative GS Methods

We already have presented random forest and SVM with linear kernel for GS. In Box 4, we introduce two linear, parametric rrBLUP [7] and BGLR [8] packages of R project. They are easy to implement, fast and can deal with the additives properties of rust resistance. From our experience [9], although these linear methods cannot reach the performance of random forest, in a very short time an anxious researcher can have the idea of the success of GS in a particular population. A key issue in BGLR is a two-level list used to specify the regression function: ETA (ETA <−list(list(X = impMarker, model = "BL"))). The options for model allows FIXED (Flat prior), BRR (Bayesian ridge regression with Gaussian prior), BayesA (scaled-t prior), BL (Double-Exponential prior), BayesB (two component mixture prior with a point of mass at zero and a scaled-t slab), and BayesC (two component mixture prior with a point of mass at zero and a Gaussian slab) (*see* **Note 6**).

Finally, in Box 5 we present an alternate protocol of classification to deal with scale problems [9]. We must recall that this last part is just to introduce the classification issue of GS, we strongly encourage the reader to go deeper into this matter (*see* for example https://mlr-org.github.io/mlr-tutorial/).

Box 5
R script for predict non-phenotyped individuals using a phenotyped training population and machine learning classification:

```
require(mlr)
require(randomForest)
require(kernlab)
p <- ncol(impMarkerTrain)
threshold <- quantile(y, probs = 0.3, na.rm = TRUE, type = 7)
yQual <- rep("res",length(y)) #resistant
yQual[which(y> threshold)] <-"sus" # susceptible
dataGG= cbind(as.data.frame(impMarkerTrain), as.data.frame(yQual))
rdesc = makeResampleDesc("CV", iters = 10L)
taski = makeClassifTask (data = dataGG, target = "yQual")
ctrl = makeTuneControlGrid()
###Random Forest Optimization is time demanding, default values may present good
enough predictions
lrner = makeLearner("classif.randomForest")
ranMtry <- round (sqrt(p)*c(0.5,0.75,1,1.25,1.5))
psRF  <- makeParamSet(
  makeDiscreteParam ("mtry", values = ranMtry),
  makeDiscreteParam("ntree", values = seq(400, 600, by = 50))
)
resRF = tuneParams( lrner, task = taski, resampling = rdesc, par.set = psRF,
control = ctrl)
myrf <- randomForest(impMarkerTrain, as.factor(yQual), ntree= resRF[3] $x
$ntree, mtry= resRF[3] $x$mtry)
#myrf <- randomForest(impMarkerTrain, as.factor(yQual))  # R Forest default,
mtry = √p ntree = 500
RFpred <- predict(myrf, impMarkerBreed)
print (RFpred)
###SVR Optimization
lrner = makeLearner("classif.ksvm")
psSVR = makeParamSet(
  makeDiscreteParam("kernel", values = "vanilladot"),
  makeDiscreteParam("C", values = 2^(-8:2))
)
resSVR = tuneParams( lrner, task = taski, resampling = rdesc,par.set = psSVR,
control = ctrl)
mySVR <- ksvm(yQual ~.,data = dataGG, kernel="vanilladot",C=resSVR[3]$x$C,
prob.model=TRUE)
SVRpred = predict(mySVR, impMarkerBreed)
print(SVRpred)
```

Box 6
R script generate an index for random partitions:

```
  setwd('.../workingDirectory/)
  datasT <- read.csv("trainingpopulation.csv",header=FALSE)
###set the size of training population for crossvalidation e.g. 90% or 80%
  p <- 0.9
  siz <- nrow(datasT)
  train <- round(p* siz)
  pred <- siz-train
  ind <- c(rep(1, train), rep(2, pred))
  index = sample(ind)
  for (fold in 2:50){
      index =cbind(index ,sample(ind))
    }
  write.table(index,"index.csv", sep = ",",row.names=FALSE,col.names=FALSE)
```

4 Notes

1. The GS packages we present in this protocol are the most recognized and tested in rust resistance. However, new techniques are under continuous development, or new R packages are adapted to nonadvanced users (e.g., NAM package). We encourage the reader to explore these new alternatives.
2. We adapted our protocol to remove markers with 50% of missing data; sometimes is better to lower this threshold to 30%.
3. Our protocols were designed for imputing unordered, low budget, markers. However if an order set of markers is available, the reader can try other algorithms such as Beagle or fastPHASE. Beagle is implemented in Synbreed.
4. An optimization of parameters is recommend in support vector machines (e.g., Cost parameter). From our experience, random forest, on the other hand, does not need a previous step of optimization (*see* Box 2).
5. Classification methods need alternative statistics to measure the success of the prediction, such as Area under the curve or Balanced accuracy (http://mlr-org.github.io/mlr-tutorial/release/html/measures/index.html.
6. Linear models of BGLR package are easy to implement and relatively fast. The time of the processing is highly dependent on burnIn, and nIter parameters. We recommend to set them not less than 2000 and 5000, respectively.

References

1. Bassi FM, Bentley AR, Charmet G, Ortiz R, Crossa J (2016) Breeding schemes for the implementation of genomic selection in wheat (*Triticum* spp). Plant Sci 242:23–36
2. R Core Team (2016) R: A language and environment for statistical computing. R foundation for statistical computing. Vienna, Austria. https://www.R-project.org/
3. Wimmer V, Albrecht T, Auinger HJ, Schoen CC (2012) Synbreed: a framework for the analysis of genomic prediction data using R. Bioinformatics 28:2086–2087
4. Stekhoven DJ, Bühlmann P (2011) MissForest – nonparametric missing value imputation for mixed-type data. Bioinformatics 28:112–118
5. Liaw A, Wiener M (2002) Classification and regression by random forest. R News 2:18–22
6. Karatzoglou A, Smola A, Hornik K, Karatzoglou MA (2015) An S4 Package for Kernel Methods in R. https://cran.r-project.org/eb/packages/kernlab/index.html
7. Endelman JB (2011) Ridge regression and other kernels for genomic selection with R package rrBLUP. Plant Genet 4:250–255
8. Pérez-Rodriguez P, de los Campos G (2014) Genome-Wide Regression & Prediction with the BGLR statistical package. Genetics 198:483–495
9. Ornella L, Pérez P, Tapia E, González-Camacho JM, Burgueño J, Zhang X et al (2014) Genomic-enabled prediction with classification algorithms. Heredity 112:616–626. doi:10.1038/hdy.2013.144
10. Ornella L, Singh S, Perez P, Burgueño J, Singh R, Tapia E, Bhavani S, Dreisigacker S, Braun H-J, Mathews K, Crossa J (2012) Genomic prediction of genetic values for resistance to wheat rusts. Plant Genome J 5:136–148

Chapter 16

Rapid Phenotyping Adult Plant Resistance to Stem Rust in Wheat Grown under Controlled Conditions

Adnan Riaz and Lee T. Hickey

Abstract

Stem rust (SR) or black rust caused by *Puccinia graminis* f. sp. *tritici* is one of the most common diseases of wheat (*Triticum aestivum* L.) crops globally. Among the various control measures, the most efficient and sustainable approach is the deployment of genetically resistant cultivars. Traditionally, wheat breeding programs deployed genetic resistance in cultivars, but unknowingly this is often underpinned by a single seedling resistance gene, which is readily overcome by the pathogen. Nowadays, adult plant resistance (APR) is a widely adopted form of rust resistance because more durable mechanisms often underpin it. However, only a handful of SR APR genes are available, so breeders currently strive to combine seedling and APR genes. Phenotyping adult wheat plants for resistance to SR typically involves evaluation in the field. But establishing a rust nursery can be challenging, and screening is limited to once a year. This slows down research efforts to isolate new APR genes and breeding of genetically resistant cultivars.

In this study, we report a protocol for rapid evaluation of adult wheat plants for resistance to stem rust. We demonstrate the technique by evaluating a panel of 16 wheat genotypes consisting of near isogenic lines (NILs) for known *Sr* genes (i.e., *Sr2*, *Sr33*, *Sr45*, *Sr50*, *Sr55*, *Sr57*, and *Sr58*) and three landraces carrying uncharacterized APR from the N. I. Vavilov Institute of Plant Genetic Resources (VIR). The method can be completed in just 10 weeks and involves two inoculations: first conducted at seedling stage and a second at the adult stage (using the same plants). The technique can detect APR, such as that conferred by APR gene *Sr2*, along with pseudo-black chaff (the morphological marker). Phenotyping can be conducted throughout the year, and is fast and resource efficient. Further, the phenotyping method can be applied to screen breeding populations or germplasm accessions using local or exotic races of SR.

Key words Wheat, Stem rust, Phenotyping, Seedling resistance, Adult plant resistance, Controlled environment, Speed breeding

1 Introduction

Stem rust (SR) caused by *Puccinia graminis* f. sp. *tritici* Pers. Eriks. & E. Henn. (*Pgt*) is one of the most destructive and geographically widespread diseases of breadwheat (*Triticum aestivum* L.) [1, 2]. *Pgt* has caused severe yield losses (50–70%) to wheat crops [3] and under favorable environmental conditions can destroy a field completely in just 3 weeks [1]. SR presents a major challenge as

Sambasivam Periyannan (ed.), *Wheat Rust Diseases: Methods and Protocols*, Methods in Molecular Biology, vol. 1659, DOI 10.1007/978-1-4939-7249-4_16, © Springer Science+Business Media LLC 2017

plant breeders strive to boost the rate of genetic improvement of wheat yield to meet the increasing food demand of growing human population [4]. The most sustainable and environmentally friendly strategy is the deployment of genetically resistant cultivars to reduce the losses caused by SR [5].

To date, 59 stem rust (*Sr*) genes including seedling resistance and adult plant resistance (APR) genes have been catalogued [6]. Seedling resistance is often underpinned by single major gene (R) and is effective against some races of the pathogen (i.e., race-specific). The deployment of R genes exerts selection pressure on the pathogen, and it undergoes genetic variation (i.e., mutation, gene flow, genetic drift, recombination, and natural selection) thus results in the breakdown of resistance [7]. For instance, to date, 13 variants of the SR race TTKSK (synonym Ug99) lineage have been detected (http://www.rusttracker.org, accessed February 23, 2017). On the other hand, multiple minor genes underpin APR where each gene adds small effect to overall partial resistance against all races of the pathogen thus, are considered more durable [8]. APR genes often influence pustule size, infection frequency, and latent period and thus do not exert selection pressure on the pathogen, hence they are referred as "slow rusting" genes. For instance, the APR gene *Sr2* provides effective resistance against SR and has been widely used in breeding programs for almost 100 years [5, 9–11]. APR genes are most effective at adult plant stage, thus plant breeders typically perform phenotyping and selection in the field. Regardless, the detection and deployment of APRs in breeding programs is challenging [12, 13].

Conventional breeding for disease resistance involves field-based phenotyping, which is limited to once a year, except shuttle breeding (allowing two field generations per year) at the International Maize and Wheat Improvement Center (CIMMYT), Mexico [12, 14]. The environmental factors, plant growth stage, the rate of epidemic progression, and unexpected pathotype or disease incursion often compromise the success of field phenotyping [12, 15]. These limitations slows down progress in breeding and rust research programs. Alternatively, various studies have reported successful evaluation of adult plants for resistance to foliar pathogens under glasshouse or controlled environmental conditions (CEC) [12, 15, 16, 20]. A plant management system called "speed breeding" uses constant light and controlled temperature to provide accelerated growth conditions (AGC); enabling up to six generations of wheat per year [17]. Moreover, phenotyping protocols have been adapted for other important traits in wheat (i.e., other than foliar diseases) [18, 19].

In this study, we demonstrate the ability to rapidly phenotype SR in the speed breeding system. The method involves two inoculations; first at seedling stage and second at the adult stage. The method has tremendous potential for screening large breeding populations or germplasm accessions using local or exotic pathotypes and could facilitate selection for resistance mechanisms that are stable across a range of temperatures.

2 Materials

2.1 Plant Materials

A panel comprising 16 wheat genotypes was assembled to test a newly developed protocol for phenotyping SR resistance at the adult stage using plants grown in the speed breeding system. The panel included: near isogenic lines (NILs) carrying R genes (*Sr33*, *Sr45*, and *Sr50*), APR genes (*Sr2*, *Sr55*, *Sr57*, and *Sr58*), and three landraces carrying uncharacterized APR (Table 1). The landraces were previously subjected to line purification by conducting one generation of single seed descent (SSD).

2.2 Pathogen Materials

A single *Pgt* pathotype (i.e., 343-1,2,3,5,6) was used for all speed breeding experiments. However, a mixture of pathotypes (343-1,2,3,5,6 and 34-2,12,13) was used to determine the field disease

Table 1
Information for the 16 wheat genotypes examined in this study

Sl. No.	Name	Gene information	Type	Source
1	Thatcher	None	Cultivar	[21]
2	Thatcher + *Sr57*	*Sr57*	Near isogenic line	
3	Thatcher + *Sr58*	*Sr58*	Near isogenic line	
4	Thatcher + *Sr55*	*Sr55*	Near isogenic line	
5	Chinese Spring	None	Cultivar	
6	Chinese Spring + *Sr2*	*Sr2*	Near isogenic line	
7	Chinese Spring + *Sr33*	*Sr33*	Near isogenic line	[22]
8	Chinese Spring + *Sr45*	*Sr45*	Near isogenic line	
9	CS1D 5405 (Chinese Spring + *Sr33*)	*Sr33*	Genetic stock	[22]
10	E6-*Sr33M*	*Sr33M*	Mutant for *Sr33*	[23]
11	Gabo-B + *Sr50*	*Sr50*	Genetic stock	[24]
12	Gabo-D + *Sr50*	*Sr50*	Genetic stock	[24]
13	M7-2-*Sr50*	*Sr50*	Mutant for *Sr50*	[24]
14	WLA-025	Uncharacterized APR	Landrace	[25]
15	WLA-026	Uncharacterized APR	Landrace	[25]
16	WLA-065	Uncharacterized APR	Landrace	[25]

response for all lines. Both 343-1,2,3,5,6 and 34-2,12,13 (commonly known as the "Satu triticale pathotype") occur throughout the eastern wheat belt of Australia. The 343-1,2,3,5,6 has virulence for *Sr5, Sr6, Sr7b, Sr8a, Sr9b, Sr11, Sr17* while 34-2,12,13 has virulence for *Sr5, Sr9g, Sr11, SrSatu,* and *Sr27* [26]. The rust cultures used in this study were maintained through single spore culture technique using the susceptible cultivar "Morocco" (*see* **Note 1**). Other materials used in this study were;

1. Petri dishes.
2. Filter paper.
3. Refrigerator.
4. Plastic pots (140 mm, 1.4 L).
5. Potting media comprising 70% composted pine bark fines and 30% coco peat.
6. Slow-release fertilizer.
7. Standard glasshouse.
8. Speed breeding facility: controlled temperature glasshouse fitted with sodium vapor lamps.
9. Ultrasonic fogger to maintain humidity.
10. Airbrush for pathogen inoculation.
11. Mineral oil.
12. Deionized water.
13. Dew chamber made of plastic.
14. Agar (15 g).
15. Cyclone spore collector.
16. Glass desiccator filled with 60% sulfuric acid.
17. −80 °C freezer.
18. 0.5 ml self-standing screw cap tubes (i.e., 0.1 g/tube).

3 Methods

Out of three experiments, two namely "Adult plant integrated" and "Adult plant independent" were conducted in the speed breeding system, while the third "Field disease response" was performed in the field. Below is the protocol for each experiment.

3.1 Adult Plant Integrated Method

The method outlined here is an adaptation of the rapid phenotyping method for APR to leaf rust in wheat reported by Riaz et al. [12]. The modifications for SR are reported below; however, the overall concept is the same. The adult plant integrated method involves two inoculations: first at seedling stage and again at the adult stage using the same set of plants.

3.1.1 Phenotyping at Seedling Stage

The wheat panel was evaluated for SR response at the seedling stage in a standard glasshouse at The University of Queensland, St Lucia, Queensland, Australia.

1. Ten seeds of each genotype were imbibed in a petri dish lined with a moist filter paper at room temperature for 24 h. To encourage uniform germination across wheat genotypes the imbibed seeds are placed in a refrigerator (4 °C) for 48 h. After the cold treatment, plants are returned to room temperature for 24 h.
2. Germinated seeds were transplanted into the pots (140 mm) filled with potting media (pH 5.5–6.5). At a rate of 2 g per pot, a slow-release fertilizer was applied.
3. Each pot contained four evenly spaced positions, where position one starts clockwise from the pot tag (i.e., 1–4). A single germinated seed was transplanted at each position. Genotypes were replicated three times adopting a completely randomized block design.
4. Plants were grown in a standard glasshouse, a temperature of 25/17 ± 2 °C (day/night) under natural diurnal photoperiod conditions (12 h daylight).
5. Twelve days after sowing (i.e., two-leaf stage) seedlings were inoculated (*see* **Note 2**). The SR urediniospores were suspended in light mineral oil at a rate of 0.005 g/ml. Inoculum at the concentration of 6×10^5 spores/ml was applied using an airbrush (*see* **Note 3**). Plants were then lightly misted with deionized water and placed in a dew chamber maintained at 100% humidity using an ultrasonic fogger (*see* **Note 4**).
6. After 16–18 h of incubation, plants were removed from the dew chamber and were returned to the standard glasshouse for disease development (*see* **Note 5**).
7. Fourteen days post-inoculation seedlings were assessed for infection type (IT) using the 0–4 Stakman scale [27] (Fig. 1).
8. To permit statistical analysis, the 0–4 Stakman scale which includes both numbers (e.g., 0, 1, …, 4) and symbols (e.g., ;, +) was converted to the 0–9 scale [12] (*see* **Note 6**).

3.1.2 Phenotyping at Adult Stage

At adult plant stage the panel was evaluated for resistance to SR in the speed breeding system: a fully enclosed temperature-controlled plant growth facility (5 m × 6 m) fitted with 20 low-pressure sodium vapor lamps. The lamps provided 400 watts each, generating 400–550 μmol M−2 S−1 of photosynthetically active radiation

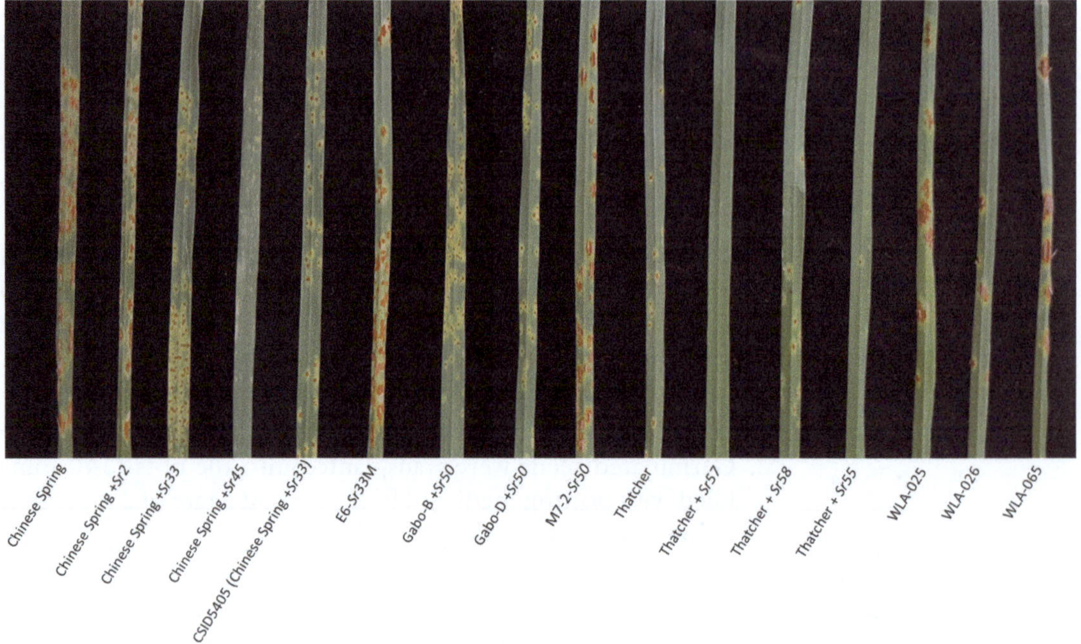

Fig. 1 Stem rust infection type on leaves of wheat seedlings grown under standard glasshouse conditions

(PAR) at pot height and 900 μmol M−2 S−1 at adult plant height (45 cm above pot level) [12].

1. Following assessment of seedlings (as described above), the same plants were re-located to the speed breeding system and subjected to 22 h light and 2 h dark period with 12 h cycling temperature regime of 22/17 ± 1 °C. Pots were positioned on the bench according to completely randomized block design.
2. Plants rapidly achieved the adult stage in just 4 weeks under AGC (*see* **Note 7**). Once all genotypes had developed a fully expanded flag leaf the plants are inoculated. Plant growth stage (GS) was recorded using Zadoks decimal code scoring system [28].
3. The plants were re-inoculated with a suspension of SR urediniospores (343-1,2,3,5,6) using the methodology, as described above. However, the elongated stems of the plants were targeted during inoculation (*see* **Note 8**). The infected plants were then sprayed lightly with mist (deionized water) and placed in a dew chamber for 12 h maintained at 100% humidity using an ultrasonic fogger.
4. Post-inoculation, infected plants relocated to the standard glasshouse with diurnal photoperiod and a 12 h cycling temperature regime at 25/17 ± 2 °C (*see* **Note 9**).

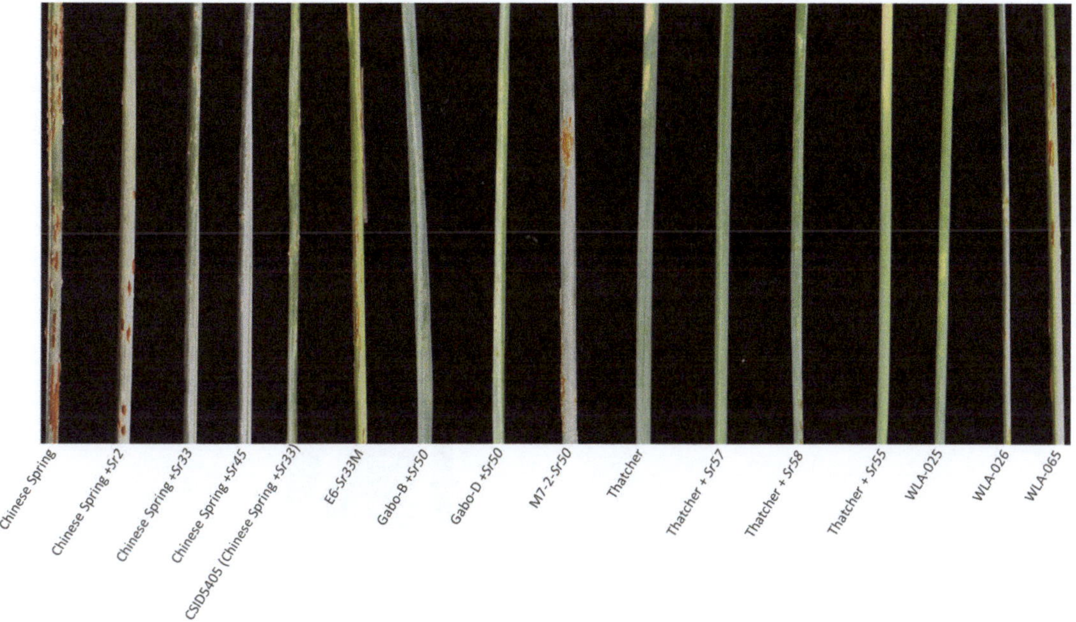

Fig. 2 Stem rust on the stems of adult wheat plants in the adult plant integrated method grown under speed breeding conditions

5. At 14 days post-inoculation, the SR response was scored using the 1–9 scale reported by Bariana et al. [29] (Fig. 2; *see* **Note 10**). The genotypes that displayed a disease response ≤5 are considered resistant. The method allowed clear detection of both seedling and APR genes for SR (Fig. 2). The method also enabled detection of pseudo-black chaff in genotype "Chinese Spring + *Sr2*," which is a morphological marker for APR gene *Sr2* (Fig. 3).
6. The adult plant integrated method can be completed within just 10 weeks. The method can be used to screen historical germplasm in genebanks to identify useful sources of resistance or evaluate breeding populations to eliminate highly susceptible genotypes in breeding populations—enabling testing of a "better" set of lines in the field (*see* **Note 11**). This technique can also accelerate the development of NILs for *Sr* genes and cloning efforts.

3.2 Adult Plant Independent Method

In this experiment, a subset of 9 genotypes from the panel was directly sown and raised in the speed breeding system and was inoculated with SR at the adult plant stage only.

1. Seeds of the wheat genotypes were germinated and transplanted as described above. The same experimental design was used and speed breeding conditions were applied, as described in the adult phenotyping section above (i.e., 22 h

Fig. 3 Completely susceptible adult SR response displayed by M7-2-*Sr50* (*left*), and restricted sporulation and pseudo-black chaff displayed by Chinese Spring + *Sr2* (*right*) obtained in the adult plant independent method

light and 2 h dark period with 12 h cycling temperature regime of 22/17 ± 1 °C). Five weeks after sowing, wheat lines achieved the adult plant stage (fully extended flag leaves) and were inoculated with SR, as outlined above. Prior to inoculation, the GS for all plants was recorded using the Zadoks scale.

2. At 14 days post-inoculation, plants were scored using the 1–9 scale reported by Bariana et al. [29] (Fig. 4).

3.3 Field Disease Response

In the field, from July to October 2016 the wheat panel was evaluated for response to SR at the Redlands Research Facility, Queensland, Australia.

1. In the field nursery, two rows of disease spreader containing the susceptible genotype "Morocco" were sown between each bay comprising two rows of hill-plots. The test material was sown as hill plots spaced 50 cm apart.
2. Three seeds of each genotype were sown as hill plots and replicated thrice in the nursery.
3. The SR epidemics were initiated by transplanting Morocco seedlings infected with 343-1,2,3,5,6 and 34-2,12,13 (as outlined above) into the field among the spreader rows about

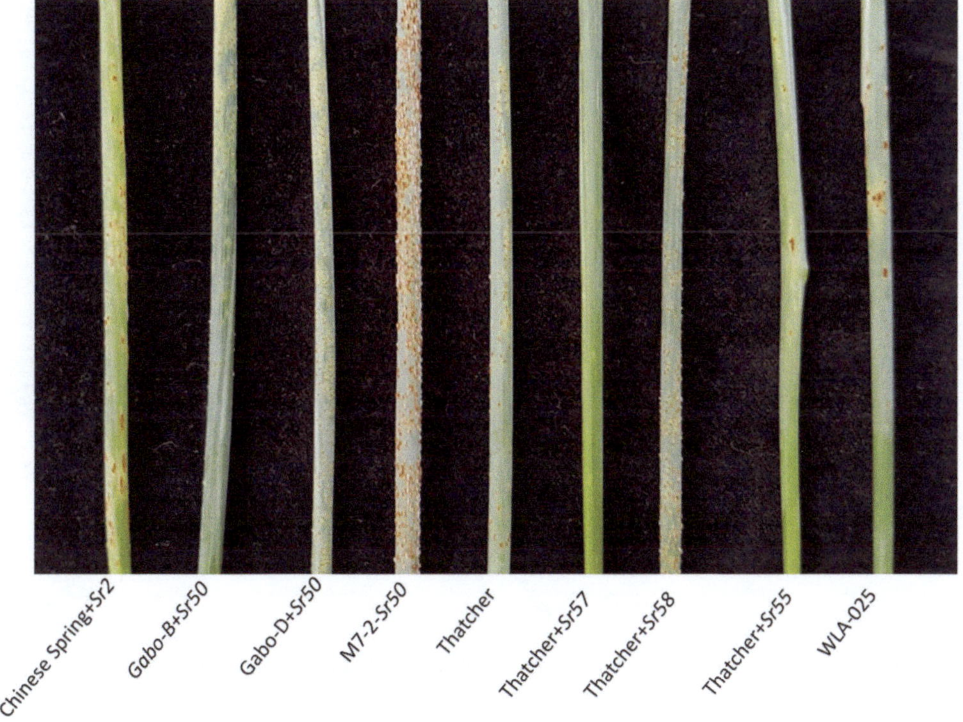

Fig. 4 The adult plant response to stem rust for the subset of nine wheat genotypes evaluated using the adult plant independent method under speed breeding conditions

8 weeks after sowing. Once the epidemic had sufficiently developed to allow a clear differentiation between susceptible and resistant wheat lines, disease response was assessed on a whole plot basis using the 1–9 scale reported by Bariana et al. [29]. Disease assessment was conducted at the grain filling stage (i.e., 113 days after sowing; DAS). The genotypes that displayed a disease response ≤ 5 are considered resistant (Fig. 5).

4. To compare disease response of the genotypes across the experiments, all the 1–9 field scores were converted to 0–9 [30]. Analysis of variance (ANOVA) was performed for each experiment. The mean disease response and standard error of the mean (SEM) for each genotype were calculated and presented in the form of a bar chart created using GraphPad Prism version 6.00 for Windows (GraphPad Software, La Jolla California USA). The means and standard error of means (SEM) are presented in Fig. 6.

4 Notes

1. *Pgt* urediniospores can be maintained and multiplied using the susceptible cultivar "Morocco." At 14 days post-inoculation, spores can be collected using a cyclone spore collector, dried in

Fig. 5 Stem rust response of wheat genotypes in the panel at adult plant stage in the field

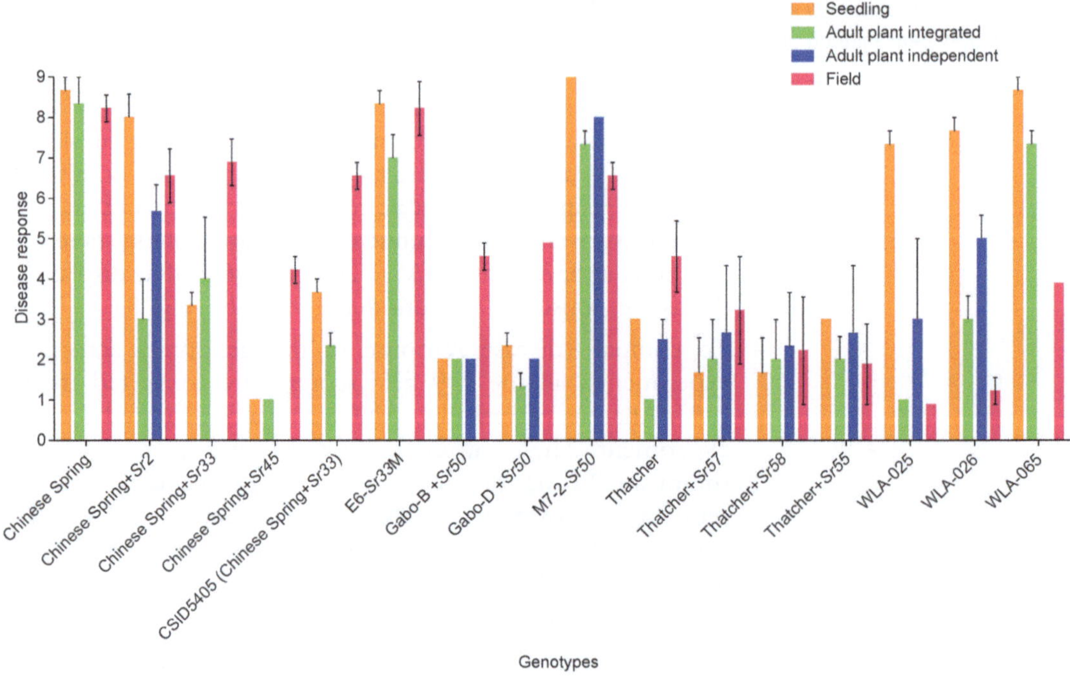

Fig. 6 Comparison of mean disease response for the 16 genotypes evaluated against stem rust (SR) in three experiments: adult plant integrated method, adult plant independent method, and in the field. The standard error of the means (SEM) are calculated for each of the three experiments

a desiccator filled with 60% sulfuric acid for 2 days and then stored in 0.5 ml self-standing tubes with screw caps (i.e., 0.1 g/tube). Tubes should be clearly labeled with the pathotype information and date of collection and can be are stored in plastic cryo boxes in the −80 °C freezer, until required.

2. Prior to inoculation, a tube containing *Pgt* spores should be removed from −80 °C and heat-shocked by submerging the tube in a water bath maintained at 42–45 °C for 3 min.
3. During inoculation of plants, a petri dish containing agar (15 g/L of distilled water) can be sprayed with the spore suspension and misted with deionized water. The petri dish is best placed inside the chamber (lid off) to expose the spores to the same conditions.
4. For small experiments, the inoculated plants can be incubated using black polyethylene garbage bags. The inside of the bag should be lightly misted with deionized water and then sealed air tight.
5. When plants are removed from the incubation chamber, the spore germination in the petri dish can be examined under a microscope to determine the germination rate.
6. The 0–4 Stakman scale can be converted to the 0–9 scale, where 0 = immune and 9 = very susceptible, using a conversion table [30]. The ITs may be converted as follows: 0;, ;n, ; , 1−, 1, 1+, 2−, 2, 2+, 2++, 3−, 3, 3+, 3++, and 4 are coded as 0, 0.5, 1, 2.5, 3, 3.5, 4, 5, 6, 6.5, 7, 8, 8.5, and 9, respectively. For heterogeneous ITs, each score is converted individually to the 0–9 scale, and the average calculated. These converted datasets are more desirable for correlation analyses to compare phenotypes across environments or QTL mapping.
7. Under the fast growth conditions in the speed breeding system, it is important to maintain plant health by applying liquid fertilizer.
8. For plants grown at high density, the elongated stems can be better targeted during inoculation by slowly rotating the pots or using your arm to gently separate groups of plants and leaning them side by side.
9. The SR inoculated plants can remain in the speed breeding system if a diurnal photoperiod is adopted to favor disease development.
10. Disease scoring can also be performed using the modified Cobb scale for severity (0–100) and infection response (Resistance, Moderately Resistance, Moderately Susceptible, and Susceptible), as described in Roelfs et al. [31]. The host

response and disease severity data are used to calculate the coefficient of infection (CI) [32].

11. The SR phenotyping method reported in this study enables efficient use of resources and time and can be performed all year round—compared to traditional field-based phenotyping. The method can be scaled up to screen large collections of historical germplasm and breeding populations. If applied to segregating populations, SR resistant plants can be identified and crossed in same plant generation. Following crossing, SR screening can be performed in parallel with SSD performed in the speed breeding system. For instance, plants can be raised at higher densities using a cell-based system (i.e., up to 900 plants per meter square; [16]). This method enables rapid development of inbred lines enriched with SR resistance. This phenotyping approach can also accelerate screening of mutant populations, as required for gene cloning techniques such as MutRenSeq [33].

Acknowledgments

We would like to acknowledge Dr. Mark Dieters and Dr. Ian Delacy at The University of Queensland for concepts relating to speed breeding. We thank Dr. Rohit Mago and Dr. Sambasivam Periyannan at CSIRO Agriculture and Food, Canberra for providing wheat genotypes examined in this study. We would also like to thank Mr. Gregory Platz at the Department of Agriculture and Fisheries Queensland, Australia for providing stem rust inoculum.

References

1. Leonard KJ, Szabo LJ (2005) Stem rust of small grains and grasses caused by *Puccinia graminis*. Mol Plant Pathol 6(2):99–111
2. Park RF (2007) Stem rust of wheat in Australia. Aust J Agric Res 58(6):558–566
3. Singh RP, Hodson DP, Huerta-Espino J, Jin Y, Bhavani S, Njau P, Herrera-Foessel S, Singh PK, Singh S, Govindan V (2011) The emergence of Ug99 races of the stem rust fungus is a threat to world wheat production. Annu Rev Phytopathol 49(1):465–481
4. Ray DK, Mueller ND, West PC, Foley JA (2013) Yield trends are insufficient to double global crop production by 2050. PLoS One 8 (6):e66428
5. Singh RP, Hodson DP, Jin Y, Lagudah ES, Ayliffe MA, Bhavani S, Rouse MN, Pretorius ZA, Szabo LJ, Huerta-Espino J, Basnet BR, Lan C, Hovmøller MS (2015) Emergence and spread of new races of wheat stem rust fungus: continued threat to food security and prospects of genetic control. Phytopathology 105(7):872–884
6. McIntosh R, Dubcovsky J, Rogers W, Morris C, Xia X (2017) Catalogue of gene symbols for wheat: 2017 Supplement. pp 9–10
7. McDonald BA, Linde C (2002) Pathogen population genetics, evolutionary potential, and durable resistance. Annu Rev Phytopathol 40 (1):349–379
8. Figueroa M, Upadhyaya NM, Sperschneider J, Park RF, Szabo LJ, Steffenson B, Ellis JG, Dodds PN (2016) Changing the game: using integrative genomics to probe virulence mechanisms of the stem rust pathogen *Puccinia graminis* f. Sp. *tritici*. Front Plant Sci 7:205
9. Moore JW, Herrera-Foessel S, Lan C, Schnippenkoetter W, Ayliffe M, Huerta-Espino J,

Lillemo M, Viccars L, Milne R, Periyannan S, Kong X, Spielmeyer W, Talbot M, Bariana H, Patrick JW, Dodds P, Singh R, Lagudah E (2015) A recently evolved hexose transporter variant confers resistance to multiple pathogens in wheat. Nat Genet 47(12):1494–1498

10. Krattinger SG, Lagudah ES, Spielmeyer W, Singh RP, Huerta-Espino J, McFadden H, Bossolini E, Selter LL, Keller B (2009) A putative ABC transporter confers durable resistance to multiple fungal pathogens in wheat. Science 323(5919):1360–1363
11. Lagudah ES (2011) Molecular genetics of race non-specific rust resistance in wheat. Euphytica 179(1):81–91
12. Riaz A, Periyannan S, Aitken E, Hickey L (2016) A rapid phenotyping method for adult plant resistance to leaf rust in wheat. Plant Methods 12(1):1–10
13. Singh RP, Huerta-Espino J, Bhavani S, Herrera-Foessel SA, Singh D, Singh PK, Velu G, Mason RE, Jin Y, Njau P, Crossa J (2011) Race non-specific resistance to rust diseases in CIMMYT spring wheats. Euphytica 179 (1):175–186
14. Ortiz R, Trethowan R, Ferrara GO, Iwanaga M, Dodds JH, Crouch JH, Crossa J, Braun H-J (2007) High yield potential, shuttle breeding, genetic diversity, and a new international wheat improvement strategy. Euphytica 157 (3):365–384
15. Bender CM, Prins R, Pretorius ZA (2016) Development of a greenhouse screening method for adult plant response in wheat to stem rust. Plant dis 100(8):1627–1633
16. Hickey LT, Wilkinson PM, Knight CR, Godwin ID, Kravchuk OY, Aitken EA, Bansal UK, Bariana HS, De Lacy IH, Dieters MJ (2012) Rapid phenotyping for adult-plant resistance to stripe rust in wheat. Plant Breed 131(1):54–61
17. Mackay M, Street K, Hickey L (2016) Toward more effective discovery and deployment of novel plant genetic variation: reflection and future directions. In: Bari A, Damania A, Mackay M, Dayanandan S (eds). CRC Press, pp 139–150
18. Hickey LT, Dieters MJ, DeLacy IH, Kravchuk OY, Mares DJ, Banks PM (2009) Grain dormancy in fixed lines of white-grained wheat (*Triticum aestivum* L.) grown under controlled environmental conditions. Euphytica 168(3):303–310
19. Richard CA, Hickey LT, Fletcher S, Jennings R, Chenu K, Christopher JT (2015) High-throughput phenotyping of seminal root traits in wheat. Plant Methods 11(1):1–11
20. Dinglasan E, Godwin ID, Mortlock MY, Hickey LT (2016) Resistance to yellow spot in wheat grown under accelerated growth conditions. Euphytica 209(3):693–707
21. Hayes HK, Ausemus E, Stakman E, Bailey C, Wilson H, Bamberg R, Markley M, Crim R, Levine M (1936) Thatcher wheat. Minn Agric Exp Stn Tech Bull 325:39
22. Jones S, Dvořák J, Knott D, Qualset C (1991) Use of double-ditelosomic and normal chromosome 1D recombinant substitution lines to map *Sr33* on chromosome arm 1DS in wheat. Genome 34(4):505–508
23. Periyannan S, Moore J, Ayliffe M, Bansal U, Wang X, Huang L, Deal K, Luo M, Kong X, Bariana H, Mago R, McIntosh R, Dodds P, Dvorak J, Lagudah E (2013) The gene *Sr33*, an ortholog of barley *Mla* genes, encodes resistance to wheat stem rust race Ug99. Science 341(6147):786–788
24. Mago R, Spielmeyer W, Lawrence GJ, Ellis JG, Pryor AJ (2004) Resistance genes for rye stem rust (*SrR*) and barley powdery mildew (*Mla*) are located in syntenic regions on short arm of chromosome. Genome 47(1):112–121
25. Riaz A, Hathorn A, Dinglasan E, Ziems L, Richard C, Singh D, Mitrofanova O, Afanasenko O, Aitken E, Godwin I, Hickey L (2016) Into the vault of the Vavilov wheats: old diversity for new alleles. Genet Resour Crop Evol 64:1–14
26. Park RF (2016) The wheat stem rust pathogen in Australia pathogenic variation and pathotype designation. Cereal rust report, plant breeding institute, University of Sydney, vol 14
27. Stakman E, Stewart D, Loegering W (1962) Identification of physiologic races of *Puccinia graminis* var. *tritici*. USDA-ARS Bull, St. Paul, p E-617
28. Zadoks JC, Chang TT, Konzak CF (1974) A decimal code for the growth stages of cereals. Weed res 14:415–421
29. Bariana HS, Miah H, Brown GN, Willey N, Lehmensiek A (2007) Molecular mapping of durable rust resistance in wheat and its implication in breeding. In: Buck HT, Nisi JE, Salomón N (eds) Wheat production in stressed environments, Developments in plant breeding, vol vol 12. Springer, the Netherlands, pp 723–728
30. Ziems LA, Hickey LT, Hunt CH, Mace ES, Platz GJ, Franckowiak JD, Jordan DR (2014) Association mapping of resistance to *Puccinia hordei* in Australian barley breeding germplasm. Theor Appl Genet 127(5):1199–1212

31. Roelfs A, Singh R, Saari E (1992) Rust diseases of wheat: concepts and methods of disease management. CIMMYT, Mexico, DF (Mexico)
32. Loegering W (1959) Methods for recording cereal rust data. USDA International Spring Wheat Nursery
33. Steuernagel B, Periyannan SK, Hernandez-Pinzon I, Witek K, Rouse MN, Yu G, Hatta A, Ayliffe M, Bariana H, Jones JDG, Lagudah ES, Wulff BBH (2016) Rapid cloning of disease-resistance genes in plants using mutagenesis and sequence capture. Nat Biotechnol 34(6):652–655

Part IV

Rust Resistance Gene Isolation and Characterization

Chapter 17

Generation of Loss-of-Function Mutants for Wheat Rust Disease Resistance Gene Cloning

Rohit Mago, Bradley Till, Sambasivam Periyannan, Guotai Yu, Brande B.H. Wulff, and Evans Lagudah

Abstract

One of the most important tools to identify and validate rust resistance gene function is by producing loss-of-function mutants. Mutants can be produced using irradiation, chemicals, and insertions. Among all the mutagens, ethyl methanesulfonate (EMS) and sodium azide are most favored because of the ease of use and generation of random point mutations in the genome. The mutants so produced facilitate the isolation, identification and cloning of rust resistance genes. In this chapter we describe a protocol for seed mutagenesis of wheat with EMS and sodium azide.

Key words Ethyl methanesulfonate (EMS), Sodium azide, Mutagenesis, Loss of function

1 Introduction

Mutagenesis is an important technique whereby DNA mutations are deliberately generated to produce mutant genes. Various constituents of a gene, such as its control elements and its gene product, may be mutated so that the functioning of a gene or protein can be examined in detail. The mutation may also produce mutant proteins with interesting properties, such as enhanced or novel functions that may be of commercial use.

Ethyl methanesulfonate (EMS) is a mutagenic and carcinogenic organic compound with formula $C_3H_8SO_3$. It produces random mutations in genetic material by nucleotide substitutions resulting in GC to AT transition. This changes the genetic information, causing loss of gene functions such the rust resistance gene becoming susceptible. Sodium azide (NaN_3) also creates mutations and has been a mutagen of choice in barley [1]. It causes AT to GC base pair transitions. Sodium azide is a respiratory inhibitor, and hence, its action may be affected by the presence of oxygen. Lack of

Sambasivam Periyannan (ed.), *Wheat Rust Diseases: Methods and Protocols*, Methods in Molecular Biology, vol. 1659,
DOI 10.1007/978-1-4939-7249-4_17, © Springer Science+Business Media LLC 2017

aeration during sodium azide treatment is shown to have physiological damage on M1 seedlings in barley [2].

Rust caused by the fungus of *Puccinia* sp. is one of the most devastating diseases of wheat and barley. The most effective and environmentally sustainable method to control rust is by using effective resistance genes present in the cereal gene pool [3]. This can be most efficiently achieved by marker-assisted breeding which requires gene specific markers obtained from cloned resistance gene. The cloned genes can also be introduced into wheat varieties by developing transgenics. Several rust resistance genes have been cloned from wheat [4–7], and mutants have been critical in confirming the identity of the resistance gene in each of the cases. More recently, new techniques namely MutRenSeq [8] and MutChromSeq [9] have been developed which use isolation and characterization mutants and new sequencing technologies for rapid cloning of rust resistance or other gene/s without the need of difficult and time consuming map-based cloning. Stem rust resistance genes *Sr22* and *Sr45* and barley Eceriferum-q gene and the wheat *Pm2* were cloned using these techniques [8, 9].

The mutants produced using EMS or sodium azide have not only been useful for resistance gene discovery but, have also been used for understanding gene function in vivo and also fundamental processes such as cell cycle control [10]. Chemical mutagenesis has led to the development of the reverse-genetic strategy known as Targeting Induced Local Lesions IN Genomes (TILLING) [11]. This along with next generation sequencing technologies have allowed the development of methods to rapidly clone EMS-induced alleles in plants, keeping forward genetic analysis an attractive approach to identify gene function [12, 13].

Here we report the protocols used for developing mutants with both EMS and sodium azide which have been used for producing mutants in wheat. The EMS treatment procedure is based on Mago et al. (2002) [14] and Jankowicz-Cieslak and Till (2016) [15]. The sodium azide treatment has been adapted from Zwar and Chandler (1995) [16].

2 Materials

2.1 Materials for EMS Treatment

1. Plant seeds.
2. Beakers 250 ml and 1000 ml.
3. Orbital shaker.
4. Fume hood.
5. Micropipettes.
6. Graduated cylinders.
7. Capped bottles.

8. 50 ml Falcon tubes.
9. Tube rack.
10. 0.1 M sodium thiosulfate (prepared from 1 M stock solution): Dissolve 158.097 g sodium thiosulfate in 1 L deionized water to make a 1 M stock and dilute it 1:10 in water to make 0.1 M solution.
11. 2% DMSO (Dimethyl sulfoxide): Mix 2 ml DMSO in 98 ml water to make a 2% solution.
12. Ethyl methanesulfonate (EMS).
13. Bin for dry hazardous materials disposal.
14. Plastic bucket with lid for hazardous liquid waste disposal.
15. Nitrile gloves.
16. Laboratory coat.
17. Goggles.
18. Deionized water.
19. Cheese cloth.

2.2 Additional Materials for Sodium Azide Treatment

1. Sodium azide.
2. Potassium hydroxide (KOH).
3. 0.1 M pH 3.0 Phosphate buffer: Add 6.8 ml of phosphoric acid to 800 ml deionized water, adjust pH to 3.0 using KOH, make up volume to 1 L.
4. Pressurized air or a small pump used for fish tanks.

3 Method

It is advisable to consult the Materials Safety Data Sheets for all chemicals used. Caution should be taken when using EMS or sodium azide. Laboratory coats, gloves, and goggles should be worn when working with hazardous chemicals. It is advised to wear double gloves so that contaminated gloves can be removed while avoiding contact of contaminated materials with skin and carry out the mutagenesis in a fume hood to avoid any risk of exposure to the toxic fumes.

A prerequisite for any mutagenesis is bulking up 2000–3000 or more seed of the required genotype, which are homozygous resistant to the desired rust isolate. It is important to start with a larger number of seeds for mutagenesis as the final dosage of mutagen chosen will only allow 50% normal plants, relative to the 0% control concentration. The seed should ideally be from the same batch of plants (grown at same time in same environment). It is also important to determine the mutagen dosage required for the seed treatment. The dose–response curve should be done for each seed batch

Fig. 1 Effect of different sodium azide concentrations on seed germination and plant growth. *Left* to *right*, 0 mM (control); 1 mM; 2 mM; 3 mM; and 5 mM

closer to full treatment and needs to be repeated in case of delay with the full treatment. Ten to fifty seeds for each mutagen concentration are enough to determine the optimum concentration required for final treatment. Fig. 1 shows the effect of various concentrations of sodium azide on seed germination and plant growth.

Development of a mutant population suitable for genotypic or phenotypic screening requires the production of M_2 or higher generations. The timing of production of a suitable population depends on the species. Thus, planning when to mutagenize seed is especially important in field propagated plants as seed are typically planted immediately after treatment with chemical mutagens. Experiments should therefore be timed to synchronize with sowing.

3.1 EMS Mutagenesis of Seeds

This protocol gives detailed instructions for EMS mutagenesis of wheat or barley seeds. This is a 3-day protocol consisting of three major steps.

3.1.1 Day 1

1. Soak seed overnight in a beaker (~16 h) in water with orbital rotation at 4 °C. Check that seeds are moving freely.

3.1.2 Day 2

1. Evaluate movement of seed under orbital rotation and adjust water if necessary and test to ensure seed are moving freely.
2. Mark the beakers at the level of the adjusted volume of liquid to estimate the volume of mutagen required for seed treatment and also the volume of wash solutions required.
3. Drain water into a graduated cylinder, and record the volume needed to reach the mark made in **step 5**. Store seeds at 4 °C for few hours and bring them to room temperature an hour before starting the EMS treatment.

4. Add the required amount of water to the seeds (as estimated in **step 4**) and add the EMS for required concentration (based on dosage curve prepared: *see* **Notes 1** and **2**).
5. Keep the seeds on shaker with gentle shaking for 12 h at room temperature.

3.1.3 Day 3

1. Gently pour of the EMS solution into a bucket for waste disposal and add same volume of 100 mM Sodium thiosulfate to the seeds. Shake for 15 min on a shaker and pour off the liquid for disposal.
2. Repeat **step** 7 twice.
3. Rinse the seeds with water 2–3 times similar to wash with Sodium thiosulfate for 15 min each.
4. Cover the beaker with a muslin cloth and secure with a rubber band/string and keep it under running water for 1 h.
5. Pour off water and spread seeds on blotting paper and leave to air-dry for 1 h.
6. Sow treated seeds in soil.
7. Let plants to maturity and harvest heads from each plant separately.
8. Use M_2 seeds from a single head of each plant to screen with a desired rust pathotype for mutants (*see* **Note 4**).

3.2 Sodium Azide Mutagenesis of Seeds

1. Soak seed overnight in a beaker (~16 h) in water with orbital rotation at 4 °C. Check that seeds are moving freely.
2. Transfer seeds to a 2 L measuring cylinder filled with water. Aerate with pressurized air for 8 h with a change of water after 4 h as lack of aeration is shown to have physiological damage in M_1 seedlings [2].
3. Replace water with a the desired concentration (as determined by dosage response curve, *see* **Note 3**) of sodium azide (in mM) prepared in 0.1 M pH 3.0 phosphate buffer and aerate for 2 h.
4. Dispose the sodium azide solution in waste and transfer seeds to a beaker.
5. Cover the beaker with a cheese cloth and secure with a rubber band/string and keep it under running water for 2 h.
6. Pour off water and spread seeds on blotting paper and leave to air-dry for at least 1 h (can be left overnight).
7. Sow treated seeds in soil.
8. Grow plants to maturity and harvest heads from each plant separately.
9. Use M_2 seeds from a single head of each plant to screen with a desired rust pathotype for mutants (*see* **Note 4**).

4 Notes

1. Making the Dosage Curve and Determining the Optimum Concentration for EMS Treatment.

   Treat 10–50 seeds each with EMS solution of the following concentrations: 0, 0.2, 0.4, 0.5, 0.6, 0.7, 0.8, and 0.9% (*see* **step 4** above: variations 1 and 2) in a 50 ml bottle or Falcon tube using the protocol described below. Sow seeds in soil and determine the EMS concentration that gives approximately 50 germination; in some cases we found no effect of EMS on germination, but reduction in plant growth. Hence, the concentration that gives 50% growth reduction was chosen for final treatment.
2. Variation 2.

   We found with certain wheat varieties that even a very low concentration of EMS was toxic, resulting in no seed germination. Alternatively, we prepare required amount of EMS in 2% DMSO solution, the rest of the protocol is same as described above. If high lethality is observed with EMS it may be advisable then to try an alternative chemical mutagen like sodium azide.
3. Making the Dosage Curve and Determining the Optimum Concentration for Sodium Azide Treatment.

   Treat 10–50 seeds each with sodium azide solution of the following concentrations: 0, 0.5 mM, 1.0 mM, 2.0 mM, 3.0 mM, 4.0 mM, and 5.0 mM in a 50 ml bottle or Falcon tube using the protocol described above (Fig. 1). Sow seeds in soil and determine the sodium azide concentration that gives approximately 50 germination; in some cases we found no effect of sodium azide concentration on germination, but reduction in plant growth. Hence, the concentration that gives 50% growth reduction was chosen for final treatment.
4. Screening for Susceptible Mutants
   This is common to both EMS and sodium azide-treated seeds. Plant 8–15 M_2 seeds from each separate spike (an M_2 family), inoculate with the desired pathogen isolate and screen for families segregating for resistant and susceptible plants. For a dominant resistance gene, the probability of obtaining at least one recessive susceptible mutant in a segregating family is 0.90 if eight seedlings are screened, and 0.97 if 12 seedlings are screened. In hexaploid wheat, a susceptible mutant is typically obtained for every 100–200 single heads screened. Obtain self-fertilized seed (M_3) from each mutant to confirm the susceptibility phenotype. Based on observations in barley [17], different tillers in wheat are likely derived from different cell lineages in the meristem, whereby individual tillers from the same M_1

plant will represent different mutagenic events. Therefore, the effective M_2 population size can be increased by harvesting multiple individual tillers from each plant. To obtain independent M_2 families will require harvesting of each spike separately. Harvest tissue from the M_2 or M_3 to extract DNA.

References

1. Olsen O, Wang X, Wettstein DV (1993) Sodium Azide mutagenesis: preferential generation of AT->GC transitions in the barley *Antl8* gene. Proc Natl Acad Sci U S A 90:8043–8047
2. Molina-Cano JL (1979) Effect of aeration during Sodium Azide treatment of barley seeds on the physiological damage in M1 seedlings. Barley Genet News Lett 9:62–65
3. Ellis JG, Lagudah ES, Spielmeyer W, Dodds PN (2014) The past, present and future of breeding rust resistant wheat. Front Plant Sci 5:641
4. Krattinger SG, Lagudah ES, Spielmeyer W, Singh RP, Huerta-Espino J, McFadden H, Bossolini E, Selter LL, Keller B (2009) A putative ABC transporter confers durable resistance to multiple fungal pathogens in wheat. Science 323:1360–1363
5. Periyannan S, Moore J, Ayliffe M, Bansal U, Wang X, Huang L, Deal K, Luo MC, Kong X, Bariana H, Mago R, McIntosh R, Dodds P, Dvorak J, Lagudah E (2013) The Ug99 effective wheat stem rust resistance gene *Sr33* is an ortholog of barley *Mla* genes. Science 341:786–788
6. Mago R, Zhang P, Vautrin S, Šimková H, Bansal U, Luo M-C, Rouse M, Karaoglu H, Periyannan S, Kolmer J, Jin Y, Ayliffe MA, Bariana H, Park RF, McIntosh R, Doležel J, Bergès H, Spielmeyer W, Lagudah ES, Ellis JG, Dodds PN (2015) The wheat *Sr50* gene reveals rich diversity at a cereal disease resistance locus. Nat Plants 1:15186
7. Moore JW, Herrera-Foessel S, Lan C, Schnippenkoetter W, Ayliffe M, Huerta-Espino J, Lillemo M, Viccars L, Milne R, Periyannan S, Kong X, Spielmeyer W, Talbot M, Bariana H, Patrick JW, Dodds P, Singh R, Lagudah E (2015) A recently evolved hexose transporter variant confers resistance to multiple pathogens in wheat. Nat Genet 47:1494–1498
8. Steuernagel B, Periyannan SK, Hernández-Pinzón I, Witek K, Rouse MN, Yu G, Hatta A, Ayliffe M, Bariana H, Jones JD, Lagudah ES, Wulff BBH (2016) Rapid cloning of disease-resistance genes in plants using mutagenesis and sequence capture. Nat Biotechnol 34:652–655
9. Sánchez-Martín J, Steuernagel B, Ghosh S, Herren G, Hurni S, Adamski N, Vrána J, Kubaláková M, Krattinger SG, Wicker T, Doležel J, Keller B, Wulff BBH (2016) Rapid gene isolation in barley and wheat by mutant chromosome sequencing. Genome Biol 17:221
10. Coen ES, Meyerowitz EM (1991) The war of the whorls: genetic interactions controlling flower development. Nature 353:31–37
11. McCallum CM, Comai L, Greene EA, Henikoff S (2000) Targeting induced local lesions IN genomes (TILLING) for plant functional genomics. Plant Physiol 123:439–442
12. Abe A, Kosugi S, Yoshida K, Natsume S, Takagi H, Kanzaki H, Matsumura C, Mitsuoka M, Tamiru M, Innan H, Cano L, Kamoun S, Terauchi R (2012) Genome sequencing reveals agronomically-important loci in rice from mutant populations. Nat Biotechnol 30:174–178
13. Schneeberger K, Ossowski S, Lanz C, Juul T, Petersen AH, Nielsen KL, Jan-Elo J, Weigel D, Andersen SU (2009) SHOREmap: simultaneous mapping and mutation identification by deep sequencing. Nat Methods 6:550–551
14. Mago R, Spielmeyer W, Lawrence GJ, Lagudah ES, Ellis JG, Pryor A (2002) Identification and mapping of molecular markers linked to rust resistance genes located on chromosome 1RS of rye using wheat-rye translocation lines. Theor Appl Genet 104:1317–1324
15. Jankowicz-Cieslak J, Till BJ (2016) Chemical mutagenesis of seed and vegetatively propagated plants using EMS. Curr Protoc Plant Biol 1:617–635
16. Zwar JA, Chandler PM (1995) α-Amylase production and leaf protein synthesis in gibberellin-responsive dwarf mutant of 'Himalaya' barley (*Hordeum vulgare* L.) Planta 197:39–48
17. Sarvella P, Nilan RA, Konzak CF (1962) Relation of embryo structure, node position, tillering and depth of planting to the effects of x-rays in barley. Radiation Bot 2:89–108

Chapter 18

Isolation of Wheat Genomic DNA for Gene Mapping and Cloning

Guotai Yu, Asyraf Hatta, Sambasivam Periyannan, Evans Lagudah, and Brande B.H. Wulff

Abstract

DNA is widely used in plant genetic and molecular biology studies. In this chapter, we describe how to extract DNA from wheat tissues. The tissue samples are ground to disrupt the cell wall. Then cetyltrimethylammonium bromide (CTAB) or sodium dodecyl sulfate (SDS) is used to disrupt the cell and nuclear membranes to release the DNA into solution. A reducing agent, β-mercaptoethanol, is added to break the disulfide bonds between the cysteine residues and to help remove the tanins and polyphenols. A high concentration of salt is employed to remove polysaccharides. Ethylenediaminetetraacetic acid (EDTA) stops DNase activity by chelating the magnesium ions. The nucleic acid solution is extracted with chloroform–isoamyl alcohol (24:1) or 6 M ammonium acetate. The DNA in aqueous phase is precipitated with ethanol or isopropanol, which makes DNA less hydrophilic in the presence of sodium ions ($Na+$).

Key words Wheat, Tissue, Leaf, Seeds, DNA extraction, Cetyltrimethylammonium bromide, Sodium dodecyl sulfate

1 Introduction

Wheat DNA extraction methods abound by the dozen. In terms of purpose, they can generally be classified into two types: (1) those laborious methods that yield large quantities of highly pure, high quality, high molecular weight DNA suitable for generating whole genome shotgun libraries for screening or next generation sequencing, and (2) those "quick-and-dirty" high-throughput methods suitable for PCR genotyping purposes. Here, we present two methods covering these two aims, including a modified CTAB method [1] for isolating large quantities of high quality DNA, and a modified SDS method [2] for rapidly isolating small quantities of DNA from large numbers of samples. The modified CTAB method does not require the use of phenol which is highly toxic. The modified SDS method requires neither phenol nor chloroform. Both methods have been tested extensively in our labs.

Sambasivam Periyannan (ed.), *Wheat Rust Diseases: Methods and Protocols*, Methods in Molecular Biology, vol. 1659, DOI 10.1007/978-1-4939-7249-4_18, © Springer Science+Business Media LLC 2017

2 Materials

2.1 Materials for Modified CTAB Method

1. H_2O: deionized water or distilled water.
2. 1 M Tris–HCl (pH 8.0): Dissolve 121.14 g Tris in 800 ml dH_2O. Adjust pH to 8.0 with the appropriate volume of concentrated HCl. Bring final volume to 1 l with deionized water.
3. 4 M NaCl: Dissolve 116.9 g of NaCl (m.w. 58.44) in 250 ml of deionized or distilled water. Then, add deionized or distilled water to make a total volume of 500 ml of solution.
4. 0.5 M EDTA: Add 18.6 g EDTA (disodium salt, m.w. 372.24) to 80 ml deionized or distilled water. Adjust to pH 8.0 by slowly adding approximately 2.2 g of sodium hydroxide pellets (m.w. 40.00).
5. Extraction buffer without β-mercaptoethanol: Transfer 65.8 ml of H_2O, 14 ml of 1 M Tris–HCl, 49.0 ml of 4 M NaCl, and 5.6 ml of 0.5 M EDTA to a 250 ml bottle and warm it up to 37 °C, and then dissolve 2.8 g of CTAB in it.
6. Extraction buffer: right before use, add 2.8 ml of β-mercaptoethanol to the bottle in a fume hood.
7. Mortar (10 cm in diameter) and pestle.
8. Sand (50–70 mesh).
9. Liquid nitrogen.

2.2 Materials for Modified SDS Method

1. H_2O: deionized water or distilled water.
2. 1 M Tris–HCl (pH 8.0): Dissolve 121.14 g Tris. Adjust pH to 8.0 with the appropriate volume of concentrated HCl. Bring final volume to 1 l with deionized water.
3. 0.5 M EDTA: Add 18.6 g EDTA (disodium salt, m.w. 372.24) to 80 ml deionized or distilled water. Adjust to pH 8.0 by slowly adding approximately 2.2 g of sodium hydroxide pellets (m.w. 40.00).
4. 10% SDS: Dissolve 10 g electrophoresis-grade SDS (m.w. 288.37) in 80 ml deionized water. Add deionized or distilled water to make 100 ml total solution.
5. Extraction buffer: transfer 270 ml of H_2O, 40 ml of 1 M Tris–HCl, 40 ml of 0.5 M EDTA, and 50 ml of 10% SDS in a 500 ml bottle.
6. 6 M ammonium acetate: Dissolve 131.2 g of CH3COONH4 in 250 ml of H_2O, and bring it up to 500 ml.
7. Isopropanol.
8. Deep-well microtiter plate and 8-cap strips.
9. Grinding beads (3 mm, Tungsten Carbide Beads).

3 Methods

3.1 Modified CTAB Method

Carry out **steps 1** and **2** in the fume hood.

1. Harvest clean, young leaf material by placing ~2.5 cm cut sections into a bag, sealing the bag, and placing into liquid nitrogen. The samples can be stored at −80 °C until required.
2. Chill a mortar and pestle in liquid nitrogen for 5 min. Place 2.0–4.0 g leaf material in the mortar, add 2.0–4.0 g grinding sand and some liquid nitrogen, and then carefully grind to a very fine powder. Transfer the powder into a 50 ml polypropylene conical tube containing 20 ml CTAB extraction buffer and immediately mix the powder with buffer using a spatula (*see* **Note 1**).
3. Place in a 65 °C water bath for 60 min. Vortex every ~10 min.
4. In the fume hood, add 20 ml of chloroform–isoamyl alcohol (24:1).
5. Mix thoroughly by inversion until a homogenous emulsion is formed.
6. Centrifuge for 15 min at 2500 × *g* in a swing-out centrifuge to separate the aequeous and chloroform phases.
7. Slowly remove the top phase (14 ml) with a disposable 25 ml pipet and transfer to a fresh 50 ml polypropylene conical tube.
8. Do this one tube at a time. Add 28 ml of absolute ethanol (99.9% v/v) down the side of each tube. Mix carefully by holding the tube flat (sideways) and rocking slowly so that the solution rocks back and forth from one end of the tube to the other.
9. Using a pipette, carefully remove DNA from the tube and place it into a fresh 15 ml tube.
10. Add 5.0 ml 1× TE buffer. Invert tube to dislodge the DNA. Gently resuspend until no more DNA is visible. At this point the samples can be placed in the fridge overnight.
11. Add 50 μl of RNase A (1000 μg/ml) to a final concentration of 10 μg/ml, mix gently, and digest at 37 °C for at least 1 h. At this point the samples can be placed in the fridge overnight.
12. Add ~5 ml of chloroform–isoamyl alcohol (24:1) and rock tubes until homogeneous.
13. Centrifuge for 15 min at 2500 × *g* to separate the phases.
14. Slowly remove 3 ml of the top phase with a disposable 1 ml pipette and transfer to a fresh 15 ml polypropylene conical tube.
15. Add 6 ml of absolute ethanol (99.9% v/v).

16. Mix carefully by holding tube flat and rocking slowly so that the solution rolls back and forth from one end of the tube to the other. When mixed well, a DNA blob should become visible.
17. Remove the DNA blob with a pipette and put into a fresh 15 ml tube.
18. Add ~3 ml 70% EtOH and wash the DNA overnight on an orbital shaker at 50–100 rpm at room temperature.
19. Carefully transfer the DNA from the 15 ml tubes into 2 ml tubes with a pipette.
20. Using a pipette tip squeeze the DNA blob. Continue to squeeze the DNA until it has reduced in size and the alcohol has mostly gone.
21. Place the tube with the DNA in a fume hood until it has completely dried or leave on counter overnight.
22. Add 500 μl of TE buffer to dissolve the DNA and close the tube. Let stand in fridge at 4 °C overnight to rehydrate it. If required, further incubate at 37 °C at 300 rpm on a thermal mixer.
23. Quantify and adjust to desired concentration (*see* **Note 2**). The method should yield 100–200 μg of DNA. The DNA on gel should look like those in the following image, i.e., with little degradation and no obvious residual RNA (Fig. 1).

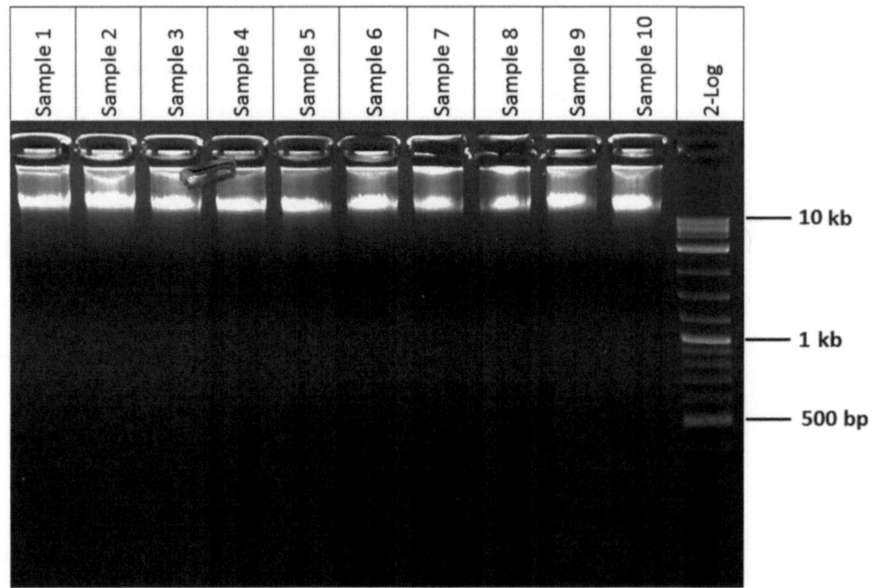

Fig. 1 10 DNA samples extracted with modified CTAB method. 1 μl of DNA prep was loaded per well. A 2-Log (NEB, N3200S) DNA ladder was loaded on the far right. Note the absence of RNA contamination in the DNA preps

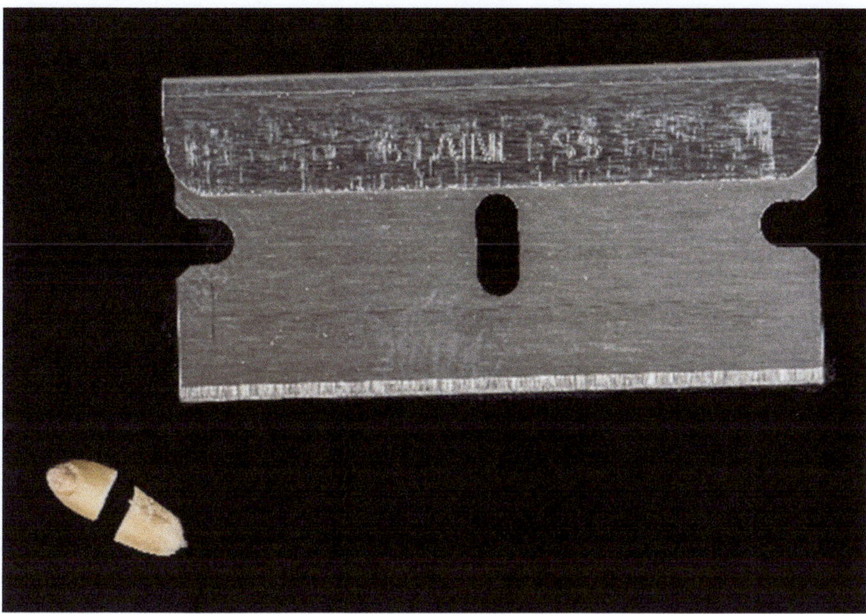

Fig. 2 Wheat grain cut in half with a razor blade. DNA for molecular analysis can be extracted from the bottom half, while the top half containing the embryo can be stored and germinated at a later date

3.2 Modified SDS Method

1. Sample leaf tissue or seed in a deep-well microtiter plate. About 3 cm of a seedling leaf (2–3 leaf stage) is enough. A half-seed may also be used (*see* Fig. 2).
2. For leaf tissue, freeze-dry overnight or longer.
3. Add 1 or 2 (*see* **Note 3**) SDS-washed-and-ethanol-rinsed grinding beads to each well. Seal plate with 8-lid strips. Shake on the Qiagen shaker at a frequency of 28/s for 2 min each side (change orientation). Store at −20 °C if required and allow plates to warm up to room temperature before resuming protocol.
4. Quick spin to settle contents down before opening lids. Then add 400 μl extraction buffer to each well and put the respective strips of lids back. Make sure they are tight.
5. Incubate at 55 or 65 °C (*see* **Note 4**) for a minimum of 1 h in an oven. Tightly seal caps and weigh down the lids to avoid cross-contamination of samples.
6. Cool plates in fridge (or freezer) to room temperature.
7. Add 200 μl 6 M ammonium acetate (*see* **Note 5**). Tightly seal caps then invert plates to mix, then leave in fridge for 15–30 min.
8. Centrifuge plates for 30 min at 2500 × *g*.
9. Recover 240 μl of supernatant into a new deep well microtiter plate containing 150 μl isopropanol and mix well by inverting

20 times. Allow DNA to precipitate for 5 min at room temperature.

10. Centrifuge for 30 min at 2500 × *g*. Immediately (within 1 min.) pour off supernatant (*see* **Note 6**).
11. Add 300 μl of 70% ethanol. Centrifuge for 30 min at 2500 × *g*. Pour off supernatant and dry off ethanol for 10 min.
12. Add 200 μl of H_2O to every well, seal plates with adhesive sheet, leave plates overnight at 4 °C for DNA pellets to dissolve.
13. Centrifuge plate for 20 min at 2500 × *g*.
14. Transfer 150 μl of supernatant to a fresh microtiter plate.
15. Test DNA concentration with Nanodrop and dilute to appropriate level for PCR. Store at 4 °C or −20 °C for long term storage.

4 Notes

1. 50 mg of polyvinyl pyrrolidone (PVP) per 50 ml of extraction buffer can be used to remove more polyphenols [1].
2. Quantification should be done with two or three methods, including NanoDrop spectrophotometry (http://www.nanodrop.com/nucleicacid.aspx), PicoGreen [3], Qubit fluorometer [4], and agarose gel. The two ratios (260/280 and 260/230) obtained from a Nanodrop reading can be used to indicate the level of contamination from protein and small organic molecules. The PicoGreen and Qubit readings give a realiable quantification of the double-stranded DNA concentration. The agarose gel is useful for estimating the degree of degradation and contamination with residual RNA.
3. Samples are ground better with 2 beads.
4. At 55 °C, the cross-contamination risk is less than at 65 °C.
5. The 6 M ammonium acetate is suitable for a robot work station due to its low corrosive activity, but it can be substituted with chloroform–isoamyl alcohol (24:1) when manually extracting the DNA.
6. If RNA removal is required, continue with the following steps:
   (a) Resuspend pellets in 100 μl of TE + 10 ng/μl RNAse. Incubate plate at 37 °C for 60 min.
   (b) Add 10 μl of 3 M CH3COONa (pH 5.2) to each well and then add 200 μl of 100% ethanol. Mix well and allow DNA to precipitate for 5 min at room temperature.
   (c) Centrifuge for 30 min at 2500 × *g*. Immediately (within 1 min) pour off supernatant.

References

1. Porebski S, Bailey LG, Baum BR (1997) Modification of a CTAB DNA extraction protocol for plants containing high polysaccharide and polyphenol components. Plant Mol Biol Rep 15:8–15
2. Kota R, Spielmeyer W, McIntosh RA, Lagudah ES (2005) Fine genetic mapping fails to dissociate durable stem rust resistance gene *Sr2* from pseudo-black chaff in common wheat (*Triticum aestivum* L.) Theor Appl Genet 112:492–499
3. Ahn SJ, Costa J, Emanuel JR (1996) PicoGreen quantitation of DNA: effective evaluation of samples pre- or post-PCR. Nucleic Acids Res 24:2623–2625
4. Simbolo M, Gottardi M, Corbo V, Fassan M, Mafficini A, Malpeli G, Lawlor RT, Scarpa A (2013) DNA Qualification workflow for next generation sequencing of histopathological samples. PLoS One 8:e62692

Chapter 19

MutRenSeq: A Method for Rapid Cloning of Plant Disease Resistance Genes

Burkhard Steuernagel, Kamil Witek, Jonathan D.G. Jones, and Brande B.H. Wulff

Abstract

MutRenSeq is a method to clone disease resistance (*R*) genes in plants. Tips and detailed experimental protocols for the pipeline, including the complexity reduction by *R* gene targeted enrichment sequencing, and computational analysis based on comparative genomics are provided in this chapter.

Key words MutRenSeq, Gene cloning, NB-LRR, Resistance gene, Target enrichment sequencing

1 Introduction

Upon infection, pathogens secrete proteins that alter the plant host in a way that is beneficial to the pathogen. These proteins are called effectors. As a countermeasure, plants have evolved *R* genes, the products of which can detect the presence of specific effectors [1]. Most *R* proteins consist of a multidomain structure encompassing a nucleotide-binding (NB) domain fused to a variable number of leucine-rich repeats (LRRs) [2]. The NB domain may be preceded by a coiled-coiled (CC) or a Toll/interleukin-1 receptor homology (TIR) domain [3]. NB-LRR genes are one the most numerous gene families in plants, typically having hundreds of members in a genome.

MutRenSeq is an *R* gene cloning pipeline, which combines sequence capture targeting *R* genes belonging to the NB-LRR structural class of genes and mutational genomics: the sequence comparison of wild-type parental with multiple independently derived mutants to identify causative mutations in a single candidate gene [4].

The MutRenSeq pipeline requires that a single *R* gene has been genetically isolated in an otherwise susceptible background. On this basis, with a certain probability, random mutations will knock

Sambasivam Periyannan (ed.), *Wheat Rust Diseases: Methods and Protocols*, Methods in Molecular Biology, vol. 1659, DOI 10.1007/978-1-4939-7249-4_19, © Springer Science+Business Media LLC 2017

out the *R* gene and instigate loss of resistance provided by that gene. Typically, loss-of-function mutants are due to a mutation directly in the *R* gene. Indeed, based on multiple *R* gene loss-of-function screens in dicots and monocots, whether diploids or polyploids, second-site suppressors in positive regulators have been found to be uncommon (typically less than 10% of suppressors; *see* Supplementary Table 1 in [4]). Nonetheless, it is prudent to perform complementation and testcross analyses to confirm that all the mutants belong to a single complementation group defining the *R* gene, particularly in diploid genomes where there is less genetic redundancy in downstream signaling components compared to a polyploid. The probability, that several independently derived mutants, e.g., six, have a mutation in the same gene is extremely low, except where these mutants have been selected for by screening for loss of resistance. Thus, sequencing of the resistant wild type and the six susceptible mutants followed by comparative genomics will reveal the target *R* gene (*see* **Note 1**).

The Triticeae genomes are very large (4–17 billion base pairs). Therefore, whole genome shotgun sequencing of a wild type and multiple mutant individuals would be cost prohibitive and impose significant computational challenges. Since we are only interested in NB-LRR genes, we can enrich DNA prior to sequencing using exome capture targeted to the *R* gene complement only (i.e., *R* gene *en*richment *seq*uencing; RenSeq [5]). Conveniently, this also reduces the amount of data to be analyzed to much less than 0.1% of the original genome size.

The method has successfully been applied to clone the stem rust resistance genes *Sr22* and *Sr45* from hexaploid wheat [4]. Here, we provide a detailed step-by-step protocol of MutRenSeq.

2 Materials

2.1 Plant Material

MutRenSeq includes the generation and screening for disease susceptible mutants followed by RenSeq. In order to obtain susceptible mutants, the disease resistance in the line used for mutagenesis must be controlled by a single gene. The majority of mutations caused by EMS are point mutations but some are deletion mutations (e.g., found in cloning of *Sr33* [6]). A wheat–alien introgression line, a near isogenic line (NIL) or a recombinant inbred line (RIL) containing a single effective *R* gene against a given pathogen isolate is usually still segregating for some unlinked loci. These segregating loci may be confounded with true EMS-derived mutations. A doubled haploid line would be ideal for EMS mutant development. A line with two or more genes conditioning disease resistance would generally not be suitable for MutRenSeq as it is difficult to obtain susceptible mutants.

2.2 Equipment and Reagents

1. Covaris sonicator S2.
2. NEBNext Ultra DNA library preparation kit.
3. AMPure XP Beads.
4. Ethanol.
5. Magnetic stand.
6. KAPA HiFi DNA PCR kit.
7. PicoGreen ds DNA Assay kit.
8. MYcroarray MYbaits kit.
9. SeqCap EZ Developer Reagent, Roche.

3 Methods

Two detailed protocols describing enrichment of short insert-size Illumina-style and long insert-size PacBio-style libraries with NB-LRR-specific baits were previously published [5, 7]. Jupe et al. 2014 described the enrichment of short insert-size Illumina libraries with the Agilent SureSelect system. Here we describe a modified Jupe et al. (2014) protocol, to carry out enrichments with the MYcroarray MYbait enrichment system.

In this protocol a sequencing library containing Illumina-style adaptors is constructed, followed by hybridization with NB-LRR-specific baits. We have successfully used barcoded libraries to carry out one enrichment on up to six multiplexed samples [4]. Perform the following library preparation protocol for the parental (non-mutated) line and each mutant separately, and then combine multiple barcoded libraries into one enrichment reaction. We usually enrich the resistant parent in a separate reaction, so that it can be sequenced with longer reads.

3.1 DNA Fragmentation

We use a Covaris sonicator S2 for genomic DNA (gDNA). We typically start with 2 μg gDNA (dissolved in 130 μl of TE buffer) and use the following settings on the Covaris device, which results in fragments of ~500 bp appropriate for both HiSeq and MiSeq libraries:

Duty cycle 5%.

Intensity 5.

Cycle/burst 200.

Time (s) 35.

Temperature of bath 4 °C.

3.2 Library Preparation

The generation of gDNA libraries prior to enrichment comprises purification of sheared gDNA, end repair, and adapter ligation. We routinely use the NEBNext Ultra DNA Library Prep Kit. However, any other Illumina library preparation method can be used (*see* **Note 2**).

3.2.1 Purification of Sheared gDNA Using AMPure Beads (See Note 3)

1. Bring the AMPure® XP beads to room temperature and homogenize by vortexing.
2. Add 80 μl of AMPure® XP beads (0.6×) to the 130 μl of sheared gDNA, mix by vortexing and incubate at room temperature for 5 min. Briefly centrifuge the tube to collect the solution from the sides of the tube. Be careful not to pellet the magnetic beads. Place the tube in a magnetic stand to separate the beads from the supernatant. Wait for 5 min for the solution to become clear. Carefully transfer the supernatant to a clean tube without disturbing the beads.
3. Add 200 μl of freshly prepared 80% ethanol to the tube with the beads whilst positioned in the magnetic stand. Incubate at room temperature for 30 s to let the beads settle down and then carefully remove and discard the supernatant. Repeat the wash with freshly prepared 80% ethanol a further two times.
4. Carefully remove the residual ethanol using a small-volume (10 μl) pipette after the third wash. Air-dry the beads for 10 min while the tube remains in the magnetic stand with the lid open.
5. Elute the DNA from the beads by adding 60 μl water. Mix by vortexing, briefly centrifuge the tube to collect the solution from the sides and lid of the tube. Place tube in the magnetic stand for 5 min to collect the beads and wait until the solution becomes clear.
6. Transfer 56 μl of the supernatant which contains the eluted DNA to a clean 1.5 ml tube.

3.2.2 End Repair of the Sheared gDNA

1. Set up the end repair reaction using the NEBNext Ultra DNA library Prep Kit (New England BioLabs, E7370S), by mixing the following components in a sterile, nuclease-free PCR tube:

   End Prep Enzyme Mix—3.0 μl.

   End Repair Reaction Buffer—6.5 μl.

   Sheared DNA—56 μl.

   Total volume—65.5 μl.
2. Mix the components by pipetting, followed by a brief centrifugation step to collect all liquid from the sides of the tube.
3. Transfer the tube to a thermocycler, with the "heated lid" option engaged and run the following program:

   20 °C for 30 min.

   60 °C for 30 min.

   Hold at 4 °C.

3.2.3 Adapter Ligation

1. Add the following components from the NEBNext Ultra DNA library Prep Kit (New England BioLabs, E7370S) directly to the 65.5 μl of the end repair reaction and mix well by pipetting:

   Blunt/TA ligase Master Mix—15 μl.

   NEBNext Adapter for Illumina—2.5 μl.

   Ligation Enhancer—1.0 μl.

   Total volume—83.5 μl.
2. Briefly centrifuge the sample to collect liquid from the sides of the tube before incubating the reaction mixture at 20 °C for 15 min in a thermal cycler with the "heated lid" option enabled.
3. Add 3 μl of USER™ enzyme to the ligation mixture, mix well by pipetting followed by a brief centrifugation to collect liquid from the sides of the tube.
4. Place the reaction in a thermal cycler and incubate at 37 °C for 15 min, with the "heated lid" option enabled.

3.2.4 Purification and Size Selection of Adapter-Ligated gDNA

At this point, the first size selection of the library is performed using a two-step AMPure purification. We usually select for fragments between 400 and 600 nucleotides, which give optimal libraries for all HiSeq and MiSeq platforms with various read lengths.

1. Transfer the 86.5 μl ligation reaction into a clean 1.5 ml tube and adjust the volume to 100 μl with water.
2. Add 40 μl of resuspended AMPure XP Beads to the 100 μl ligation reaction. Mix well by pipetting up and down at least 10 times.
3. Incubate for 5 min at room temperature.
4. Spin the tube briefly and place the tube on an appropriate magnetic stand to separate the beads from the supernatant. After the solution is clear (about 5 min), carefully transfer the supernatant containing your DNA to a new tube **(Caution: do not discard the supernatant)**. Discard the beads that contain the unwanted large fragments.
5. Add 20 μl resuspended AMPure XP Beads to the supernatant, mix well and incubate for 5 min at room temperature.
6. Add 200 μl of freshly prepared 80% ethanol to the tube with the beads whilst positioned in the magnetic stand. Incubate at room temperature for 30 s to let the beads settle down and then carefully remove and discard the supernatant. Repeat the wash with freshly prepared 80% ethanol a further two times.
7. After the third wash, carefully remove the residual ethanol using a small-volume (10 μl) pipette after the third wash. Air-- dry the beads for 10 min while the tube remains in the magnetic stand with the lid open.

8. Elute the DNA from the beads by adding 25 μl water. Mix by vortexing, then briefly centrifuge the tube to collect the solution from the sides and lid of the tube. Place tube in the magnetic stand for 5 min to collect the beads and wait until the solution becomes clear.
9. Transfer 22 μl to a clean PCR tube for amplification without disturbing the beads.

3.2.5 PCR Amplification of the Purified Adapter-Ligated DNA

1. Mix the following components in the PCR tube:

   Adapter-ligated DNA fragments—10 μl.

   2× KAPA HiFi HotStart ReadyMix—25 μl.

   Index Primer—1 μl.

   Universal PCR Primer—1 μl.

   Water—13 μl.
2. Run the PCR using the cycling conditions:

98 °C 30 s	
98 °C 10 s	8–15 cycles
65 °C 30 s	
72 °C 30 s	
72 °C 5 min	
4 °C ∞	

First, perform 8 cycles and run 5 μl of the reaction in a 1% agarose gel next to 1 μl of unamplified library (as a control of input DNA). *See* Fig. 1 for a typical result. In case there is not enough DNA, put the reaction back in the thermocycler for an additional 2–5 cycles. Do not amplify more than 15 cycles. If there is no product after 15 cycles, the library preparation failed and you need to start again.

3.2.6 Purification and Quantification of the Amplified Library

1. The ampified library is size-selected with AMPure® XP beads (45 μl) as described in Subheading (3.1.2). DNA should be eluted in 15 μl of water.
2. Quantify precisely the concentration of the purified libraries using a fluorometric method such as a Qubit or the PicoGreen® ds DNA Assay Kit (ThermoFisher Scientific).
3. Combine equimolar amounts of each barcoded library. You will need at least 800 ng of combined library for enrichment.

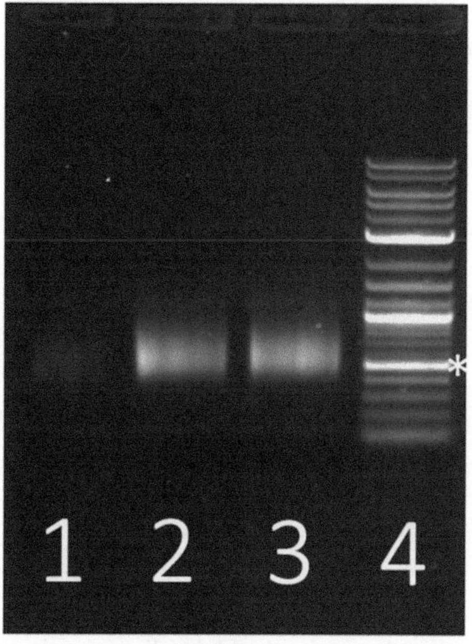

Fig. 1 Typical Illumina library constructed with NEBNext Ultra DNA library Prep Kit. A comparison of an adaptor ligated, size selected (PCR input, Subheading 3.2.4) gDNA (lane 1) with PCR amplified library (lanes 2 and 3). 1 μg of 2-log DNA ladder (NEB, N3200S) was loaded on the agarose gel (lane 4). The asterisk indicates a 500 bp fragment with a DNA mass of 124 ng. The next (above) band with increased intensity is a 1 kb fragment with a DNA mass of 122 ng

3.3 Targeted Enrichment Using the MYcroarray MYbaits Kit

The enrichment of target DNA fragments is achieved through hybridization of the PCR amplified genomic libraries generated in Subheading 3.2 with complementary RNA (cRNA) baits. All reagents, unless stated otherwise, are part of the MYcroarray MYbaits kit.

Perform two enrichments, one for the combined susceptible mutants and a separate one for the resistant parent (*see* **Note 4**).

3.3.1 Library Master Mix Preparation

In a sterile, nuclease-free PCR tube, mix:

SeqCAP (from Roche, not in MYbaits kit)—5 μl.

Block #3—0.6 μl.

Illumina library 500–800 ng—max volume 7 μl.

Total volume—12.6 μl.

Transfer the tube containing the Library Master Mix to the thermocycler and incubate at 95 °C for 5 min and then hold at 65 °C, with the "heated lid" option enabled. While Library Master Mix is incubated in the thermocycler, proceed to Hybridization Master Mix preparation.

3.3.2 Hybridization Master Mix Preparation

In a sterile, nuclease-free PCR tube, mix:

Hyb #1—9 μl.

Hyb #2—0.5 μl.

Hyb #3—2.5 μl.

Hyb #4—0.5 μl.

RNase Block—1 μl.

Baits—5 μl.

Total volume—18.5 μl.

1. Mix the components by vortexing, followed by a quick spin to collect all liquid from the sides of the tube.
2. Transfer the tube containing the Hybridization Master Mix to the thermocycler and incubate at 65 °C for 5 min.
3. While keeping the tube at 65 °C, transfer 9 μl of Hybridization Master Mix to the Library Master Mix and mix by pipetting.
4. Hybridize solution at 65 °C for 16–24 h.

3.3.3 Recovery of Captured Targets

Before starting, preheat the Wash Buffer 2 to 65 °C for at least 1 h, then prepare Wash Buffer 2.2 as follows:

1. Combine 400 μl HYB #4, 39.6 ml nuclease-free water and 10 ml Wash Buffer 2 (to create Wash Buffer 2.2).
2. Heat the Wash Buffer 2.2 to 65 °C for at least 45 min before use.
   The prepared volume of Wash Buffer 2.2 is sufficient for washing 33 samples. It can be stored at 4 °C for up to 6 weeks.

3.3.4 Capture and Washing

1. Transfer 30 μl of MyOne Streptavidin C1 magnetic beads to a sterile, nuclease-free microcentrifuge tube.
2. Place the tube on a magnetic stand to separate beads from supernatant. After the solution is clear, carefully remove and discard the supernatant.
3. Add 200 μl Binding Buffer to wash the beads. Vortex the tube for 5–10 s, place on the magnetic stand for 2 min, and then carefully remove and discard the supernatant.
4. Repeat wash step twice for a total of three washes.
5. Resuspend the beads in 70 μl Binding Buffer and incubate at 65 °C for 2 min.
6. Transfer the hybridization solution to the Binding Buffer/ Beads and incubate for 45 min at 65 °C, mixing the solution every 5–10 min.
7. Pellet the beads on the magnetic stand for 2 min and then carefully remove and discard the supernatant.

8. Add 500 μl of the Wash Buffer 2.2 from 65 °C to the beads and mix by pipetting. Incubate for 10 min at 65 °C in a thermal block. Flick the tube occasionally to resuspend the beads. Pellet the beads on the magnetic stand for 2 min and carefully remove and discard the supernatant.
9. Repeat washing step twice for a total of three 65 °C washes. After the third wash, make sure that all additional buffer is removed by giving the tube a quick spin after the supernatant has been removed, and repelleting the beads with the magnetic stand.
10. Resuspend the beads in 30 μl molecular biology grade water.

3.3.5 Amplification of the Captured Library

This step consists of amplifying the captured DNA while it is still attached to the streptavidin beads. It is important to limit the number of cycles to get just enough material for sequencing while minimizing PCR amplification bias. For this step, KAPA HiFi DNA Polymerase is used, which compares favorably to other available DNA polymerases.

1. Mix the following components in the PCR tube:

   Captured Library—10 μl.

   2× KAPA HiFi HotStart ReadyMix—25 μl.

   P5 primer (10 μM)—1.5 μl.

   P7 primer (10 μM)—1.5 μl.

   Water—12 μl.

   Total volume—50 μl.
2. Run the PCR using the following cycling conditions:

98 °C 30 s	
98 °C 10 s	14–18 cycles
65 °C 30 s	
72 °C 30 s	
72 °C 5 min	
4 °C ∞	

Start with 14 cycles and check 5 μl of PCR product on the 1% agarose gel. Size and amout of the amplified library should be similar to the one in Fig. 1 (lanes 2 and 3). If amplification is weak (e.g., library not visible on gel), perform 2–4 additional cycles. (*see* **Note 5**).

3.3.6 Purification and Size Selection of Enriched and Amplified Library

1. Perform a two-step size selection and purification with AMPure XP beads as described in Subheading (3.2.4).
2. Additional QC steps like Qubit and qPCR quantification may be necessary, depending on the requirements of your sequencing service provider.

3.4 Sequencing

The goal is to create a de novo assembly based on wild type RenSeq data. Then, map RenSeq data from the susceptible mutants to the assembly (*see* **Note 6**).

3.4.1 Preprocessing of Read Data

You need to discuss with your sequencing service provider how much effort has to be put into further processing of read data. Adapters have to be removed. Programs such as fastqc (http://www.bioinformatics.babraham.ac.uk/) help to assess read data quality. Some de novo assemblers require quality trimming of reads. In such a case, sickle (downloadable from https://github.com/najoshi/sickle) can be used.

3.4.2 De Novo Assembly of Wild Type RenSeq Data

We have tested the MutRenSeq pipeline with the commercial CLC assembly cell (http://www.clcbio.com/products/clc-assembly-cell/) and with the free software MaSuRCA [8]. Assemblies from CLC assembly cell were slightly better, but since MaSuRCA is free for scientific use, it is detailed below.

1. Reads do NOT have to be trimmed.
2. Generate a configuration file for MaSuRCA:

```
masurca -g config.txt
```

3. Edit config file. Set parameter USE_LINKING_MATES = 1. Add the correct location of your raw reads. Choose mean read distance and standard deviation of read distances. This information can be obtained from your sequencing service provider.
4. Generate assembly script and run assembly:

```
masurca config.txt
./assemble.sh
```

5. MaSuRCA creates a lot of files. The output file that is the downstream reference.fasta is in the directory CA/10-gap-close/genome.scf.fasta.

   Some tools are not compatible with the naming convention for scaffolds/contigs of MaSuRCA. To avoid this, it might make sense to rename contigs to conitg_1 .. contig_n.

3.4.3 Generation of Mapping and Pileup Files for Each Mutant and Wild Type

The mapping of wild type data to the wild type de novo assembly is used as a positive control and essential for running MutantHunter (*see* below). This step requires the mapping software BWA [9] and SAMtools [10]. The following steps are shown here as an example for wild type reads. The same has to be repeated for reads of each mutant.

1. Create BWA index and fasta index (required for step d.).

```
bwa index reference.fasta
samtools faidx reference.fasta
```

2. Create mappings for separate reads

```
bwa aln reference.fasta wt_R1.fastq > wt_R1.aln
bwa aln reference.fasta wt_R2.fastq > wt_R2.aln
```

(*see* **Note 7**).

3. Combine mappings and create SAM file

```
bwa sampe reference.fasta wt_R1.aln wt_R2.aln wt_R1.fastq
wt_R2.fastq > wt.raw.sam
```

4. Convert SAM file to BAM, sort BAM and remove redundancy. "SAM" stands for sequence alignment/map format. "BAM" is the binary version. For subsequent steps the BAM file needs to be sorted. During the PCR amplification steps of enrichment the generation of PCR-duplicates is very likely and should be removed before further processing.

```
samtools view -f2 -Shub -o wt.raw.bam wt.raw.sam
samtools sort wt.raw.sam wt.sorted
samtools rmdup wt.sorted.bam wt.rmdup.bam
```

(*see* **Note 8**)

5. Generate mpileup format from mapping. Mapping files are read-centric. In SAM format, one line corresponds to one read and it includes information about the mapping and the position where the read maps within the reference. The mpileup format is reference-centric. Each line corresponds to a position in the reference.

```
samtools mpileup -f reference.fasta -BQ0 wt.rmdup.bam >
wt.pileup
```

3.4.4 Run NLR-Parser on the Reference File

The NLR-Parser can be downloaded from github/steuernb/MutantHunter/. You will need to install MAST [11] from the meme suite, version 4.9.1 (download from http://meme-suite.org/) and Java Runtime Environment 1.6 or higher for running NLR-Parser.

```
java -jar NLR-Parser.jar -x meme.xml -y path/to/meme/bin/mast
-i reference.fasta -c nlr.xml -b nlr.bed
```

The nlr.xml file is used in **step 7** and nlr.bed is used in **step 10**.

This step extracts all information from mpileup files that is necessary for downstream analysis, thus reducing the data input for the

3.4.5 Convert mpileup Files to an Internal XML Format

MutantHunter. The program Pileup2XML.jar can be downloaded from github/steuernb/MutantHunter/. When performing this step on wild type data, the argument (-w) needs to be added.

```
java -jar Pileup2XML.jar -m nlr.xml -i wt.pileup -o wt.xml -w
java -jar Pileup2XML.jar -m nlr.xml -i m1.pileup -o m1.xml
```

3.4.6 Define On-Target Regions in Your Assembly

Use a set of NB-LRR sequences or the bait-sequence file as a reference. Essentially, only those regions in the reference will be used for subsequent analysis that have a BLAST hit to your reference file. This step requires NCBI BLAST [12].

```
makeblastdb -in baits.fasta -dbtype nucl
blastn -query assembly.fasta -db baits.fasta -outfmt 5 -out
blastn.xml
```

3.4.7 Run MutantHunter

The program MutantHunter.jar can be downloaded from github/steuernb/MutantHunter/.

```
java -jar MutantHunter.jar -w wt.xml -m m1.xml m2.xml [...] -b
blastn.xml -o output.txt
```

The program has a couple of additional parameters to adjust the sensitivity:

1. Maximum reference allele frequency of a SNP to be reported (-a).

   The default value is 0.1. This means that if less than 1 out of 10 reads support the same allele as the reference at a position this position is reported as a SNP.
2. Minimum coverage for a position to be regarded (-c).

   The default value is 10, which means if less than 10 reads map at a position, no SNP will be called.
3. Number of coherent positions with zero coverage to call a deletion mutation (-z).
   The default value is 50. If a section of 50 or more bp in the reference are not covered by reads, this is regarded as a deletion mutation. This will create a lot of noise if the sequenced material is not isogenic. Increasing this parameter reduces noise but decreases sensitivity to call deletion mutations.

3.4.8 Manual Inspection and Validation of Candidate Contigs

Depending on the level of noise in the data, false positive candidates can exist. The reference file as well as the nonredundant BAM files can be loaded into a genome viewer, such as Savant [13] for manual inspection (Fig. 2) (*see* **Note 9**). The nlr.bed can be loaded to add information about the position of the NB-LRR gene. Further information about motifs can be found in Jupe et al. [14]. Briefly, motifs 17 and 16 mark a coil-coil domain of an NB-LRR gene. Motifs 1, 6, 4, 5, 10, 3, 12, and 2 represent parts of an NB-domain, and motifs 9, 11, and 19 mark LRR regions.

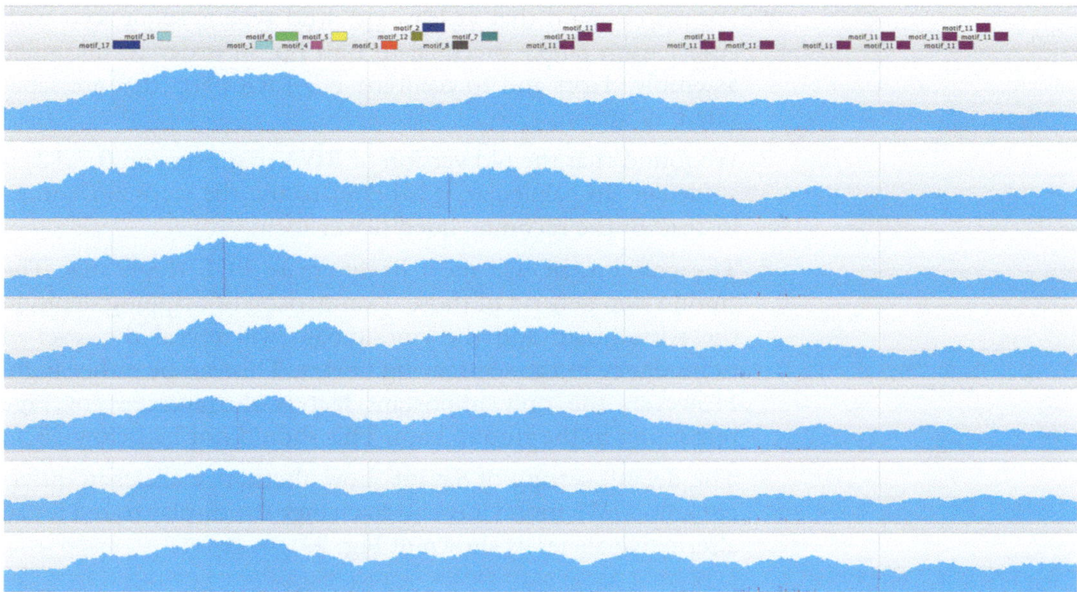

Fig. 2 An example of a target contig visualized with Savant. The bed file derived from NLR-Parser can be loaded (first track) as well as bam file from mapped RenSeq data. Here the second track is from mapped wild type data and the following are data from individual EMS mutants (*see* **Note 9**)

4 Notes

1. The more isogenic the wild type and mutant genomes are the better. Therefore, the source seed used for mutagenesis should be in a homozygous state as residual heterogeneity complicates the downstream comparative genomics analysis.
2. Make sure that blocking oligo reagent (Block #3) is compatible with the adaptors you are using.
3. During this purification, a simple size selection step is performed. Adding 0.6× volume of AMPure beads to sheared DNA results in binding fragments above 400 bp. This step can be adjusted by changing the ratio of AMPure beads to volume of sheared DNA (http://core-genomics.blogspot.co.uk/2012/04/how-do-spri-beads-work.html).
4. To save reagent, you can use a half-reaction volume for the resistant parent.
5. It is highly recommended to check the performance of the enrichment by qPCR. Primer pairs should be designed for a few genes targeted in the enrichment. The difference between enriched and nonenriched hexaploid wheat samples should oscillate between 8 and 11 cycles.

6. We recommend that at least the wild type is sequenced with read length larger or equal to 250 bp, in order to minimize assembly errors and to combine NB-LRR parts that are separated by introns.
7. We found that the old version of BWA (e.g., version 0.7.12) is sufficient. In particular, it is better to use the steps aln/sampe instead of the recommended newer mem.
8. The enrichment will most likely create PCR duplicates. The rmdup step gets rid of those. Comparison of mapping depth in the sorted.bam and the rmdup.bam will reveal the level of redundancy in the sequencing library. This can be rather high. However, the only important fact is the average coverage remaining in the rmdup.bam. This should not be below 25.
9. The default setting of Savant show alternative alleles in different colors. We found it useful to change the display to red for all alternative alleles as shown in Fig. 2.

Acknowledgements

The authors thank the support by the Gatsby Charitable Foundation, UK, the Two Blades Foundation, USA and the Biotechnology and Biological Sciences Research Council, UK, Grant numbers BB/H019820/1, BB/L009293/1 and BB/L011794/1.

References

1. Dodds PN, Rathjen JP (2010) Plant immunity: towards an integrated view of plant-pathogen interactions. Nat Rev Genet 11:539–548
2. Dangl JL, Horvath DM, Staskawicz BJ (2013) Pivoting the plant immune system from dissection to deployment. Science 341:746–751
3. Takken FLW, Goverse A (2012) How to build a pathogen detector: structural basis of NB-LRR function. Curr Opin Plant Biol 15:375–384
4. Steuernagel B, Periyannan SK, Hernandez-Pinzon I, Witek K, Rouse MN, Yu G, Hatta A, Ayliffe M, Bariana H, Jones JD et al (2016) Rapid cloning of disease-resistance genes in plants using mutagenesis and sequence capture. Nat Biotechnol 34:652–655
5. Jupe F, Chen X, Verweij W, Witek K, Jones JD, Hein I (2014) Genomic DNA library preparation for resistance gene enrichment and sequencing (RenSeq) in plants. Methods Mol Biol 1127:291–303
6. Periyannan S, Moore J, Ayliffe M, Bansal U, Wang X, Huang L, Deal K, Luo M, Kong X, Bariana H et al (2013) The gene Sr33, an ortholog of barley Mla genes, encodes resistance to wheat stem rust race Ug99. Science 341:786–788
7. Witek K, Jupe F, Witek AI, Baker D, Clark MD, Jones JDG (2016) Accelerated cloning of a potato late blight-resistance gene using RenSeq and SMRT sequencing. Nat Biotechnol 34:656–660
8. Zimin AV, Marcais G, Puiu D, Roberts M, Salzberg SL, Yorke JA (2013) The MaSuRCA genome assembler. Bioinformatics 29:2669–2677
9. Li H, Durbin R (2009) Fast and accurate short read alignment with burrows-wheeler transform. Bioinformatics 25:1754–1760
10. Li H, Handsaker B, Wysoker A, Fennell T, Ruan J, Homer N, Marth G, Abecasis G, Durbin R, GPDPS (2009) The sequence alignment/map format and SAMtools. Bioinformatics 25:2078–2079
11. Bailey TL, Gribskov M (1998) Combining evidence using p-values: application to sequence homology searches. Bioinformatics 14:48–54
12. Zhang Z, Schwartz S, Wagner L, Miller W (2000) A greedy algorithm for aligning DNA sequences. J Comput Biol 7:203–214

13. Fiume M, Smith EJ, Brook A, Strbenac D, Turner B, Mezlini AM, Robinson MD, Wodak SJ, Brudno M (2012) Savant genome browser 2: visualization and analysis for population-scale genomics. Nucleic Acids Res 40: W615–W621

14. Jupe F, Pritchard L, Etherington GJ, Mackenzie K, Cock PJ, Wright F, Sharma SK, Bolser D, Bryan GJ, Jones JD, Hein I (2012) Identification and localisation of the NB-LRR gene family within the potato genome. BMC Genomics 13:75

Chapter 20

Rapid Gene Isolation Using MutChromSeq

Burkhard Steuernagel, Jan Vrána, Miroslava Karafiátová, Brande B.H. Wulff, and Jaroslav Doležel

Abstract

MutChromSeq is an approach for isolation of genes and DNA sequences controlling gene expression in plants with complex and polyploid genomes. It involves a lossless complexity reduction by flow cytometric chromosome sorting and shotgun sequencing DNA from isolated chromosomes. Comparison of sequences from wild-type parental chromosome with chromosomes from multiple independently derived mutants identifies causative mutations in a single candidate gene or a noncoding sequence. MutChromSeq does not rely on recombination-based genetic mapping and does not exclude any DNA sequence from being targeted.

Key words Chromosome isolation, FISHIS, Flow cytometry and chromosome sorting, Mutational genomics, Gene cloning

1 Introduction

The knowledge of DNA sequences controlling traits of interest is needed to fully exploit the potential of molecular techniques in plant breeding and deliver a new generation of agricultural crops with improved yield, quality, and resistance to pests, diseases and adverse environmental conditions. Current evidence in human shows that in addition to coding sequences, noncoding sequences, including cell type- and stimulus-specific enhancer regions may be involved in the control of phenotypic expression [1, 2]. The advances in DNA sequencing technologies made the identification of DNA loci controlling various traits a realistic goal in plants with small genomes such as *Arabidopsis* and rice, where whole genome sequencing is technically and economically feasible. Thus, it has been possible to sequence genomes of individuals from mapping populations to aid in gene mapping and cloning, or genomes of

---

Burkhard Steuernagel and Jan Vrána contributed equally to this work.

Sambasivam Periyannan (ed.), *Wheat Rust Diseases: Methods and Protocols*, Methods in Molecular Biology, vol. 1659, DOI 10.1007/978-1-4939-7249-4_20, © Springer Science+Business Media LLC 2017

independently derived mutants with the same phenotype for direct gene identification [3, 4].

Unfortunately, such mutational genomics approaches remain prohibitively expensive and impose significant computational challenges in a number of important crops such as those from the tribe Triticeae, which includes barley, rye, and wheat. Genomes of these species are huge; their size exceeds many giga base pairs, and some of them are allopolyploid. Consequently, only a limited number of their genes have been cloned to date [5–9]. Most of these genes were identified by map-based cloning, an approach that is costly, time-consuming and hampered by low recombination regions which span almost half of the length of each chromosome. The advent of next generation sequencing technologies has allowed new gene cloning approaches to be explored. These are typically characterized by reduction of DNA sequence complexity to decrease the sequencing costs, for example by methylation filtration [10, 11], duplex-specific nuclease digestion [12], and more recently transcriptome sequencing and exome capture sequencing [13, 14]. A common disadvantage of these approaches is that they do not capture all potentially relevant sequences.

To avoid any bias in complexity reduction, the MutChromSeq (Mutant Chromosome Sequencing) gene cloning approach was developed. It combines (1) mutagenesis and screening for mutants with (2) a lossless complexity reduction based on flow cytometric chromosome sorting, followed by (3) sequencing and sequence analysis. The sequence comparison of the wild-type parental chromosome with chromosomes from multiple independently derived mutants allows the identification of causative mutations in a single candidate gene or a noncoding sequence [15]. The extent of complexity reduction depends on the number of chromosomes of a species and in wheat and barley it is 21-fold and 7-fold, respectively. MutChromSeq does not exclude any DNA sequence from being targeted and, as it does not rely on recombination-based genetic mapping, is particularly suited to complex genomes with large chromosome regions practically devoid of recombination.

MutChromSeq can be applied to any plant species that is amenable to mutagenesis and from which chromosomes can be flow-sorted. Moreover, the target DNA sequence must be associated with a distinct phenotype and the chromosome to which the phenotype maps must be known. The method has successfully been applied to reclone the *Eceriferum-q* gene in barley and clone de novo the *Pm2* gene from hexaploid wheat [15]. Here, we provide a detailed step-by-step protocol of MutChromSeq.

2 Materials

2.1 Plant Material

Dried vernalized seeds of a wild type and a set of mutants (*see* **Note 1**).

2.2 Reagents and Solutions for Preparation of Chromosome Suspensions

1. Hoagland's stock solution (10×): 4.7 g $Ca(NO_3)_2{\cdot}4H_2O$ (40 mM), 2.6 g $MgSO_4{\cdot}7H_2O$ (20 mM), 3.3 g KNO_3 (65 mM), 0.6 g $NH_4H_2PO_4$ (10 mM), 5 mL solution A, and 0.5 mL solution B, in deionized water. Adjust volume to 500 mL. Prepare just before use.
   (a) Solution A: 45 mM H_3BO_3 (280 mg), 20 mM $MnSO_4{\cdot}H_2O$ (340 mg), 0.4 mM $CuSO_4{\cdot}5H_2O$ (10 mg), 0.8 mM $ZnSO_4{\cdot}7H_2O$ (22 mg) and 0.08 mM $(NH_4)_6Mo_7O_{24}{\cdot}4H_2O$ (10 mg) in deionized water (100 mL). Store at 4 °C.
   (b) Solution B: 0.05 mM concentrated H_2SO_4 (0.5 mL) in deionized water (100 mL). Store at 4 °C.
   (c) Solution C: 18 mM Na_2EDTA (3.36 g) and 20 mM 2.79 g $FeSO_4$ (20 mM) in deionized water. Heat the solution to 70 °C while stirring until the color turns yellow-brown. Cool down, adjust the volume with deionized water (500 mL) and store at 4 °C.
2. Hoagland's nutrient solution (0.1×): 100 mL Hoagland's stock solution (10×) and 0.5 mL solution C in deionized water. Adjust volume to 1000 mL. Prepare just before use.
3. 2 mM HU solution: dissolve 121.6 mg hydroxyurea in 800 mL 0.1× Hoagland's nutrient solution. Prepare just before use.
4. Amiprophos methyl (APM) stock solution (20 mM): dissolve 60.86 mg APM in 10 mL ice-cold acetone and store at −20 °C, in 1 mL aliquots.
5. APM working solution (2.5 μM): 101.3 μL APM stock solution in 800 mL deionized water. Prepare just before use.
6. Tris buffer: 10 mM Tris (606 mg), 10 mM Na_2EDTA (1.861 g), 100 mM NaCl (2.922 g) in deionized water (500 mL). Adjust pH to 7.5 using 1 N NaOH.
7. Formaldehyde 2% fixative: 13.5 mL formaldehyde in Tris buffer. Adjust volume to 250 mL. Prepare just before use.
8. LB01 buffer [16]: 15 mM Tris (0.363 g), 2 mM Na_2EDTA (0.149 g), 0.5 mM spermine·4HCl (0.0348 g), 80 mM KCl (1.193 g), 20 mM NaCl (0.234 g), 0.1% (v/v) Triton X-100 (200 μL) in deionized water (200 mL). Adjust pH to 9. Filter through a 0.22 μm filter to remove small particles. Add 220 μL β-mercaptoethanol and mix well. Store at −20 °C, in 8 mL aliquots.

2.3 Reagents and Solutions for Chromosome Sorting

1. 4′,6-diamidino-2-phenylindole (DAPI) stock solution (0.1 mg/mL): dissolve DAPI in deionized water by stirring. Filter through a 0.22 μm filter to remove small particles. Store at −20 °C, in 0.5 mL aliquots.
2. 10 M NaOH: dissolve solid NaOH in deionized water. Store at room temperature.
3. 1 M Tris–HCl: dissolve Tris in deionized water by stirring; adjust the pH to 7.5 using 1 N HCl. Store at 4 °C.
4. $(GAA)_7$ microsatellite probe labeled with FITC: Dissolve the probe to 100 μM concentration with 2× SSC (see recipe below) according to manufacturer's instructions. Prepare working solution by adding 2× SSC to a final concentration of 80 ng/μL. Store in the dark at −20 °C.

2.4 Reagents and Solutions for Fluorescence In Situ Hybridization (FISH)

1. P5 buffer: 10 mM Tris (30.28 mg), 50 mM KCl (93.2 mg), 2 mM $MgCl_2 \cdot 6H_2O$ (10.17 mg) and 5% sucrose (1.25 g) in deionized H_2O (25 mL). Adjust pH to 8 using 1 N HCl. Store at −20 °C, in 1 mL aliquots.
2. 20× SSC stock solution: 3 M NaCl (175.3 g) and 300 mM $Na_3C_6H_5O_7 \cdot 2H_2O$ (88.2 g) in deionized H_2O (1000 mL). Adjust pH to 7. Sterilize by autoclaving. Store at room temperature.
3. 4× SSC washing buffer: 20× SSC (200 mL) and 0.2% Tween 20 in deionized H_2O (1000 mL).
4. 2× SSC washing buffer: 20× SSC (100 mL) in deionized H_2O (1000 mL). Prepare just before use.
5. 0.1× SSC stringent washing buffer: 20× SSC (5 mL), 0.1% Tween 20 and 2 mM $MgCl_2 \cdot 6H_2O$ in deionized H_2O (1000 mL). Prepare just before use.
6. Hybridization mix: 40% formamide (10 μL), 20x SSC (1.25 μL), 0.625 μL salmon sperm DNA (250 ng/μL), labeled DNA probe(s) (1 ng/μL). Add 5% dextrane sulfate (final volume 25 μL). Prepare just before use. Labeled DNA probes (either directly labeled with fluorescent probes, or labeled by digoxigenin or biotin) may be prepared using PCR [17].
7. Detection of digoxigenin-labeled probes: FITC-labeled anti-digoxigenin antibody raised in sheep.
8. Detection of biotin-labeled probes: Cy3-labeled streptavidin antibody.
9. Blocking solution: dissolve 0.5 g blocking reagent in 50 mL 4× SSC. Autoclave. Store at −20 °C, in 1 mL aliquots.
10. Vectashield antifade solution containing DAPI (Vector Laboratories, Inc., Burlingame, CA, USA).

2.5 Reagents and Materials for DNA Amplification

1. Proteinase K buffer (40×): 100 μL 1 M Tris–Cl (pH 8.0), 100 μL 0.5 M EDTA (pH 8.0), 500 μL 10% SDS (Sigma-Aldrich). Adjust volume to 1 mL with sterile deionized H_2O. Store up to 3 months at room temperature.
2. Proteinase K (10 mg/mL): 1 mg proteinase K (Roche). Dissolve in 100 μL sterile deionized H_2O. Store up to 1 week at 4 °C.
3. Illustra GenomiPhi V2 DNA Amplification kit (GE Healthcare).

3 Methods

3.1 Preparation of Chromosome Suspensions (See Note 2)

1. For better germination results, soak the seeds in a beaker filled with deionized H_2O at room temperature for about 15 min (*see* **Note 3**).
2. Spread the seeds evenly on a layer of wet paper towel sandwiched by two layers of filter paper in a glass petri dish and germinate in the dark at 25 °C, until optimal root length is achieved (typically 3 cm).
3. For easier manipulation with the seedlings in subsequent treatments, thread their roots through the holes of a plastic cover.
4. To synchronize the cell cycles of root meristems, transfer the plastic cover with seedlings onto a plastic box filled with 2 mM HU solution and incubate for 18 h by aerating in the dark at 25 °C.
5. Remove the seedlings from the HU solution and transfer them onto a plastic box filled with 0.1× Hoagland's solution. Incubate by aerating in the dark at 25 °C for 5.5 h (wheat) or 6.5 h (barley).
6. To accumulate cells in metaphase, transfer the cover with seedlings onto a box filled with 2.5 μM solution of APM and incubate in the dark at 25 °C by aerating for 2 h.
7. Remove the cover with seedlings and put it into a container filled with ice water containing ice cubes and keep it overnight in a refrigerator (*see* **Note 4**). Make sure that all roots are fully immersed.
8. Cut 100 roots (approx. 1 cm from apex) and transfer them into a beaker filled with deionized H_2O.
9. Place the roots into the beaker containing 2% formaldehyde fixative solution and incubate at 5 °C for 20 min (*see* **Note 5**).
10. Wash the roots three times in Tris buffer at 5 °C for 5 min. Keep the roots in the Tris buffer on ice after the last wash. Process fixed roots within a few hours.

11. Cut the root apices (1–2 mm long) and transfer them into a 5 mL polystyrene tube containing 1 mL LB01 buffer. Grind the root tips using a blender using the following settings: 15,000 RPM for 13 s (barley), 20,000 RPM for 13 s (wheat).
12. Filter the crude suspension through 50 μm nylon mesh into a new 5 mL polystyrene tube.
13. Keep the suspension on ice until FISHIS labeling.

3.2 Chromosome Labeling Using FISH in Suspension (FISHIS [18]) (See Note 6)

1. Filter 1 mL of chromosome suspension (*see* **Note 7**) through 20 μm nylon mesh into 1.5 mL tube.
2. Add 10 M NaOH to reach pH range of 12.8–13.3.
3. Incubate the sample for 15 min on ice.
4. Adjust the pH in the range of 8.5–9.1 using Tris–HCl and keep on ice for 1 min.
5. Add $(GAA)_7$ probe working solution to final concentration of 4.6 ng/μL and let the suspension incubate for 1 h in the dark at room temperature.
6. Keep the suspension on ice until the flow cytometric experiments.

3.3 Chromosome Sorting Using Flow Cytometry

1. Start up the flow sorter. Make sure that optical path alignment and sorting precision are in peak condition in order to get best results. If not, follow manufacturer's instructions to improve it.
2. Filter the sample through 20 μm nylon mesh.
3. Add DAPI to a final concentration 2 μg/mL (for 1 mL sample use 20 μL of DAPI stock solution).
4. In acquisition software of the flow sorter, open or create the appropriate histograms and dot plots. First, use dot plot FSC vs. DAPI to visualize populations representing chromosomes. Create a region surrounding the chromosomes and use this gating on dot plot FITC vs. DAPI and use it for sorting chromosome(s) of interest.
5. Run the sample and adjust instrument settings for each parameter so that the populations corresponding to chromosomes are in the field. Analyze at least 20,000 chromosomes and save the data.
6. Create sorting region surrounding the population of chromosome(s) of interest (*see* Fig. 1 and **Note 8**).

3.4 Estimation of Purity in Sorted Fractions (See Note 9)

1. Sort approximately 2000 chromosomes of interest into a 5 μL drop of P5 buffer on a microscope slide. Leave to air-dry and keep in the dark at room temperature until use.
2. Add 25 μL of hybridization mix, place a coverslip and seal with rubber cement.

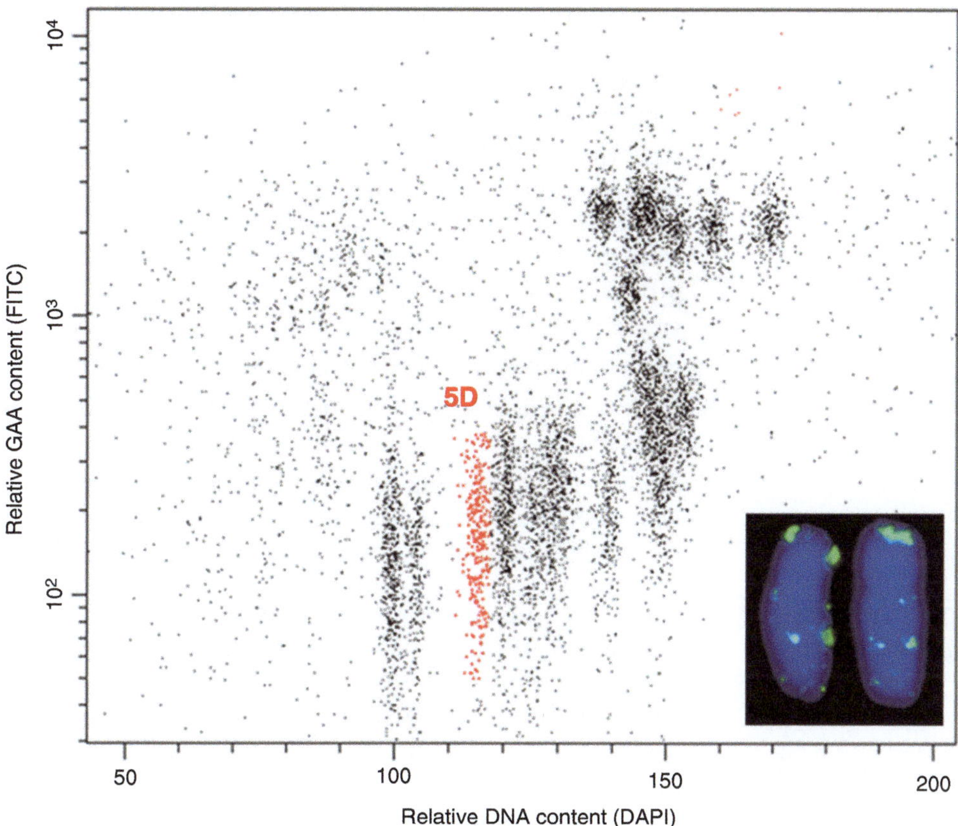

Fig. 1 Biparametric dot plot obtained after flow cytometric analysis of chromosome suspension of bread wheat *T. aestivum* cv. Chancellor after FISHIS with a FITC-labeled probe for GAA microsatellite. The region highlighted in *red* corresponds to chromosome 5D, which was sorted in this experiment. *X axis*: DAPI fluorescence intensity (linear scale); *Y axis*: FITC fluorescence intensity (logarithmic scale). *Inset*: chromosome-specific distribution of GAA microsatellite on chromosome 5D, which served to identify the chromosome and estimate the purity in the sorted chromosome 5D fraction. FISH was done with FITC-labeled probe for GAA microsatellite (*green*) and the chromosomes were counterstained by DAPI (*blue*)

3. Denature chromosomal DNA at 80 °C for 40 s on a hot plate.
4. Transfer the slide into a humidity chamber and incubate overnight at 37 °C.
5. Transfer the slide into container filled with preheated (42 °C) 2× SSC and carefully remove the coverslip using tweezers. Wash for 10 min at 42 °C.
6. Wash in preheated 0.1× SSC for 5 min at 42 °C.
7. Incubate in preheated 2× SSC for 10 min at 42 °C.
8. Remove the container from incubator. Replace the solution with preheated (42 °C) 2× SSC solution, and incubate the slide for 10 min at room temperature.
9. Wash in 4× SSC (RT) for 10 min at room temperature.

10. Remove the slide from the container and put 60 μL of 1% blocking solution over the area with the chromosomes. Cover the slide with parafilm and incubate for 10 min at room temperature.
11. Add the solution of fluorescently labeled antibody (follow manufacturer's instructions regarding the concentration) in 60 μL of 1% blocking solution, and incubate for 1 h at 37 °C. This step is omitted in the case of directly labeled fluorescent probes.
12. Wash the slide three times in heated (42 °C) 4× SSC solution for 10 min at 42 °C. Let the slides air-dry.
13. Add Vectashield solution containing DAPI and cover with a coverslip.
14. Analyze the slide using a fluorescence microscope. Evaluate at least 100 chromosomes on three different slides with the same-sorted chromosome fraction (*see* the inset of Fig. 1 and **Note 10**).

3.5 DNA Purification and Amplification [19]

1. Sort chromosomes of interest into three 0.5-mL PCR tubes containing 40 μL sterile, deionized H_2O (*see* **Note 11**). The number of chromosomes to be sorted into each tube should correspond to 50 ng DNA.
2. Add proteinase K stock solution and 40x proteinase buffer so that the final concentration of proteinase K reaches 60 ng/μL and that of proteinase buffer reaches 1×. Incubate for 20 h at 50 °C on PCR cycler.
3. Add half amount of proteinase K stock solution used in **step 2** and incubate for another 20 h at 50 °C on PCR cycler.
4. To remove proteinase K and buffer, use a Vivacon 500 column (Sartorius). Add deionized H_2O to the sample in the column to reach 500 μL volume and centrifuge for approximately 10 min at 3000 × *g* at 24 °C.
5. Repeat **step 4** three times. Do not let the sample dry. The remaining volume should be about 10 μL.
6. Turn the column bottom-up and transfer the sample into a 1.5-mL tube by centrifuging for 3 min at 1000 × *g* at 24 °C.
7. Estimate the concentration of purified DNA using fluorimeter (*see* **Note 12**).
8. Use 10 ng of purified DNA as a template for amplification. Reduce the volume to 1 μL by overnight evaporation at 4 °C.
9. Amplify the DNA using Illustra GenomiPhi V2 DNA Amplification kit. Follow the manufacturer's instructions.
10. Merge three independent amplification products (*see* **Note 13**).

3.6 Sequencing

The goal is to create a de novo assembly of the wild-type target chromosome. Subsequently, the data from the mutant chromosomes is mapped to the assembly, and the chromosomes are compared to each other to identify induced, causative mutations in a candidate gene. We have found that 35 Gb of Illumina HiSeq 150 bp paired-end reads per chromosome is sufficient.

3.7 Bioinformatics

1. Preprocessing of read data. You need to discuss with your sequence service provider how much effort has to be put into further processing of read data. Adapters have to be removed. Programs such as FastQC (http://www.bioinformatics.babraham.ac.uk/) can be used to assess read quality. Some de novo assemblers require quality trimming of reads. In such a case, sickle (downloadable from https://github.com/najoshi/sickle) can be used.
2. De novo assembly of wild-type MutChromSeq data. We have tested the MutChromSeq pipeline with the commercial CLC assembly cell (http://www.clcbio.com/products/clc-assembly-cell/) and default parameters. Other softwares might be sufficient or superior. The better the de novo assembly the easier are the subsequent steps.
3. Mask repetitive sequences in the de novo assembly. Since we are not interested in mutations within repetitive parts of the genome we can reduce complexity further by masking transposable elements. Any software identifying repeats is sufficient. We use RepeatMasker (www.repeatmasker.org) with the TriticeaeRepeat Database (TREP) [20]. Download the nonredundant nucleotide sequences from http://botserv2.uzh.ch/kelldata/trep-db/downloadFiles.html. The program call for RepeatMasker is as follows:

   ```
   RepeatMasker -lib trep-db_nr_Rel-XX.fasta wt_assembly.fasta
   ```

   Among the result files of RepeatMasker a file

   ```
   wt_assembly.fasta.masked
   ```

   can be used for subsequent analysis.
4. Generate mapping and pileup files for each mutant and the wild type. The mapping of wild-type data to the wild-type de novo assembly is used as a positive control and essential for running MutantHunter (*see* below). This step requires the mapping software BWA [21] and samtools [22]. The following steps are shown here as an example for wild-type reads (abbreviated as wt in the code below). The same has to be repeated for reads of each mutant.

   (a) Create BWA index and fasta index (required for step d).

```
bwa index wt_assembly.fasta.masked
samtools faidx wt_assembly.fasta.masked
```

(b) Create mappings for separate read files

```
bwa aln wt_assembly.fasta.masked wt_R1.fastq > wt_R1.
aln
bwa aln wt_assembly.fasta.masked wt_R2.fastq > wt_R2.
aln
bwa aln wt_assembly.fasta.masked MT1_R1.fastq >
mt1_R1.aln
...
bwa aln wt_assembly.fasta.masked MTx_R2.fastq >
MTx_R2.aln
```

(c) Combine mappings and create SAM file

```
bwa sampe wt_assembly.fasta.masked wt_R1.aln wt_R2.
aln wt_R1.fastq wt_R2.fastq > wt.raw.sam
```

Repeat this command for each mutant.

(d) Convert SAM file to BAM, sort BAM and remove redundancy. “SAM” stands for sequence alignment/map format. “BAM” is the binary version. For subsequent steps the BAM file needs to be sorted. During the PCR of multiple displacement amplification and Ilumina library preparation the generation of duplicates is very likely and should be removed before further processing.

```
samtools view -f2 -Shub -o wt.raw.bam wt.raw.sam;
samtools sort wt.raw.sam wt.sorted;
samtools rmdup wt.sorted.bam wt.rmdup.bam;
```

Repeat this command for each mutant.

5. Generate mpileup format from mapping. Mapping files are read-centric. In SAM format, one line corresponds to one read and it includes information about the mapping and the position where the read maps within the reference. The mpileup format is reference-centric. Each line corresponds to a position in the reference.

```
samtools mpileup -f wt_assembly.fasta.masked -BQ0 wt.
rmdup.bam > wt.pileup
```

Repeat this command for each mutant.

6. Run Pileup2XML

This step is part of the MutChromSeq pipeline [15] and preprocesses a pileup file screens for single nucleotide variations (SNV). It has to be executed for the pileup generated from wild-type mapping as well as for every pileup file from each individual mutant line. For running the Pileup2XML on the wild-type data the parameter -w needs to be added.

Download latest release of Pileup2XML.jar from Github (github.com/steuernb/MutChromSeq/releases).

```
java -jar Pileup2XML.jar -i wt.pileup -o wt.xml -f refer-
ence.fasta -a 0.1 -c 10 -w
java -jar Pileup2XML.jar -i MT1.pileup -o MT.xml -f
reference.fasta -a 0.1 -c 10
```

Repeat the command for the other mutants.

The parameter `-a` is the reference allele frequency. A perfect SNV would have 0.0 reference allele frequency. Sensitivity of the pipeline increases with increasing this parameter to up to 0.7. This will, however influence the file-sizes, runtime and false positive rate. The parameter `-c` is the minimum coverage to consider a position for SNV calling.

7. Run MutChromSeq

   This step combines the individual xml output files from **step 6** and calls candidate contigs.

   Download latest release of MutChromSeq.jar from Github (github.com/steuernb/MutChromSeq/releases).

```
java -jar MutChromSeq.jar -w wildtype.pileup.xml -m mu-
tant1.pileup.xml mutant2.pileup.xml [...] -o output.txt -n
6 -c 10 -a 0.1 -z 2
```

   Parameter -n denotes the minimum number of mutants that need to have a mutation in a contig to be reported as a candidate. Parameter `-z` denotes the maximum number of mutants that are allowed to have a mutation at the same position. Parameters `-c` and `-a` are the same as in **step 6** (*see* **Note 14**).

8. Manual validation

   Load the mapping data (.bam files from **step 4d**) into a genome browser. For the large chromosome data sets we suggest IGV [23] as a browser.

4 Notes

1. Several rounds of backcrosses are recommended to recover homozygous material. If heterozygous material is used, the sorted fraction consists of two chromosome types, which may significantly complicate sequence analysis.
2. This protocol has been optimized for barley and wheat. For protocols suitable for some other species, *see* Vrána et al. [24].
3. For better results, always use viable and healthy seeds. Check the germination on a small sample of seeds before the experiment. Thirty seeds of both barley and wheat are necessary for preparation of 1 mL sample.

4. This treatment helps to reduce the frequency of chromosome clumps for better yields of metaphase chromosomes.
5. As formaldehyde is harmful, always wear protective laboratory gloves and work in a biosafety hood.
6. The FISHIS labeling procedure may be omitted if peaks of chromosome(s) of interest are well resolved according to DAPI fluorescence.
7. When using FISHIS to label chromosome repeats, it is advisable to double the amount of root tips per sample due to dilution of the chromosome suspension.
8. If not sure which population corresponds to the chromosome of interest, sort small number of chromosomes from several populations separately onto a microscope slide and inspect them microscopically after FISH with appropriate probe(s).
9. Estimation of purity in sorted chromosome population using FISH relies on the availability of cytogenetic markers providing chromosome-specific labeling patterns. This is a critical step as it allows estimation of the frequency and identity of contaminating chromosomes.
10. Mutagenesis may change chromosome morphology, e.g., due to deletions and translocations, and thus the hybridization pattern of the probe on mutated chromosome may differ from that on wild-type chromosome.
11. If the sorted fraction is not processed immediately, it can be stored for up to 6 months at −20 °C.
12. The yield of purified chromosomal DNA typically reaches about 50% of the original DNA amount.
13. The final product of DNA amplification can be stored up to 1 year at −20 °C before further processing.
14. The redundancy of parameters Pileup2XML and MutChromSeq gives the opportunity to apply very sensitive values in Pileup2XML and vary the strictness of thresholds in the MutChromSeq step. Pileup2XML is the long running data reduction step. But this can be executed in parallel for each mutant separately.

Acknowledgements

This work was supported by the JIC Innovation Fund (IF2015BW22) and the Biotechnological and Biological Sciences Research Council (BBSRC)-UK and grant LO1204 from the Czech National Program of Sustainability I. We also thank Ms. Zdeňka Dubská and Ms. Romana Šperková for excellent technical assistance.

References

1. Piraino SW, Furney SJ (2016) Beyond the exome: the role of non-coding somatic mutations in cancer. Ann Oncol 27:240–248
2. Spitz F (2016) Gene regulation at a distance: from remote enhancers to 3D regulatory ensembles. Semin Cell Dev Biol 57:57–67
3. Fekih R, Takagi H, Tamiru M, Abe A, Natsume S, Yaegashi H, Sharma S, Sharma S, Kanzaki H, Matsumura H et al (2013) MutMap+: genetic mapping and mutant identification without crossing in rice. PLoS One 8:e68529
4. Thole JM, Strader LC (2015) Next-generation sequencing as a tool to quickly identify causative EMS-generated mutations. Plant Signal Behav 10:e1000167
5. Feuillet C, Travella S, Stein N, Albar L, Nublat A, Keller B (2003) Map-based isolation of the leaf rust disease resistance gene *Lr10* from the hexaploid wheat (*Triticum aestivum* L.) genome. Proc Natl Acad Sci USA 100:15253–15258
6. Huang L, Brooks SA, Li W, Fellers JP, Trick HN, Gill BS (2003) Map-based cloning of leaf rust resistance gene *Lr21* from the large and polyploid genome of bread wheat. Genetics 164:655–664
7. Uauy C, Distelfeld A, Fahima T, Blechl A, Dubcovsky J (2006) A NAC Gene regulating senescence improves grain protein, zinc, and iron content in wheat. Science 314:1298–1301
8. Krattinger SG, Lagudah ES, Spielmeyer W, Singh RP, Huerta-Espino J, McFadden H, Bossolini E, Selter LL, Keller B (2009) A putative ABC transporter confers durable resistance to multiple fungal pathogens in wheat. Science 323:1360–1363
9. Rawat N, Pumphrey MO, Liu S, Zhang X, Tiwari VK, Ando K, Trick HN, Bockus WW, Akhunov E, Anderson JA, Gill BS (2016) Wheat Fhb1 encodes a chimeric lectin with agglutinin domains and a pore-forming toxin-like domain conferring resistance to Fusarium head blight. Nat Genet 48:1576–1580
10. Rabinowicz PD (2003) Constructing gene-enriched plant genomic libraries using methylation filtration technology. Methods Mol Biol 236:21–36
11. Peterson DG, Schulze SR, Sciara EB, Lee SA, Bowers JE, Nagel A, Jiang N, Tibbitts DC, Wessler SR, Paterson AH (2002) Integration of cot analysis, DNA cloning, and high-throughput sequencing facilitates genome characterization and gene discovery. Genome Res 12:795–807
12. Shagina I, Bogdanova E, Mamedov IZ, Lebedev Y, Lukyanov S, Shagin D (2010) Normalization of genomic DNA using duplex-specific nuclease. BioTechniques 48:455–459
13. Oono Y, Kobayashi F, Kawahara Y, Yazawa T, Handa H, Itoh T, Matsumoto T (2013) Characterisation of the wheat (*Triticum aestivum* L.) transcriptome by de novo assembly for the discovery of phosphate starvation-responsive genes: gene expression in pi-stressed wheat. BMC Genomics 214:77
14. Steuernagel B, Periyannan SK, Hernández-Pinzón I, Witek K, Rouse MN, Yu G, Hatta A, Ayliffe M, Bariana H, Jones JDG et al (2016) Rapid cloning of disease-resistance genes in plants using mutagenesis and sequence capture. Nat Biotechnol 34:652–655
15. Sanchez-Martin J, Steuernagel B, Ghosh S, Herren G, Hurni S, Adamski N, Vrána J, Kubaláková M, Krattinger SG, Wicker T et al (2016) Rapid gene isolation in barley and wheat by mutant chromosome sequencing. Genome Biol 17:221
16. Doležel J, Binarová P, Lucretti S (1989) Analysis of nuclear DNA content in plant cells by flow cytometry. Biol Plant 31:113–120
17. Vrána J, Šimková H, Kubaláková M, Číhalíková J, Doležel J (2012) Flow cytometric chromosome sorting in plants: the next generation. Methods 57:331–337
18. Giorgi D, Farina A, Grosso V, Gennaro A, Ceoloni C, Lucretti S (2013) FISHIS: fluorescence in situ hybridization in suspension and chromosome flow sorting made easy. PLoS One 8:e57994
19. Šimková H, Svensson JT, Condamine P, Hřibová E, Suchánková P, Bhat PR, Bartoš J, Šafář J, Close TJ, Doležel J (2008) Coupling amplified DNA from flow-sorted chromosomes to high-density SNP mapping in barley. BMC Genomics 9:294
20. Wicker T, Matthews DE, Keller B (2002) TREP: a database for Triticeae repetitive elements. Trends Plant Sci 7:561–562
21. Li H, Durbin R (2009) Fast and accurate short read alignment with burrows-wheeler transform. Bioinformatics 25:1754–1760
22. Li H, Handsaker B, Wysoker A, Fennell T, Ruan J, Homer N, Marth G, Abecasis G, Durbin R, GPDPS (2009) The sequence alignment/map format and SAMtools. Bioinformatics 25:2078–2079
23. Robinson JT, Thorvaldsdottir H, Winckler W, Guttman M, Lander ES, Getz G, Mesirov JP (2011) Integrative genomics viewer. Nat Biotechnol 29:24–26
24. Vrána J, Cápal P, Šimková H, Karafiátová M, Čízková J, Doležel J (2016) Flow analysis and sorting of plant chromosomes. Curr Protoc Cytom 78:531–534

Chapter 21

Rapid Identification of Rust Resistance Genes Through Cultivar-Specific De Novo Chromosome Assemblies

Anupriya K. Thind, Thomas Wicker, and Simon G. Krattinger

Abstract

"Map-based cloning" is a frequently used approach to isolate rust resistance genes. A critical step during map-based cloning is the transition from genetic information, i.e., a genetic map, to physical sequence information. Bacterial artificial chromosome clones are often used to establish sequence information spanning a genetic interval. However, a major limitation of BAC clones consists in their small insert size of 100–200 kb. Targeted chromosome-based cloning via long-range assembly (TACCA) is a method that can replace BAC library screening. This approach involves chromosome flow-sorting and the establishment of a long-range de novo assembly. This chapter provides an overview of TACCA as well as a detailed description of sequence analyses, molecular marker development, and candidate gene identification.

Key words Map-based cloning, Long-range scaffolding, Chromosome flow-sorting, Molecular marker, Candidate gene

1 Introduction

"Map-based cloning" is a widely used method for the isolation of genes, including the rust resistance genes *Lr1, Lr10, Lr34, Sr33*, and *Sr35* [1–5]. This gene cloning strategy involves the generation of a mapping population, most often a biparental population, from a cross between parents that differ for the trait of interest. Genotyping and phenotyping of a large number of progenies allows determining the genetic position of a specific gene, i.e., a genetic interval flanked by molecular markers. One of the major challenges in map-based cloning consists in the establishment of continuous sequence information spanning the genetic interval with the gene of interest. Often, this is achieved through multiple rounds of bacterial artificial chromosome (BAC) library screening, an approach referred to as "chromosome walking" [6]. The insert size of BAC clones however is limited to 100–200 kb. Chromosome walking is therefore tedious in hexaploid wheat with its large 17 Gb and repeat-rich genome. Often, wheat BAC clones are void of genes. The polyploid nature of

Sambasivam Periyannan (ed.), *Wheat Rust Diseases: Methods and Protocols*, Methods in Molecular Biology, vol. 1659, DOI 10.1007/978-1-4939-7249-4_21, © Springer Science+Business Media LLC 2017

the wheat genome further complicates the design of specific probes for BAC library screening. Gene order and content can differ between different wheat cultivars, and the gene of interest can be missing in the reference cultivar Chinese Spring [7, 8], and it is thus essential that the physical interval is established from a wheat line that carries the gene of interest. To solve the problem of limited BAC insert sizes, we developed targeted chromosome-based cloning via long-range assembly (TACCA). This method includes the generation of a high-quality de novo assembly from single chromosomes that are isolated through flow cytometry from a wheat cultivar carrying the gene of interest. TACCA eliminates the need for chromosome walking and allows for a rapid and inexpensive establishment of physical intervals for leaf rust resistance genes. In this chapter, we describe the principle of TACCA. In Subheading 3.1 we give an overview on the long-range de novo chromosome scaffolding. Subheadings 3.2 and 3.3 describe the scaffold annotation and marker development in detail.

We used TACCA to clone the broad-spectrum leaf rust resistance gene *Lr22a* [9]. *Lr22a* was crossed into hexaploid wheat from the diploid wild wheat progenitor *Aegilops tauschii* in the 1960s [10], and the gene was subsequently mapped to the short arm of chromosome 2D [11]. The protocols in Subheadings 3.1, 3.2, and 3.3 will be described in the context of *Lr22a* as an example.

2 Materials

2.1 Plant Material

To isolate high-molecular weight (HMW) DNA of isolated chromosomes, 100 g of seeds of the *Lr22a*-containing Swiss spring wheat line "CH Campala *Lr22a*" [12] are used.

2.2 Computer Setup and Databases Required

1. Computer with a Linux operating system is required for the bioinformatics analyses.
2. CLC Main workbench (Qiagen) is required for the assembly.
3. *Brachypodium distachyon* predicted coding sequences and genes can be obtained from http://plants.ensembl.org/index.html. For our analysis, we use version 1.0 [13].
4. Repeat sequences for Triticeae can be obtained from the TREP database (botinst.uzh.ch/en/research/genetics/thomas-Wicker/trep-db.html) [14].

3 Methods

3.1 Generating of a Long-Range De Novo Chromosome Assembly

Given the huge size of the wheat genome it is not yet feasible to obtain complete high-quality genome sequences for specific wheat cultivars for map-based cloning projects. A genome complexity reduction is thus an essential step to obtain a fraction of the wheat genome that can be sequenced cost-effectively. Complexity reduction for TACCA is achieved through the isolation of single chromosomes. Chromosome flow-sorting is a well-established method in wheat and single chromosomes can be isolated by flow cytometry with purities of >90%. Chromosome sorting requires special equipment and experience. It is thus advisable that this step is done in collaboration with a laboratory specialized in chromosome flow-sorting. The isolation of HMW DNA for the cloning of *Lr22a* was done in collaboration with Prof. Jaroslav Doležel's group at the Institute of Experimental Botany, Olomouc, Czech Republic. ~640 ng of HMW DNA of chromosome 2D was isolated from the Swiss spring wheat line "CH Campala *Lr22a*." An essential consideration for the generation of a de novo chromosome assembly is the amount of input DNA that is required. This is because the amount of DNA defines the time needed for chromosome sorting. It was thus important to find a technology that allows generation of high-quality genome assemblies from small amounts (<1 μg) of DNA. For the cloning of *Lr22a*, Chicago long range linkage offered by Dovetail Genomics was chosen [15]. Other long-range scaffolding or long-read sequencing technologies that work with comparable amounts of DNA however can also be incorporated into TACCA. Chicago combines short read Illumina sequencing with proximity ligation of in vitro reconstituted chromosomes. The preparation of libraries, shotgun sequencing and scaffolding is provided as a full service by Dovetail Genomics (Santa Cruz, CA). A Chicago library of isolated chromosome 2D from "CH Campala *Lr22a*" was prepared from 250 ng HMW DNA (mean fragment length ~ 100 kb). Three paired-end libraries are produced for shotgun sequencing, two prepared from 50 ng of chromosomal DNA and one prepared from 150 ng of chromosomal DNA. Libraries are sequenced on an Illumina HiSeq 2500 (rapid run mode) and resulted in 145 million 150 bp paired-end reads for the Chicago library and 709 million 150 bp paired-end reads for the shotgun libraries, respectively. The shotgun reads are trimmed for quality, sequencing adapters, and mate pair adapters using Trimmomatic [16], and a de novo assembly of shotgun reads was performed with Meraculous 2 (2.2.2.3) [17]. The input de novo assembly, shotgun reads, and Chicago library reads are assembled with the HiRise software pipeline that was specifically designed for Chicago data [15]. The resulting "CH Campala *Lr22a*" assembly consisted of 10,344 scaffolds and had an estimated physical coverage of 37×. The total length of the assembly was 567.2 Mb with a scaffold N50 length of 9.76 Mb and a scaffold N90 length of 1.93 Mb. The longest scaffold was 36.4 Mb in size and covered 6.4% of the entire assembly.

The following two sections provide a detailed description of the assembly analysis, candidate gene identification and the development of molecular makers.

3.2 Bioinformatics Analysis of the Chromosome Assembly

3.2.1 Identification of the Physical Region Between Flanking Markers

For the identification of a Chicago scaffold(s) spanning a defined genetic interval, the sequence information of the flanking markers is used. The nucleotide or primer sequences of simple sequence repeat (SSR) markers for example can be downloaded from GrainGenes (https://wheat.pw.usda.gov/GG3/). These sequences are then used for a blast search against the Chicago scaffolds. For *Lr22a*, the SSR markers *gwm455*, *gwm296*, *wmc25*, and *wmc503* are identified as being linked to the gene [9, 11]. The protocol to identify the physical interval containing these SSR markers is as follows:

1. Create a database with the scaffolds of the Chicago assembly.
2. Obtain the primer sequence of the flanking SSR markers from GrainGenes.
3. Perform a blast search of the marker sequences against the scaffold database using blastn with a cutoff of 30 bp and 96% identity.
4. Download the scaffold(s) with the best hit and the lowest E-value. Check for the alignment identity and the position of the hit on the scaffold from the output file. The forward and reverse primer for a given SSR marker should be located at a distance of 100–200 bp.
5. Repeat the above steps for all the mapped markers.
6. Note down the position of the flanking markers and splice out the region between the markers to get the target region for gene identification and annotation.

In the case of *Lr22a*, both flanking markers are located on a single scaffold of 6.39 Mb in size.

3.2.2 Identification of Genic Sequences in the Target Interval

To identify genes in the physical candidate region identified in Subheading 3.2.1, the *Brachypodium* protein database can be used because *Brachypodium distachyon* is a well annotated model species for cereals. In the case of *Lr22a*, a 438-kb region flanked by upper and lower flanking markers on a single scaffold was annotated using the *Brachypodium* protein database. For the gene identification the following steps can be followed:

1. Perform a blast search with the nucleotide sequence of the identified target region (*see* Subheading 3.2.1) against the *Brachypodium* protein database using blastx with an E-value cutoff of 10E-10.
2. Open the output file and download the coding sequence (CDS) from the *Brachypodium* CDS database of the corresponding *Brachypodium* protein hit with the lowest E-value and make an annotation file as follows:

(Start position) (stop position) (gene name);orientation of gene (forward or reverse);(gene or pseudogene);(gene Id).
Example: 670,711 673,437 Leucine-rich repeat;for;gene; Bradi4g06970

3. Mask the positions annotated in the previous step using a perl script.

   This script replaces the bases of the previously annotated region with "x" in the sequence. The masked bases will be ignored during the next blast search. Repeat the above steps for the remaining sequence until no further genes can be identified. For the 438-kb region of *Lr22a* target interval, six rounds of annotations are performed.

   Example of a masked sequence:

   TTGCTAAGAGACAAGCAACACGAGAATGATACTGCATT
   CGGACGCCCTCGTCTCGTCTGGACTGACTAGAGGAG
   GAAGAAGACGGGGAGGGAGGGAGGGAGAAAAATGG
   GGTCTTTGGCTGGTTCCGTTACXXXXXXXXXXXXX-
   XXXXXXXXXXXXXXXXXXXXXXXXXXXXXXXXXXX-
   XXXXXXXXXXXXXXXXXXXXXXXXXXXXXXXXXXX-
   XXXXXXXXXXXXXXXXXXXXXXXXXXXXXXXXXXX-
   XXXXXXXXXXXXXXXXXXXXXXCTCGAACCTCTCCCTC
   TCTACCACCGCCGCCGGCTCCACCCCTGCTGCTCTA
   ACTTGCTGGCTCAGAGTCTCCGGGGAGAGAAAGAAT
   GGCGCCGGCTCCTCTACTTATATGATAAGTATGATG
   CTCTGCGGCTGCTGCTTCTTCGTCCTGCCA

3.2.3 Manual Annotation of Candidate Genes

After the identification of genic sequences in the target region, the next step is to perform a manual annotation of these sequences to identify complete genes and pseudogenes. This will result in the final list of candidate genes that are considered for gene validation. For this, dot plots are generated between the identified genes in the spliced target region and the corresponding *Brachypodium* CDS to obtain the positions of start and stop codons. This CDS is then translated to check whether the gene is full length or a pseudogene. For *Lr22a*, we identified nine full length and two pseudogenes in the target interval. The identification of the full length gene can be done using the following steps:

1. Splice out the region between the putative start and stop positions of the hit mentioned in the annotation file of the chicago scaffolds (*see* **Note 1**).
   Example: 670711 673437 Leucine-rich repeat;for;gene; Bradi4g06970.
2. Download the CDS of the corresponding *Brachypodium* gene from the *Brachypodium* CDS database.
3. Make a dot plot between the spliced region of the Chicago scaffold and the *Brachypodium* CDS (e.g., Bradi4g06970).

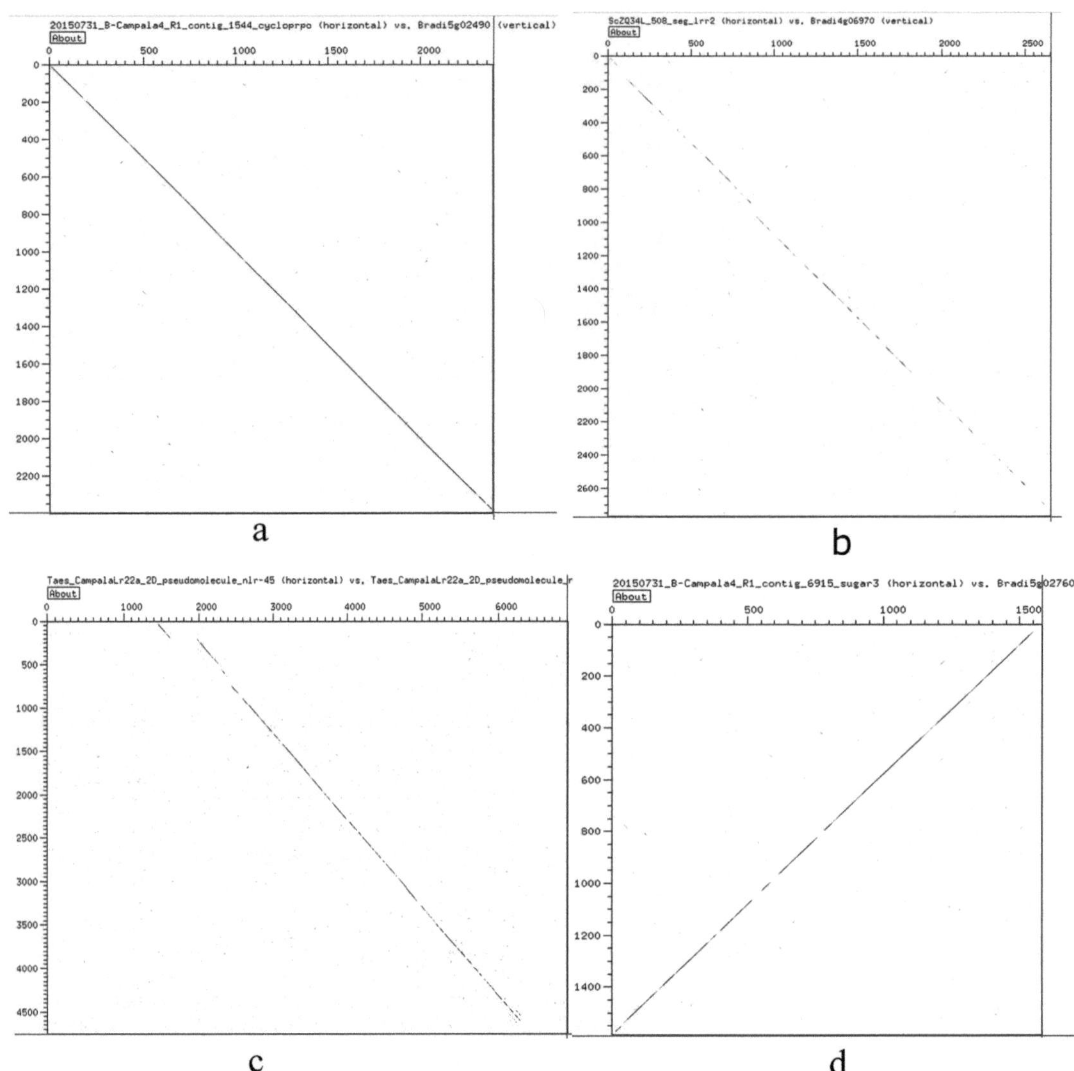

Fig. 1 Dot plot for gene annotation (**a**) Example of a dot plot of a Brachypodium CDS (*vertical*) with a spliced Chicago scaffold (*horizontal*). The gene is complete and the start and stop codon can easily be identified. The gene has no introns. (**b**) Example of a gene with a frame shift which can be identified by the break in the line. (**c**) Example of a gene with an intron at spanning positions 1600–1950. (**d**) Example of an intronless gene in reverse orientation

4. Identify the start and the stop codon positions based on the dot plot (*see* **Note 2**).

Apart from this, dot plots also provide information about the orientation of the gene (forward or reverse), frameshifts, and the intron–exon junctions (Fig. 1).

5. Splice out the region from start to stop codon and translate it using the ExPASy translate tool (http://web.expasy.org/

translate/) to check for internal stop codon that disrupt the gene (*see* **Note 3**).

6. Genes that translate into full-length proteins will be considered as candidate genes.

3.3 Marker Development

This section describes the identification of polymorphisms and the design of molecular markers based on the "CH Campala *Lr22a*" assembly. In addition to "CH Campala *Lr22a*," chromosome 2D was also isolated from the susceptible near isogenic line "CH Campala" and amplified by multiple displacement amplification [18]. In contrast to the isolation of HMW DNA for the Chicago scaffolding, multiple displacement amplification is not suitable for long-range chromosome scaffolding, but it requires a lower number of chromosomes that need to be isolated (30,000 for multiple displacement amplification and 1.5 million copies for long-range scaffolding). The isolated chromosomes of "CH Campala" are sequenced on a lane of Illumina HiSeq 2500 with 125 bp paired-end reads and the "CH Campala *Lr22a*" Chicago scaffolds are used to anchor the "CH Campala" reads. SNPs and InDel markers in gene-containing contigs are subsequently identified and locus-specific markers are developed.

3.3.1 De Novo Assembly Using CLC Main Workbench

This section describes the assembly of Illumina raw reads from isolated chromosome 2D amplified by multiple displacement amplification. For this, the CLC Main workbench 7 (Qiagen) is used. The protocol is detailed below:

1. Go to the De novo sequencing in the tool box option which opens a drop-down menu and select De novo assembly.
2. Select the files obtained from the Illumina Hiseq 2500 with 125-bp paired-end read.

   Example:

   CH_Campala_R1.fastq

   CH_Campala_R2.fastq
3. Graph parameters can be set to automatic and minimum contig length should be set to 500 bp.
4. For the paired reads parameters, select auto-detect paired distances and perform scaffolding.
5. For the mapping options, select map reads back to contig (slow) and set the following parameters as follows:

   Mismatch cost—2.

   Insertion cost—3.

   Deletion cost—3.

   Length fraction—0.95.

   Similarity fraction—0.9.

Select the following options:

Global alignment.

Update contigs.

Create list of un-mapped reads.

6. Result handling: select create report and give a destination folder to save results.
7. CLC de novo assembly tool creates five output files.

   CH_Campala_R1(paired).clc.

   CH_Campala_R1(paired) assembly.clc.

   CH_Campala_R1(paired) assembly summary report.clc.

   CH_Campala_R1(paired) un-mapped reads [CH_Campala_R1] (paired).clc.

   CH_Campala_R1(paired) un-mapped reads [CH_Campala_R1] (single).clc.

3.3.2 Filter De Novo Assembly for Gene-Containing Contigs

To eliminate contigs only consisting of repeats, a filtering step for gene-containing contigs is introduced because transposable elements will hinder marker development and further mapping of the markers. Gene-containing contigs for of the Illumina contigs obtained from the susceptible cultivar and Chicago scaffolds from the resistant cultivar will be used for the designing of SNP marker and insertion/deletion (InDel) markers. For working with large data sets, linux system works best to perform such blast analysis.

1. Develop a perl script for blast searches of multiple sequences against any database. This is a simple perl script that loops through flat file (i.e., concatenated fasta sequences) and uses each sequence for a blast search against the respective database. If a sequences has a blast hit stronger than the user-defined cutoff, it is stored in a second flat file (sequences without hits are not stored).
2. Perform a blast search with blastx using the above designed script for the assembly generated from CLC against the *Brachypodium* protein database to identify scaffolds that contain gene sequences. Usually, an E-value cutoff of 10E-10 is used.

   File to be used:

   CH_Campala_R1(paired) assembly.clc.
3. This creates an output file with only gene containing contigs Example: CH_Campala_R1_assembly_x_Bdis_proteins_hits.

3.3.3 Design of SNP or InDel Markers

The goal of this section is to show how the Illumina contigs of the susceptible cultivar "CH Campala" are aligned to the Chicago scaffold of the resistant parent "CH Campala *Lr22a*" to identify polymorphisms.

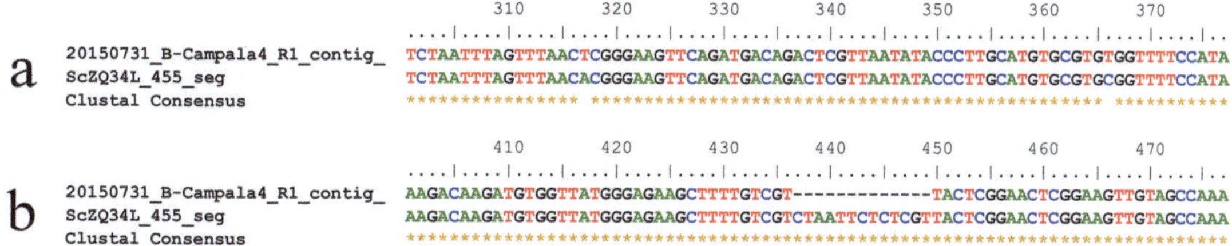

Fig. 2 Development of SNP or InDel marker. Alignment of a sequence from the resistant and the susceptible cultivar where (**a**) shows the possible positions such as 317 and 366 to design SNP based marker and (**b**) shows region with deletion in one of the cultivar which can be used for the designing of InDel marker

1. Create a database of the gene containing contigs of the Illumina contigs from the susceptible cultivar.

   Campala_R1_assembly_x_Bdis_proteins_hits
2. Perform blast search of the nucleotide sequence of the putative candidate genes identified in Subheading 3.2.3 against the gene-containing contigs database of the susceptible cultivar.
3. Extract the corresponding Illumina contigs from the susceptible cultivar and the Chicago scaffold from the resistant cultivar and align them using the clustal omega tool (http://www.ebi.ac.uk/Tools/msa/clustalo/).
4. Check for SNPs, insertions or deletions within the gene or in the 5′ or 3′ region of the gene which could be used to design the SNP or InDel markers (Fig. 2) (*see* **Note 4**).
5. Design primers manually for 500–600 bp amplicon size from the region outside of the SNP or InDel (*see* **Note 5**).

3.3.4 PCR Amplification of the Designed SNP and InDel Markers

Set up a gradient polymerase chain reaction (PCR) on the DNA of resistant and susceptible cultivar to check if the specific SNP or the InDel for which the primer is designed can be recovered. In case if the deletion is more than 10 bp in one of the cultivars, it can be resolved on an agarose gel and can be used as codominant marker. However, for SNP based markers, Sanger sequencing or other SNP genotyping technologies can be used. The PCR amplifications can be done as follows:

1. Mix the Following Components in PCR Tube for One Sample (20 μl Reaction Volume)

   DNA (65 ng/μl)—1 μl.

   dNTPs (2.5 mM)—1 μl.

   Primer forward (10 μM)—1 μl.

   Primer reverse (10 μM)—1 μl.

   Sigma Buffer (10×)—1 μl.

   Sigma Taq Polymerase—0.25 μl.

   Milli Q water—13.75 μl.

2. Run the gradient PCR using the cycling conditions:

95 °C 4 min	
95 °C 30 s	35 cycles (**steps 2–4**)
50–55–60 °C 30 s	
72 °C 2 min	
72 °C 5 min	
10 °C ∞	

3. Run 1% agarose gel and elute the bands from the gel and purify using GenElute™ Gel Extraction Kit (Sigma) using the manufacturer's protocol.
4. Sequence the extracted bands using Sanger sequencing.
5. Align the sequences using clustal omega to confirm the SNP or InDel.
6. Ones the SNP or InDels are confirmed, map the marker on the biparental population and look for critical recombination between the mapped markers which would further narrow down the candidate gene list.

4 Notes

1. In case the start and the stop codons are missed in the spliced region, increase the length of the segment to be spliced by ~2 Kb upstream and downstream of start and stop positions, respectively.

   For example: splice out the region from 2 kb upstream of 670,711 to 2 kb downstream of 673,437.
2. For some of the dot plots with a *Brachypodium* gene, it can be difficult to visualize the start and/or stop codon. In such cases, try to find an ortholog in rice, sorghum or *Arabidopsis* for the dot plot.
3. The splicing of intron–exon junction has to be very precise. Otherwise a gene can be mistakenly classified as a pseudogene.
4. While designing the markers, perform a blast search with the extracted scaffold sequence or contigs against the repeat database to check for transposable elements.
5. For the SNP or InDel markers, design forward and reverse primer approximately 100 bp before and after the SNP, respectively.

Acknowledgments

We are grateful to the staff at Dovetail Genomics for constructing the "*Lr22a* Campala" scaffolds and to Jaroslav Doležel and Hana Šimková for the isolation of chromosome 2D. This work was financed by an Ambizione fellowship of the Swiss National Science Foundation.

References

1. Cloutier S, McCallum BD, Loutre C et al (2007) Leaf rust resistance gene *Lr1*, isolated from bread wheat (*Triticum aestivum* L.) is a member of the large psr567 gene family. Plant Mol Biol 65:93–106
2. Feuillet C, Travella S, Stein N et al (2003) Map-based isolation of the leaf rust disease resistance gene *Lr10* from the hexaploid wheat (*Tritium aestivum* L.) genome. Proc Natl Acad Sci U S A 100:15253–15258
3. Krattinger SG, Lagudah ES, Spielmeyer W et al (2009) A putative ABC transporter confers durable resistance to multiple fungal pathogens in wheat. Science 323:1360–1363
4. Periyannan S, Moore J, Ayliffe M et al (2013) The gene *Sr33,* an ortholog of barley *Mla* genes, encodes resistance to wheat stem rust race Ug99. Science 341:786–788
5. Saintenac C, Zhang W, Salcedo A et al (2013) Identification of wheat gene *Sr35* that confers resistance to Ug99 stem rust race group. Science 314:783–786
6. Stein N, Feuillet C, Wicker T et al (2000) Subgenome chromosome walking in wheat: a 450-kb physical contig in *Triticum monococcum* L. spans the *Lr10* resistance locus in hexaploid wheat (*Triticum aestivum* L.) Proc Natl Acad Sci U S A 97:13436–13441
7. Rawat N, Pumphrey MO, Liu S et al (2016) Wheat *Fhb1* encodes a chimeric lectin with agglutinin domains and a pore-forming toxin-like domain conferring resistance to Fusarium head blight. Nat Genet 48:1576–1580
8. Mago R, Tabe L, Vautrin S et al (2014) Major haplotype divergence including multiple germin-like protein genes, at the wheat *Sr2* adult plant stem rust resistance locus. BMC Plant Biol 14:379–390
9. Thind AK, Wicker T, Šimková H et al (2017) Rapid gene isolation in hexaploid wheat through a cultivar-specific long-range chromosome assembly. Nat Biotechnol. (Accepted). doi: 10.1038/nbt.3877
10. Dyck PL, Kerber ER (1970) Inheritance in hexaploid wheat of adult-plant leaf rust resistance derived from *Aegilops squarrosa*. Can J Genet Cytol 12:175–180
11. Hiebert CW, Thomas JB, Somers DJ et al (2007) Microsatellite mapping of adult-plant leaf rust resistance gene *Lr22a* in wheat. Theor Appl Genet 115:877–884
12. Moullet O, Schori A (2014) Maintaining the efficiency of MAS method in cereals while reducing the costs. J Plant Breed Genet 2:97–100
13. The International Brachypodium Initiative (2010) Genome sequencing and analysis of the model grass *Brachypodium distachyon*. Nature 463:763–768
14. Wicker T, Matthews DE, Keller B (2002) TREP: a database for Triticeae repetitive elements. Trends Plant Sci 7:561–562
15. Putnam NH, O'Connell BL, Stites JC et al (2016) Chromosome-scale shotgun assembly using an in vitro method for long-range linkage. Genome Res 26:342–350
16. Bolger AM, Lohse M, Usadel B (2014) Trimmomatic: a flexible trimmer for Illumina sequence data. Bioinformatics 30:2114–2120
17. Chapman JA, Ho I, Sunkara S et al (2011) Meraculous: *de novo* genome assembly with short paired-end reads. PLoS One 6:e23501
18. Šimková H, Svensson JT, Condamine P et al (2008) Coupling amplified DNA from flow-sorted chromosomes to high-density SNP mapping in barley. BMC Genomics 9:294

Chapter 22

BSMV-Induced Gene Silencing Assay for Functional Analysis of Wheat Rust Resistance

Li Huang

Abstract

Virus-induced gene silencing (VIGS) is a widely used reverse genetics tool to knock down genes in plants transiently without transformation. The assay has been successfully used to downregulate the transcript abundance of a target gene at almost any plant developmental stages in any tissues. Here, we describe the VIGS assay using a barley stripe mosaic virus (BSMV) for functional genomics analysis in wheat with the focus on genes involved in rust resistance.

Key words VIGS, Reverse genetics, Gene silence, BSMV, Rust resistance, Functional analysis, Wheat

1 Introduction

Virus-induced gene silence in plants is an antiviral defense response from the hosts [1–3]. Double-stranded RNAs (dsRNAs) generated from either RNA viruses during replication or DNA viruses during infection trigger homology-dependent posttranscriptional gene silencing in the host [4, 5]. Once the host silencing machinery is activated, the viral RNAs become the targets for degradation. Host endogenous gene transcripts can also become degradation targets if a sequence with sufficient homology to the targets is engineered into the virus genome. Therefore, VIGS has been developed as a reverse genetics tool for functional analysis in plants [6, 7].

Barley stripe mosaic virus (BSMV) is a tripartite, positive-sense RNA virus [8] and can infect many agriculturally important monocots including barley (*Horedum vulgare*), wheat (*Triticum aestivum*), maize (*Zea mays*), rice (*Oryza sativa*), sorghum (*Sorghum bicolor*), and millet (*Pennisetum glaucum*) [9]. Three DNAs corresponding to the three sub-genomes (α, β, and γ) of the BSMV strain ND-18 have been cloned into plasmid DNA vectors [10]. BSMV induced gene silencing as a functional genomics tool was first established in barley using a *phytoene desaturase* (*PDS*) as a

Sambasivam Periyannan (ed.), *Wheat Rust Diseases: Methods and Protocols*, Methods in Molecular Biology, vol. 1659, DOI 10.1007/978-1-4939-7249-4_22, © Springer Science+Business Media LLC 2017

report gene [11]. The assay was soon adapted in wheat for dissecting genetic components in the defense response to leaf rust pathogen [12]. The BSMV-VIGS system was further optimized for rapidly cloning of a target gene fragment using a modified γ vector for directly PCR product cloning and silencing multiple genes simultaneously [13].

The entire BSMV-VIGS assay for functional analysis of wheat rust resistance includes five major steps: (1) construct silencing vector by cloning a target fragment into the γPCR vector; (2) prepare BSMV RNAs for virus inoculation; (3) inoculate BSMV with in vitro synthesized viral RNAs; (4) measure the silencing level; and (5) inoculate rust spores on silenced wheat plants. A videotaped protocol of BSMV-VIGS can be viewed at the link of https://vimeo.com/48603055.

2 Materials

Prepare all solutions using nuclease-free water. Inoculate 20 wheat plants for assaying one target gene.

2.1 VIGS

1. PCR-ready cloning vector kit: pγPCR vector and *Xcm*I enzyme, *Taq* polymerase, dNTPs, T_4 ligase and buffer, *E. coli* competent cells, LB broth, LB broth + Agar, and carbenicillin antibiotic.
2. Plasmid kit for BSMV RNA cloning: pα, pβ, pγ, and pγ*PDS* plasmid DNA vectors, a plasmid DNA extraction kit, and *Mlu*I, *Spe*I, and *Bss*HII restriction enzymes.
3. 10× cap/rNTPs buffer: Mix 10 μL H_2O with 20 μL of 100 mM rATP, 20 μL of 100 mM rUTP, 20 μL of 100 mM rCTP, 10 μL of 100 mM rGTP plus 10 μL of 25 A_{260} units cap analog and make up to 200 μL using H_2O.
4. T_7 RNA polymerase and buffer.
5. RNase inhibitor.
6. Agarose.
7. GP buffer: Mix 18.77 g glycine and 26.13 g K_2HPO_4 dibasic in 500 mL H_2O.
8. FES buffer: Mix 100 mL GP buffer, 5 g sodium pyrophosphate decahydrate, 5 g bentonite, 5 g celite and bring the final volume to 500 mL using H_2O. Sterilize through autoclaving.
9. Pipette tips: 200 μL wide-bore pipette tips and regular 10 μL, 20 μL, 200 μL, and 1000 μL pipette tips.
10. RNase AWAY.

2.2 Silencing Level Test

1. RNA extraction kit.
2. *iTaq* universal SYBR 1-step quantitative real-time PCR kit.

2.3 Rust Inoculation

1. Soltrol170 oil.
2. Airbrush gun.
3. Rust spores.

3 Methods

3.1 Clone a Target Gene Fragment into γPCR Vector

1. Prepare γ PCR-ready cloning vector by mixing 2 μL of 10× buffer, 1 μL of *Xcm*I restriction enzyme, 2 μg of γPCR vector plasmid and bring the final volume to 20 μL using H_2O.
2. Incubate the above solution at 37 °C for 2 h and take 1 μL to check on a 1% w/v agarose gel along with an uncut γPCR vector as a control. Heat-inactivate the enzyme at 65 °C for 20 min if the digestion is completed (Fig. 1).
3. Prepare a target insert having ≥120 bp and ≤400 bp in length and containing ≥25 bp identical base pairs with the target gene either by direct DNA synthesize or through PCR amplification from wheat cDNA (*see* **Note 1**).
4. Ligate the target gene fragment into the γPCR ready cloning vector by incubating a mixture of 1 μL of 10× buffer, 1 μL of γPCR (cutted by *Xcm*I from above), 1 μL of 200 ng conc. DNA Insert, 3 μL of H_2O at 16 °C for 5 h. Then heat-inactivate the enzyme at 65 °C for 20 min. Transform to *E. coli* and select colonies on a LB + carbenicillin plate. Select positive bacterial colonies with target gene-specific primers and confirm the accuracy of the insert through sequencing.

3.2 Prepare BSMV RNAs

1. Linearize pα, pβ, pγ, and pγ*PDS* plasmid DNA vectors by preparing a mixture with 2.5 μg plasmid DNA, 2 μL 10× buffer, 6 units of restriction enzyme and nuclease-free H_2O to make it up to 20 μL. Restriction enzyme and optimal incubation temperature for each plasmid DNA of the BSMV is indicated as follows:

Plasmid	Enzyme	Incubation temp (°C)
pα	*Mlu*I	37
pβ	*Spe*I	37
pγ	*Mlu*I	37
pγPDS	*Bss*HII	50
pγtarget	*Bss*HII	50

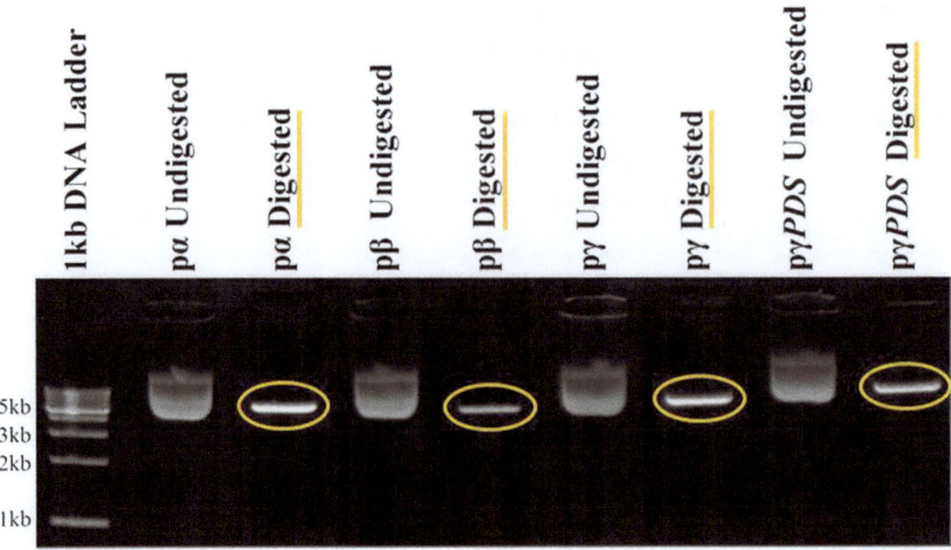

Fig. 1 Digestion of pα, pβ, pγ, and pγ*PDS* plasmid DNA vectors. Bands *encircled* indicate the digested product

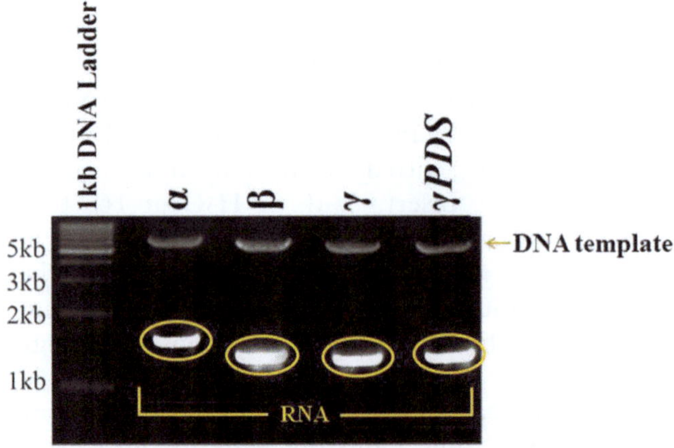

Fig. 2 In vitro synthesis of pα, pβ, pγ, and pγ*PDS* RNA. Bands *encircled* are RNAs

Incubate at corresponding optimal temperature for 5 h, and check 1 μL of the digestion on a 1% w/v agarose gel using each uncut plasmid DNA as controls. Complete digestion is shown in Fig. 1. Heat-inactivate *Mlu*I and *Spe*I at 65 °C for 20 min, and *Bss*HII at 80 °C for 20 min. Add 1 μL of 40 units RNase inhibitor per 20 μL linearization reaction (*see* **Note 2**).

2. In vitro-synthesize RNA of α, β, γ or γtarget by preparing a mixture with 2 μL 10× buffer, 2 μL 10× cap/rNTP buffer, 1 μL T_7 RNA polymerase, 7 μL linearized plasmid, and 8 μL H_2O. Incubate at 37 °C for 2 h, and take 1 μL from each reaction to check on a 1% w/v agarose gel. A sufficient RNA synthesis is shown in Fig. 2.

3.3 BSMV Inoculation

1. Combine RNAs of α:β:γ or α:β:γtarget in 1:1:1 ratio or 10:10:10 μL ratio per tube. Keep the reaction in ice, or store at −80 °C if the RNAs are not used immediately. In every experiment, 20 plants inoculated with only BSMV α:β:γ (BSMV:00) should be included as a control.
2. Add 220 μL FES buffer per 30 μL of combined α + β + γ RNAs per tube immediately before the inoculation and keep the tube in ice. Put on a new pair of gloves, spayed gloved hands with RNase AWAY, and let them air dry. Mix the BSMV RNAs in FES buffer thoroughly each time with a wide bore pipette tip and pipette 25 μL the first and second time and 22 μL after that onto the index finger and hold with the thumb. Use the other hand to hold the stem of a selected plant, and rub the RNAs onto a selected leaf from the base to tip by gently squeezing the leaf between the index finger and the thumb. In general, 20 plants should be inoculated for one target gene (*see* **Note 3**).
3. Keep the BSMV inoculated plants at 25–27 °C and water only from roots by placing the pots of the plants in a big tray with sufficient water (*see* **Note 4**).

3.4 Measure the Silencing Level

1. Viral symptoms of yellow mosaic should be visible on the newly emerged leaves 5 days post a successful BSMV inoculation. At 8–9 days post inoculation, symptom-free (or virus-free) segments or leaves appear following the viral mosaic leaves. As illustrated in Fig. 3 when the *PDS* gene was silenced in the plant inoculated with α + β + γPDS, photobleach areas are

Fig. 3 Progress of BSMV-induced gene silencing. (**A**): BSMV-inoculated leaf; (**B**): BSMV symptoms; (**C**): Photobleached stem resulted from *PDS* silence; (**D**): Photobleached leaf; (**E**): Photobleached flag leaf. The photo was taken 20 days post BSMV inoculation

corresponding to symptom-free regions if silencing of the target does not cause photobleaching. Sample a one-inch virus-free leaf segment each from the plants inoculated with the only α:β:γ and with α:β:γtarget. Place the leaf tissues into liquid N immediately.

2. Extract RNA from each sample using an RNA extraction kit. Check the RNA quality by a 1% w/v agarose gel, and measure the concentration via a Nanodrop.
3. Prepare a 60 μL reaction per RNA sample, mix thoroughly and set up as triplets with 20 μL per tube for real-time PCR using the following recipe:

   30 μL of 2× SYBR Green Supermix.

   30 μL of 5 mM forward primer.

   30 μL of 5 mM reverse primer.

   0.75 μL Reverse transcriptase.

   6 μL of 37.5 ng/μL conc. RNA.

   19.5 μL nuclease-free H_2O.

For relative expression analysis, include one or two housekeeping genes as internal standards for quantifying RNA levels. For functional analysis of genes involved in wheat rust resistance via BSMV-VIGS, glyceraldehyde 3-phosphate dehydrogenase (GAPDH) and β-actin are two commonly selected housekeeping genes for internal references because the expression levels of these genes remain unchanged during BSMV and rust infections based on the RNAseq data. To estimate silencing level, each target gene will be measured both in target-silenced plants and BSMV:00 inoculated plants. Primer sequences of the reference genes are as follows:

GAPDH-F: GACTGTTAGACTTGCGAAGC.

GAPDH-R: CATCAACGTAACCCAAAATG.

Actin-F: CCAGCAATGTATGTCGCAATCC.

Actin-R: CCAGCAAGGTCCAAACGAAGG.

Relative expression is calculated using the ΔΔCt method as described in the CFX96 manual (Bio-Rad, Hercules, CA), where fold change = $2^{-\Delta\Delta Ct}$. Silencing efficiency of a target gene is calculated as the expression level in the target-silenced plants relative to that in the BSMV:00 inoculated plants.

3.5 Rust Inoculation

1. Rust inoculation is done when the symptom-free (or virus-free) segments or leaves appear after the viral mosaic leaves at about 9–14 days post BSMV inoculation. Mark the borders of those areas with a sharper.
2. Prepare a dew chamber to achieve air temperature of 15–16 °C for leaf rust, 20–22 °C for stem rust, and 13–15 °C for stripe

rust. For details, visit the website at https://vimeo.com/48603055 to view a videotaped protocol for rust inoculations.

3. Prepare an inoculum immediately prior to use by suspending urediospores in Soltrol170 isoparaffin oil at a concentration of ~6×10^6 spores/mL (40 mg spores/8 mL oil). Spray the inoculum with an airbrush gun about 6 in. away onto the plants with one pass from front and back, making sure that the sharper-marked areas get the inoculum.
4. Incubate the plants in the dew chamber for ~18 h and then transfer the plants to a greenhouse under 16 h of photoperiod, and ~22 °C for leaf rust, ~20 °C for stripe rust, and ~25 °C for stem rust.
5. Disease scoring is done when the susceptible controls are fully susceptible. Under the abovementioned condition, leaf rust is scored at 8–10 days post inoculation (dpi), stem rust is scored at 14–16 dpi and stripe rust is scored at 12–16 dpi. The function of the target gene in the rust resistance is based on the infection types and silencing level on the symptom-free areas relative to the controls (*see* **Note 5**).

4 Notes

1. The DNA fragment inserted into the ***γPCR*** vector should be between 120 and 600 bp in size, contain no region >25 bp identical to any nontarget genes and no *Bss*HII restriction site.
2. Linearization of the plasmids completely before in vitro transcription is a crucial step.
3. Appropriate amount of initial BSMV RNAs inoculum is important to trigger the post transcriptional gene silence but not overwhelm the plant defense.
4. Plants post BSMV inoculation should be kept at 25 °C for optimal silence.
5. Phenotyping the target genes should be done in the silenced leaf segments, which are the viral symptom-free segments that appeared after the leaves with viral symptoms.

References

1. Wingard SA (1928) Hosts and symptoms of ring spot, a virus disease of plants. J Agric Res 37(3):127–153
2. Lindbo JA, Silva-Rosales L, Proebsting WM, Dougherty WG (1993) Induction of a highly specific antiviral state in transgenic plants: implications for regulation of gene expression and virus resistance. Plant Cell 5:1749–1759
3. Vance V, Vaucheret H (2001) RNA silencing in plants-defense and counter defense. Science 292:2277–2280
4. Fire A (1999) RNA-triggered gene silencing. Trends Genet 15:358–363
5. Chellappan P, Vanitharani R, Fauquet CM (2004) Short interfering RNA accumulation correlates with host recovery in DNA virus-

infected hosts, and gene silencing targets specific viral sequences. J Virol 78:7465–7477

6. Burch-Smith TM, Anderson JC, Martin GB, Dinesh-Kumar SP (2004) Applications and advantages of virus-induced gene silencing for gene function studies in plants. Plant J 39:734–746
7. Becker A, Lange M (2010) VIGS-genomics goes functional. Trends Plant Sci 15:1–4
8. Jackson AO, Lane LC, Brakke MK (1972) Characterization of multiple RNA components from purified barley stripe mosaic virus. Phytopathology 62:767
9. McKinney HH, Greeley LW (1965) Biological characteristics of barley stripe mosaic virus strains and their evolution. Technical Bulletin, USDA, Washington, DC
10. Petty IT, Hunter BG, Wei N, Jackson AO (1989) Infectious barley stripe mosaic virus RNA transcribed *in vitro* from full-length cDNA clone. Virology 171:342–349
11. Holzberg S, Brosio P, Gross C, Pogue GP (2002) Barley stripe mosaic virus-induced gene silencing in a monocot plant. Plant J 30:315–327
12. Scofield SR, Huang L, Brandt AS, Gill BS (2005) Development of a virus-induced gene silencing system for hexaploid wheat and its use in functional analysis of the *Lr21*-mediated leaf rust resistance pathway. Plant Physiol 138:2165–2173
13. Campbell J, Huang L (2010) Silencing of multiple genes in wheat using barley stripe mosaic virus. J Biotech Res 2:12–20

Chapter 23

Yeast as a Heterologous System to Functionally Characterize a Multiple Rust Resistance Gene that Encodes a Hexose Transporter

Ricky J. Milne, Katherine E. Dibley, and Evans S. Lagudah

Abstract

Recently, the *Lr67* resistance gene was identified as a hexose transporter variant which confers adult plant rust and mildew resistance in wheat. Methodologies used to characterize the protein encoded by *Lr67* may be of use to non-transporter experts conducting similar experiments with other hexose transporters. Hence, in this chapter, we detail a protocol for the functional characterization of hexose transporter proteins in the *Saccharomyces cerevisiae* expression system. We also provide guidance on the use of metabolic inhibitors and competing sugars to probe transporter structural features, energization, and specificity.

Key words Hexose transporter, Rust resistance, Heterologous expression, Yeast uptake, Transporter characterization

1 Introduction

Plant fungal pathogens cause economically significant diseases in a range of cereals and other crops. Pathogen resistance genes typically confer race-specific disease resistance in plants and are utilized to breed disease-resistant lines. However, a second, smaller class of adult plant partial resistance genes which confer resistance to multiple pathogens, has been employed in hexaploid wheat breeding for many decades [1, 2]. The recent cloning of one of these genes, *Lr67* (*Lr67/Yr46/Sr55/Pm46*), showed the putative resistance gene belongs to the sugar transport protein 13 (STP13) clade of hexose transporters [3]. Previous studies of other STP13 transporters in plants have implicated them in pathogen response, such as *AtSTP13* and *VvHT5*, both of which are induced during pathogen infection [4, 5].

There are numerous challenges in measuring the functional properties of such sugar transporters *in planta*, due to the presence and action of endogenous sugar transport, signaling and

Sambasivam Periyannan (ed.), *Wheat Rust Diseases: Methods and Protocols*, Methods in Molecular Biology, vol. 1659, DOI 10.1007/978-1-4939-7249-4_23, © Springer Science+Business Media LLC 2017

metabolism. Heterologous expression systems can be used to characterize sugar transporters, the most widely used being deficient yeast strains and *Xenopus laevis* oocytes. Oocytes of the South African clawed frog (*X. laevis*) represent a versatile system which may be used to characterize sugar transporters [6]. The lack of proton (H^+) coupled transport activity associated with oocytes [7] is an advantage for studying hexose-proton symporters including STPs, as the only proton coupled transport present will be due to the heterologous hexose transporter. *Xenopus* oocytes allow analysis of biophysical properties that cannot be undertaken in yeast by examining the proton motive force (*pmf*). *Xenopus* studies allow examination of effects of ΔpH and membrane potential on transporter activity, along with the stoichiometric relationship between proton and hexose transport. Multiple experiments can be carried out on the same oocytes [7]. Despite its advantages, the *Xenopus* system is complex and requires extensive maintenance and care of frogs for oocyte production, along with specialized equipment to measure transporter activity. The yeast system represents a simple solution which is able to be set up in most molecular biology laboratories.

The yeast strain utilized is unable to perform hexose uptake as a result of concurrent knockout of 23 hexose transporters and related genes [8]. The basis of the yeast system is to complement this phenotype with a functional hexose transporter and measure its transport properties using radiolabeled substrates. Yeast can be used to examine biochemical properties of transporters, such as pH dependence, apparent K_m values, substrate specificity, and effects of inhibitors. A disadvantage of characterizing transporters via heterologous expression systems is whether the characteristics observed are relevant, as transporters may behave differently in plant cells. However, measurement of transporter characteristics gives a point of comparison between different transporters, and also gives an indication into the optimal conditions the transporter may function under in planta. The yeast system is much more straightforward than the oocyte system, as no highly specialized equipment is required, and will be the focus of this chapter.

The transport characteristics of a number of STP13 clade members including AtSTP13, OsMST4, LeHT2, and VvHT5 have been described using the above systems [9–12]. STP13 transporters generally exhibit a high glucose affinity in the micromolar range, lower fructose affinity and an acidic pH optimum. The STPs are hexose-proton symporters that co-transport a proton and hexose in a 1:1 ratio into cells with the proton gradient. Studies using compounds which disrupt proton gradients have also provided evidence that STPs function by secondary active transport. Protonophore compounds such as cyanide m-chlorophenyl hydrazone (CCCP), and 2,4-dinitrophenol (2,4-DNP) have been used to dissipate H^+ gradients and reduce hexose uptake in experiments with STPs expressed in yeast [3, 10, 12] and *Xenopus* oocytes [11]. Secondary active transport systems are unable to function in the absence of a *pmf*. Further evidence was added by studies using antimycin to inhibit ATP generation and thereby prevent active transport. In contrast, a phylogenetically related group of sugar transporters known as SWEETs are pH and energy independent, and may facilitate diffusion of sucrose/hexoses driven by a concentration gradient across plasma membranes of cells [13]. Sugar transport mediated by SWEETs is unaffected by protonophores and ATP inhibitors.

The functional characterization system described in this chapter was used to characterize the Lr67sus and Lr67res alleles of wheat [3]. Interestingly, the hexose transporter variant which conferred multipathogen resistance (Lr67res) lacked hexose transport ability, indicating that the pathogen may depend upon the functional hexose transporter activity either directly (through substrate transport) or indirectly (for example through sugar signaling mechanisms) for successful infection of the host.

2 Materials

1. Yeast expression vector (preferably pDR196, *see* **Note 1**).
2. Hexose uptake deficient yeast (EBY.VW4000, *see* **Note 2**) transformed with transporter of interest.
3. Media for growth and selection of yeast: synthetic dropout lacking uracil (SDura$^-$; 1.72 g/L yeast nitrogen base, 5 g/L ammonium sulfate, 2% w/v maltose, 1.92 g/L yeast synthetic drop out medium supplement without uracil). Maltose should be filter-sterilized rather than autoclaved. Supplement with 15 g/L agar for solid media.
4. 50 mL screw cap tubes.
5. MES-HEPES solution: 25 mM MES + 25 mM HEPES mixed to obtain appropriate pH (*see* **Note 3**). Use Tris to reach pH above ~5.2.
6. Ethanol solution (1.1 M) for energization of yeast prior to uptake, diluted in appropriate pH buffer.
7. Sugar solutions in appropriate pH buffer.
8. [^{14}C] glucose or [^{14}C] fructose.
9. Whatman GF/C glass fiber filters.
10. Conical flask filtering apparatus.
11. Vacuum (vacuum line, vacuum pump or tap vacuum attachment may be used).
12. Scintillation vials.

13. Scintillant.
14. Liquid scintillation counter.
15. Spectrophotometer and cuvettes for determining OD values of yeast cultures.
16. Haemocytometer and microscope with 20× objective for determining number of yeast cells per reaction.

3 Methods

S. cerevisiae strain EBY.VW4000 [8] can be transformed [14] with a vector such as pDR196 [15] containing a hexose transporter such as *Lr67* [3] or STPs from other species (e.g., tomato [10] or grape [9]). From the transformation plate, each colony selected is considered a biological replicate. A minimum of three biological replicates are recommended for each experiment. Glycerol stocks should be prepared and plasmid rescue performed [16] to confirm positive by restriction digest and sequencing.

3.1 Yeast Culture and Preparation

1. Streak yeast from glycerol stock onto SDura$^-$ solid media, grow at 28–30°C for 2–3 days.
2. Inoculate starter culture with a single colony in 5 mL SDura$^-$ liquid media in a 50 mL screw cap tube, shake at 200 rpm, 28–30°C for ~16 h until OD_{600} = ~1.0 (*see* **Note 4**). Starter culture may be stored at 4°C and should remain viable for approximately two weeks.
3. Inoculate 20 mL SDura$^-$ liquid media using 100 μL of starter culture in a 250 mL conical flask.
4. Harvest culture at an OD_{600} of 0.6–1.0 by centrifugation at 4000 × *g* for 10 min at 4°C. Discard supernatant, wash pellet with 0.5 volumes MES-HEPES. Recentrifuge and resuspend to OD_{600} of 2 in MES-HEPES solution. Store on ice.

3.2 Uptake Procedures

1. Divide yeast solution into two or three technical replicates of 500 μL in 50 mL screw cap tubes.
2. Shake replicates at 150 rpm, 30°C for 10 min prior to uptake.
3. Make uptake solution by diluting [^{14}C]glucose in unlabeled glucose (begin with 1/10 dilution, *see* **Note 5**) to achieve a final concentration of 100 μM. This concentration should ideally be at or near the transporter K_m. This may be empirically determined, or estimated from affinities of closely related transporters.
4. Energize yeast with 50 μL of 100 mM ethanol for 1 min prior to addition of 50 μL uptake solution.

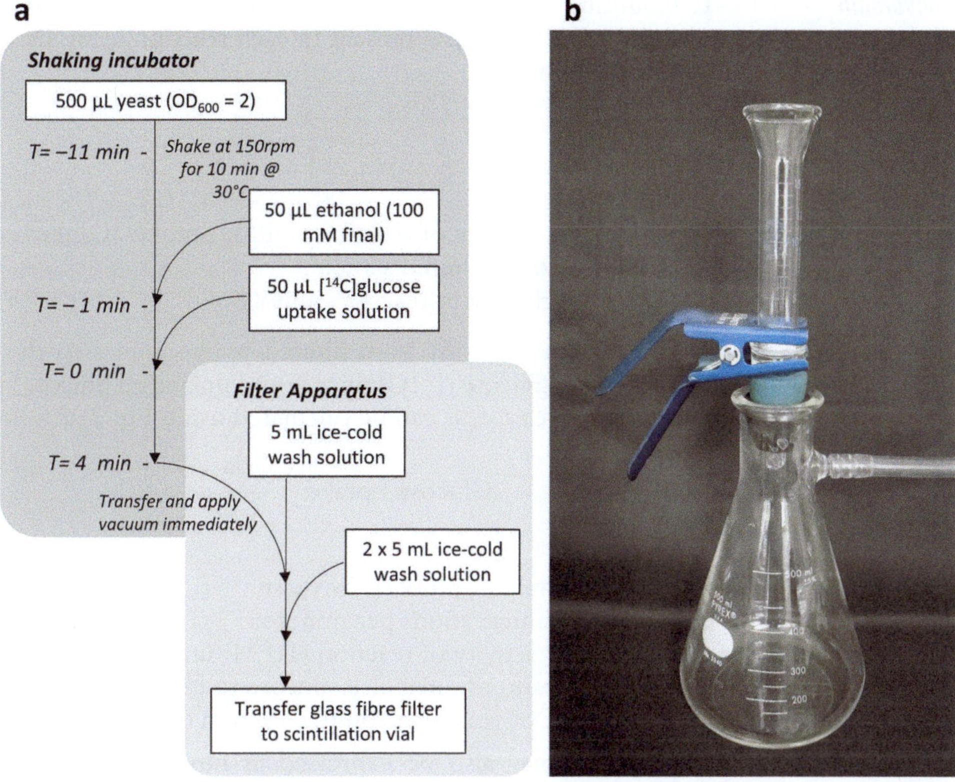

Fig. 1 (**a**) Flow diagram of uptake procedure and (**b**) photograph of filtration apparatus

5. Aliquot 5 mL of ice cold wash solution (100 μM, equal concentration to uptake solution) in filtering apparatus (Fig. 1). Aliquot yeast replicate into wash solution at set time interval and immediately apply vacuum. Wash yeast twice more with 5 mL wash solution.
6. Harvest further replicates at set time intervals for time course uptake.
7. Place glass fibre filter containing washed yeast into scintillation vial with 4 mL scintillant and incubate at RT overnight in the dark.
8. Count disintegrations per min (DPM) on scintillation counter for at least 2 min per sample.
9. Assay 10 μL of unfiltered yeast containing [^{14}C]glucose, by adding to 4 mL of scintillant in scintillation vial and counting DPM. This allows determination of the specific activity of each reaction.
10. Weigh filters containing yeast cells without [^{14}C]glucose and compare to wet filter (washed three times without yeast).
11. Yeast culture and reaction volume can be scaled up or down as required. Smaller volumes may be desirable for single time point experiments, e.g., 150 μL yeast + 15 μL ethanol + 15 μL uptake solution.

3.3 Conversion of Disintegrations per min (DPM) to Amount of Glucose

1. Counting unfiltered yeast allows determination of DPM per μmol of [^{14}C]glucose present in each reaction using the equation as follows:

$$S = \frac{X}{C} \quad (1)$$

Where:

S = specific activity of reaction (DPM/μmol [^{14}C]glucose).
X = DPM of unfiltered sample.
C = [^{14}C]glucose in reaction (μmol).

2. Conversion of DPM from filtered yeast samples to the total amount of glucose ([^{14}C]glucose and unlabeled glucose) taken up by the yeast cells can be calculated using Eq. 2 as follows:

$$\text{Glucose uptake} = \frac{\frac{\text{DPM}}{S} \times r}{w} \quad (2)$$

Where:

Glucose uptake = μmol/g fw yeast.
DPM = disintegrations per minute.
S = specific activity of reaction (DPM/μmol [^{14}C]glucose).
r = molar ratio of unlabeled glucose to [^{14}C]glucose.
w = weight of yeast per 100 μL reaction (g).

The data may also be expressed as nmol glucose/10^8 cells (*see* **Note 6**).

3.4 Determination of Michaelis-Menten Constant (K_m)

1. Concentration dependent uptakes are used to estimate a transporter K_m. Generally, three concentrations below and three concentrations above the K_m can be used as a minimum. These may be estimated using the K_m of closely related transporters, or empirically determined.
2. Perform uptake on transporter of interest and empty vector at a single time point where uptake remains linear. This can be determined from time course uptake (4-min uptake is recommended).
3. Calculate amount of glucose taken up by yeast per minute (Eqs. 1 and 2), subtract uptake by empty vector control.
4. Organize data on Eadie-Hofstee plot (glucose uptake/substrate concentration vs glucose uptake). The data should resemble a straight line where the negative of the slope of the line equals the K_m. Alternatively the data may be fitted to the Michaelis-Menten equation using software such as Prism (GraphPad Software, Inc). Example data is shown in Fig. 2.

3.5 Inhibitors of Hexose Uptake

1. Treatment of yeast using inhibitors enables examination of transporter energy dependence, such as reliance on ATP and a proton gradient. Perform single time point uptake where

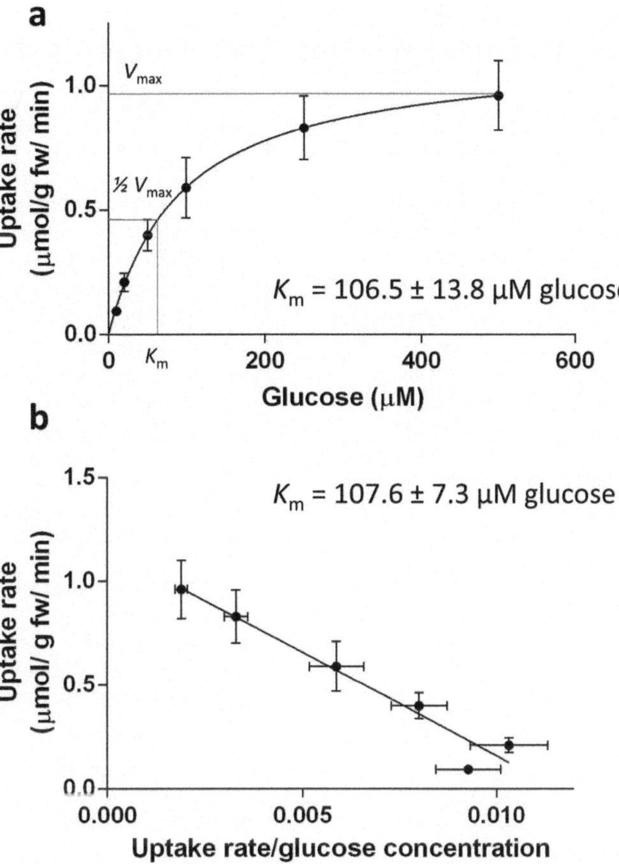

Fig. 2 Example data demonstrating calculation of Michaelis-Menten Constant (K_m). (**a**) Concentration dependent uptake by a hexose transporter. *Upper red line* indicates V_{max} (maximal velocity of reaction). The concentration at half V_{max} represents the K_m. Data was fitted to Michaelis–Menten equation using Graphpad Prism software package. (**b**) Eadie-Hofstee plot representing an alternative method used to calculate K_m. Negative slope of the line represents the K_m

uptake remains linear (4-min uptake is recommended) using a glucose concentration at the transporter K_m.

2. Add the inhibitor to yeast 30 s prior to addition of uptake solution. Inhibitors and concentrations used in previous STP13 studies are shown in Table 1.
3. Perform a control uptake without inhibitor and compare. A number of water-insoluble inhibitors need to be dissolved in ethanol or DMSO. If this is the case, the same concentration of solvent (minus inhibitor) should be added to yeast 30 s prior to addition of uptake solution as a control.
4. Express data as percentage uptake compared to control treatment.

Table 1
Commonly used inhibitors to probe energization and binding of transporter proteins

Transport relies on:	Inhibitor(s)	Action	Suggested concentration	Example study refs
A transporter protein	p-chloromercuribenzene sulfonic acid (PCMBS)	Thiol modifying agent- binds with cysteine residues on outer face of transporter. Membrane impermeable	100–1000 μM	[3, 19]
	N-ethylmaleimide (NEM)	Thiol modifying agent Membrane permeable	2 mM	[3, 19]
	Diethyl pyrocarbonate (DEPC)	Affects histidine residues on outer face of transporter. Membrane impermeable	500 μM	[3, 19]
Cellular ATP	Antimycin A	Blocks ATP-generating glycolysis/respiration systems in endomembranes	2 μM	[19]
Transmembrane proton gradient	Carbonyl cyanide m-chlorophenyl hydrazine (CCCP)	Protonophore, collapses H^+ gradient. Membrane permeable and affects endomembranes (eg.mitochondria) at high concentrations and/or long incubations	50–200 μM	[9, 11, 19][a]
	2,4-dinitrophenol (DNP)	Protonophore, collapses H^+ gradient	100–400 μM	[3, 11, 19][a]
A sugar transporter	Phlorizin	Competitive binding with glucose moiety to transporter	1–2 mM	[3, 19]

[a]**Note**—[11] studies carried out using *Xenopus* oocyte system

3.6 Competing Sugar Uptake

1. Testing competing sugars allows assessment of which sugars bind to, or may be transported by a hexose transporter (*see* **Note 7**). Perform single time point uptake where uptake remains linear (4-min uptake is recommended) using a glucose concentration at the transporter K_m.
2. Add competing sugars at ten times the concentration of glucose, in the same solution.
3. Perform a control uptake without competing sugar and compare. A reduction in glucose uptake in the presence of another sugar indicates it may compete with glucose for binding/ transport.

4 Notes

1. Higher transporter expression is likely to be achieved using the pDR196 vector, which contains the yeast plasma membrane ATPase 1 (PMA1) promoter, in comparison to vectors containing alcohol dehydrogenase (ADH) or O-acetyl homoserine sulfhydrylase (MET25) promoters [17, 18].
2. EBY.VW4000 is auxotrophic for uracil, histidine, leucine and tryptophan, so each may be used as a selectable marker. The pDR196 vector uses uracil selection, and this is incorporated into the protocol described in this chapter. For co-transformation of two transporters, each transporter can be transformed into yeast in separate vectors with different selectable markers and grown on media lacking both selectable markers (e.g., media lacking uracil and tryptophan).
3. Dependence on pH can be determined using MES-HEPES buffers to wash and resuspend yeast at appropriate pH (pH 4, 5, 6, 7, and 8 recommended). Use single time point uptake. It is preferable to test other transport properties at the pH optimum.
4. Amount of starter culture to inoculate 20 mL culture may vary depending on transporter, selectable marker, and co-expression of vectors.
5. Substitution of [^{14}C]glucose for [^{14}C]fructose may require a higher final concentration of uptake solution, as the fructose affinity of hexose transporters is generally lower. Additionally, 1/5 or 1/10 dilution of [^{14}C]fructose may be required to achieve DPM >1000. Be sure to account for dilution of uptake solution when labeled and unlabeled hexoses are combined, and for dilution when adding to yeast for uptake.
6. Expressing uptake data as nmol glucose/10^8 cells—calculate the specific activity of the uptake solution using Eq. 1, but express the concentration of glucose as nmol. Calculate the number of cells per 100 μL reaction using a haemocytometer. Divide the number of cells by 1×10^8 and use this number in place of w in Eq. 2.
7. Previous competing sugar studies show that D-enantiomers but not L-enantiomers of hexoses generally compete with glucose for STP13 binding. Pentoses are occasionally able to compete with glucose, but disaccharides and trisaccharides do not [3, 9, 11, 19].

Acknowledgments

Funding support was provided by Grains Research and Development Corporation, Australia; The Bill and Melinda Gates Foundation; and the Office of the Chief Executive (OCE) Post-Doctoral Fellowship of the Commonwealth Scientific and Industrial Research Organisation (CSIRO), Australia.

References

1. Ellis JG, Lagudah ES, Spielmeyer W et al (2014) The past, present and future of breeding rust resistant wheat. Front Plant Sci 5:641
2. Hulbert S, Pumphrey M (2014) A time for more booms and fewer busts? Unraveling cereal-rust interactions. Mol Plant-Microbe Interact 27:207–214
3. Moore JW, Herrera-Foessel S, Lan C et al (2015) A recently evolved hexose transporter variant confers resistance to multiple pathogens in wheat. Nat Genet 47:1494–1498
4. Hayes MA, Feechan A, Dry IB (2010) Involvement of Abscisic acid in the coordinated regulation of a stress-inducible hexose transporter (VvHT5) and a cell wall invertase in grapevine in response to Biotrophic fungal infection. Plant Physiol 153:211–221
5. Lemonnier P, Gaillard C, Veillet F et al (2014) Expression of Arabidopsis sugar transport protein STP13 differentially affects glucose transport activity and basal resistance to *Botrytis cinerea*. Plant Mol Biol 85:473–484
6. Ward JM (1997) Patch-clamping and other molecular approaches for the study of plasma membrane transporters demystified. Plant Physiol 114:1151–1159
7. Rentsch D, Boorer KJ, Frommer WB (1998) Structure and function of plasma membrane amino acid, oligopeptide and sucrose transporters from higher plants. J Membr Biol 162:177–190
8. Wieczorke R, Krampe S, Weierstall T et al (1999) Concurrent knock-out of at least 20 transporter genes is required to block uptake of hexoses in *Saccharomyces cerevisiae*. FEBS Lett 464:123–128
9. Hayes MA, Davies C, Dry IB (2007) Isolation, functional characterization, and expression analysis of grapevine (*Vitis vinifera* L.) hexose transporters: differential roles in sink and source tissues. J Exp Bot 58:1985–1997
10. Mccurdy DW, Dibley S, Cahyanegara R et al (2010) Functional characterization and RNAi-mediated suppression reveals roles for hexose transporters in sugar accumulation by tomato fruit. Mol Plant 3:1049–1063
11. Nørholm MHH, Nour-Eldin HH, Brodersen P et al (2006) Expression of the Arabidopsis high-affinity hexose transporter STP13 correlates with programmed cell death. FEBS Lett 580:2381–2387
12. Toyofuku K, Kasahara M, Yamaguchi J (2000) Characterization and expression of monosaccharide transporters (*OsMSTs*) in rice. Plant Cell Physiol 41:940–947
13. Chen L-Q, Hou B-H, Lalonde S et al (2010) Sugar transporters for intercellular exchange and nutrition of pathogens. Nature 468:527–532
14. Dohmen RJ, Strasser AW, Honer CB et al (1991) An efficient transformation procedure enabling long-term storage of competent cells of various yeast genera. Yeast 7:691–692
15. Rentsch D, Laloi M, Rouhara I et al (1995) NTR1 encodes a high affinity oligopeptide transporter in Arabidopsis. FEBS Lett 370:264–268
16. Hoffman CS, Winston F (1987) A ten-minute DNA preparation from yeast efficiently releases autonomous plasmids for transformation of *Escherichia coli*. Gene 57:267–272
17. Mumberg D, Muller R, Funk M (1994) Regulatable promoters of *Saccharomyces cerevisiae*: comparison of transcriptional activity and their use for heterologous expression. Nucleic Acids Res 22:5767–5768
18. Mumberg D, Muller R, Funk M (1995) Yeast vectors for the controlled expression of heterologous proteins in different genetic backgrounds. Gene 156:119–122
19. Gear ML, Mcphillips ML, Patrick JW et al (2000) Hexose transporters of tomato: molecular cloning, expression analysis and functional characterization. Plant Mol Biol 44:687–697

Part V

Biocontrol Agents for Rust Management

Chapter 24

Biocontrol Agents for Controlling Wheat Rust

Siliang Huang and Fahu Pang

Abstract

Bacterial endophytes are potential biocontrol agents of wheat rusts. Apart from disease control, these bacterial endophytes have growth-promoting efficacies which differ significantly from one isolate to another. Here, we describe the procedure for isolation, screening, and identification of endophytic bacterial isolates with high capacities to suppress strip rust infection and better ability to enhance wheat yields.

Key words Wheat rust, Biocontrol agent, Bacterial endophyte

1 Introduction

Endophytic microorganisms are important bioresources which have the potential for use as biocontrol agents for disease management. Bacterial endophytes are the predominant endophytic microorganisms of wheat plants which have been used as biocontrol agents for wheat rust [1–3] as well as for other wheat diseases caused by phytopathogenic fungi [4–12]. Considerable biodiversity exists particularly in the endophytes that reside in the roots of the wheat plant [13]. Apart from disease suppression, they also promote growth of wheat plants [14]. However, the efficacy of bacterial endophytes in promoting wheat growth and controlling wheat rust significantly varies from one isolate to another. For these reasons, isolation, screening and identification of bacterial endophytes with higher disease suppression as well as growth-promoting and yield-enhancing capacity is a key factor for its successful use in wheat cultivation.

2 Materials

Prepare double distilled water (dd-water) and analytical grade reagents. Store them at ambient temperature unless otherwise specified.

Sambasivam Periyannan (ed.), *Wheat Rust Diseases: Methods and Protocols*, Methods in Molecular Biology, vol. 1659, DOI 10.1007/978-1-4939-7249-4_24, © Springer Science+Business Media LLC 2017

2.1 Wheat Samples

1. Collect the whole healthy wheat plants from the field and gently wash the plants with tap water to remove dusts and soils from the plants.
2. Spread the plants on a washing basket to drain and air-dry superfluous water from the surface of the plants. Snip the plants into three parts (leaves, stems and roots) with sterilized scissors after air-drying.
3. Cut each part of the plants separately into small tissue pieces (5 mm length) in a sterilized Petri dish (7–9 mm in diameter and varies with the amounts of samples) for the isolation of bacterial endophytes.

2.2 NA Media

Weigh 3 g beef extract, 5 g peptone, 3 g yeast extract, 10 g sucrose, and 15 g agar powder, separately, and mix together in a 1000-mL glass beaker. Add water to a volume of 1000 mL (natural pH). Warm the beaker in boiling water and blend the medium with a glass rod while heating. After the agar melts completely, dispense 150 mL of the medium into a 300-mL conical flask for NA plates, and 5 mL into a 10 (diameter) × 100 (length) mm test tube for NA slants (*see* **Note 1**). Plug the flasks/test tubes with respective silica gel stoppers. Seal the stoppers with kraft papers or aluminum foil. Transfer the flasks/test tubes in an autoclave and keep the pressure at 1.15 bar under wet heat condition for 15 min to sterilize the media. Lay the tubes obliquely on a long, square, woody club (about 1.5×1.5 cm^2 in cross section) to shape up the media in the tubes to slants immediately after autoclaving.

2.3 1× TAE Buffer

1. Measure 242 g tris hydroxy methyl aminomethan (Tris) and ethylene diamine tetra acetic acid (EDTA) separately, and transfer them into a 1000-mL glass beaker.
2. Add 800 mL dd-water into the beaker and blend with a glass rod.
3. Add 57.1 mL glacial acetic acid while blending the solution.
4. Add dd-water to 1000 mL, and store as 50× TAE buffer at ambient temperature.
5. Dilute the TAE buffer 50 times with dd-water as 1× TAE buffer before being used.

2.4 Agarose Gel

1. Weigh 0.2 g agarose in a 50-mL glass flask.
2. Add 20 mL of 1× TAE buffer and 2 μL of commercial 10,000× Gel Red into the flask.
3. Cover the flask with a piece of food packaging film and gently shake the flask to mix solution uniformly. Heat the solution in a microwave oven until the agar completely melts.

4. After the temperature lowers down to about 50–60 °C, gently pour the melting agarose onto a gel caster (6×6 cm^2) to prepare 1% agarose gel plate (about 5 mm thickness) for electrophoresis of PCR products.

3 Methods

Carry out all procedures in a clean bench at ambient temperature for the isolation of bacterial endophytes unless otherwise specified.

3.1 Isolation of Bacterial Endophytes

1. Place a sterilized glass funnel resting on a metal support and insert a sponge plug into the top of the tunnel. Connect one side of a 10–15 mm latex tube to the base of the funnel and the other end to a 250-mL conical flask. Set a spring water stopper on the middle of the latex tube (Fig. 1).
2. Transfer wheat tissue pieces to the funnel. Transfer 10–20 mL of 75% ethanol into the funnel and drain away the ethanol by loosening the spring water stopper immediately after the elapse of 10 s. Transfer 20–30 mL of 0.1% mercuric chloride into the funnel and drain away the mercuric chloride solution by loosening the spring water stopper immediately after the elapse of 3–5 min (3 min for the leaf samples, 4 min for the stem samples and 5 min for the root samples). Rinse the samples 5–6 times with sterile water. Collect 5 mL of the last rinsing fluid in a

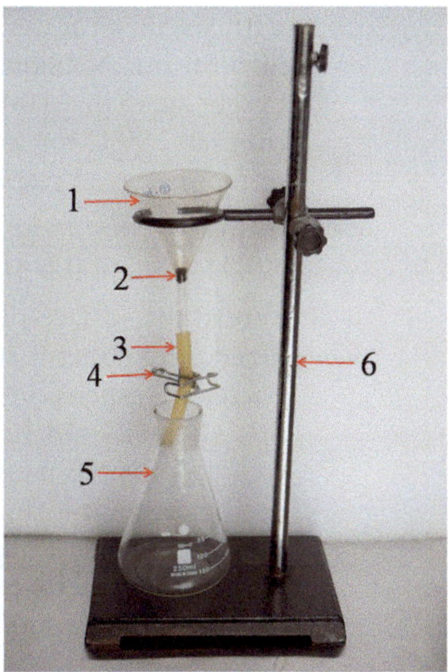

Fig. 1 A simple device for efficient surface disinfection of wheat samples. (*1*) glass funnel, (*2*) sponge plug, (*3*) latex tube, (*4*) spring water stopper, (*5*) conical flask, and (*6*) metal support

sterilized 50-mL flask for subsequent tests to evaluate whether the bacteria isolated from the wheat samples are endophytic or not (*see* **Note 2**).

3. Transfer approximately 0.1–0.2 g surface-disinfected wheat tissue pieces to a sterile mortar, add 5 mL of sterile water into the mortar and grind the samples with a sterile pestle to prepare tissue and cell (TC) suspensions. Dilute the TC suspensions 10–20 times with sterile water. Transfer 200 μL of the diluted TC suspensions to a NA plate and spread them uniformly with a sterile glass/stainless steel spreader.
4. Incubate the plates in an incubator at 28 °C for about 2–3 days. Select the plates with bacterial colonies clearly separating from each other to isolate individual colonies. Perform three successive purifications for a colony by streaking a loopful of the colony on a NA plate with a sterilized inoculating loop. Incubate the plate at 28 °C for 3–4 days for each of purifications. Transfer the purified colonies on NA slants and stored them at 4 °C before being further tested.
5. For a long period of storage, transfer a loopful of bacterial colony in a 1.5-mL Eppendorf (EP) tube consisting of 1 mL of 15% sterilized glycerol solution and gently mix the solution with the inoculating loop. Close the cover lid and cover the lid with Parafilm. Store the tube at −80 to −20 °C.

3.2 Tests on Growth Promotion

1. Disinfect the seeds with 75% ethanol for 30 s, followed by a 7-min soak in 0.1% mercuric chloride and rinsing 5–6 times with sterilized water. Inoculate a loopful of a bacterial colony into 50 mL NB liquid medium (the same components as NA except agar) in a 100-mL conical flask, and incubate the bacterium at 28 °C for 24 h on a rotational shaker at 160 rpm to obtain bacterial suspension.
2. Adjust the resultant bacterial suspension to an optical density (OD) value of 0.6–0.7 using a spectrophotometer at 600 nm.
3. Spread the surface-disinfected wheat seeds uniformly on a sterilized filter paper matting on a 9-cm petri dish and add 20 mL bacterial suspension to the filter paper. Incubate the seeds at 30 °C for 24–48 h. Use the surface-disinfected wheat seeds treated with NB medium as control. Set up three replicates for each treatment including the control.
4. As the seeds show white sprouts, sow them in a flowerpot (up diameter 25 cm, bottom diameter 15 cm, depth 20 cm) filled with sterilized soil at a density of 30 germinating seeds per pot and a depth of approximately 1 cm.
5. Transfer the potted plants to a flat ground with full sunlight during daytime and watered twice a day (in the morning and in the afternoon) to prevent excessive dehydration of the plants

Fig. 2 Two bacterial endophytes (JD204 and SB127) show growth promoting activities on a wheat cultivar (Pumai-9)

due to strong solar irradiation if necessary. Randomly sample 20 plants per pot after 25–30 days of growth. Gently wash the plants with tap water to remove soil and dusts from the plants. Measure the plant heights with a ruler and record tiller numbers per plant. To evaluate the influence of a bacterial strain on the dry weight of different parts of a wheat plant, snip the plant into three parts (stems, leaves, and roots). Separately package each part of the plant with absorbent papers and dry the plant samples in a blast drier at 102 °C for 24 h. Measure the dried wheat samples separately with an electronic balance (precision = 0.001 g). Evaluate the growth-promoting activity of a bacterial strain based on its enhanced rates of growth parameters (plant height, dry weight, and tiller number per plant). Select the bacterial strains with higher growth-promoting activities as candidates for field trials on biocontrol of wheat rusts (Fig. 2).

3.3 Field Screening

1. Use the selected bacterial strains with higher growth-promoting activities for field trials on wheat rust control. Follow the procedures of Subheading 3.2 to sterilize wheat seeds (*see* **Note 3**) and prepare bacterial suspensions.
2. Inoculate sterilized wheat seeds separately with bacterial suspensions of selected strains in a ratio of 1 mL of bacterial suspension for 4 g of wheat seeds in a beaker. Incubate the wheat seeds soaking in the bacterial suspension for 24 h at 28–30 °C.

3. Use randomized split plot design for the field trials with at least three replications per treatment. Use the wheat seeds soaked with sterile water as control. Perform fertilizer application by using 500–600 kg per ha of compound fertilizer (18% nitrogen, 18% P_2O_5, and 9% K_2O) and 160–210 kg per ha of urea shortly before rotary tillage of the wheat field soil based on soil fertility. The seeding rate is 150–240 kg per ha according to the agronomic traits of a cultivar, seed germination rates and soil fertility as well as the date of seeding (*see* **Note 4**).
4. Investigate severity of wheat rusts 2–3 times with 7–10 days interval during the occurrence of the diseases. Use a five-point sampling method to survey disease severity on the wheat plants. Investigate 20 plants for each sampling point. The investigation parameters include disease incidence and disease severity (DS) which is expressed as a percentage of infected area over the total leaf area. Divide the DS into six ratings: 0 = no symptom, 1 = <5% leaf area infected, 3 = 6–15% leaf area infected, 5 = 26–50% leaf area infected, 7 = 51–75% leaf area infected, and 9 = more than 75% leaf area infected. Record the DS on each of the top-three leaves per plant. Calculate disease incidence, disease index, and control efficacy using the following formulae:

$$\text{Disease incidence (\%)} = 100 \times (\text{No. of affected leaves/No. of leaves investigated})$$

$$\text{Disease index} = 100 \times \sum (\text{No. of affected leaves} \times \text{corresponding DS}) / (\text{No. of total leaves} \times 9)$$

$$\text{Control efficacy (\%)} = 100 \times (\text{disease index of control} - \text{disease index of bacterial endophyte treatment}) / \text{disease index of control}$$

5. Harvest all of the spikes of five sampling points (20 plants per point) and enclose them into a nylon mesh bag (one plot per bag). Thresh the spikes of each plot and collect all of the wheat kernels from the same plot into the same bag. Hang the bags uniformly on a rope and dry them under sunlight for more than 5 days until the water amount of the kernels is lower than 12.5% (*see* **Note 5**).
6. Weigh the kernels of each plot separately on an electronic balance to obtain data for evaluating yield performance of the plots. Randomly select 1000 kernels from each plot, and measure the thousand kernel weights (TKW) of each plot separately to evaluate the effect of a bacterial endophyte on the weight of kernels.

7. Evaluate yield-enhanced performance of a bacterial endophyte by calculating a PEY (percent enhanced yield compared with control) value using the following formula:

$$\text{PEY}(\%) = 100 \times (\text{yield of bacterial endophyte treatment—yield of control})/\text{yield of control}.$$

8. Perform analyses of variance on the experimental data using a statistical software such as SPSS (version 16.0; IBM corporation, New York, USA). Compare means of experimental data using the least significant difference (LSD) at $P = 0.05$ between treatments/cultivars.

3.4 Verification Field Trial

1. Select wheat cultivar–bacterial endophyte combination(s), with significantly higher rust control efficacy and/or significantly enhanced yield compared to the control for a large scale verification field trial.
2. Follow Subheading 3.2 to prepare bacterial suspension.
3. Inoculate wheat seeds in a bucket with bacterial suspension for 24 h. Remove superfluous water from the seeds, and spread the seeds uniformly on a nylon film (approximately 20–30 μm thickness) lying on the ground overnight to air-dry the seeds.
4. Select at least three experimental locations with different soil fertility for the verification field trial. Load the endophyte-treated seeds in a seeding machine and sow the seeds based on a routine mechanical sowing method. Set up a control plot in the same location in which the seeds are sown without being treated by the endophyte.
5. Investigate rust severity, control efficacy and yield performance on the endophyte-treated and control wheat plants based on the methods described in Subheading 3.3.

3.5 Identification of Bacterial Endophytes

1. Streaking a loopful of a bacterial colony on a NA plate with the inoculating loop and incubate the plate at 28 °C for 48 h. Inoculate a loopful of a single colony to 100 mL NB liquid medium in a 300-mL conical flask, and incubate at 30 °C for 24 h on a rotary shaker at 160 rpm.
2. Transfer 1 mL of the resultant bacterial suspension into 1.5-mL sterilized EP tubes with a pipette and centrifuge at 18,514 × g for 10 min. Remove the supernatant away from the tubes. Collect the bacterial cells into a new EP tube and add 600 μL cell lysate (2.29 g NaCl, 0.78 g EDTA, 0.3 g Tris, 1 g sodium dodecyl sulfonate, 50 mL dd H_2O).
3. Vortex the tube thoroughly and heat by water bath at 60 °C for 30 min. Centrifuge the tube at 18,514 × g at 4 °C for 10 min. Collect the supernatant to a new EP tube, and add equal

volume of phenol in the tube. Vortex and centrifuge the tube at 12,857 × *g* and 4 °C for 10 min.

4. Collect the supernatant in another EP tube and add equal volume of chloroform into the tube. Vortex and centrifuge the tube at 12,857 × *g* and 4 °C for 10 min. Repeat this procedure twice. Collect the supernatant in a new EP tube after the last chloroform extraction.
5. Add two volume of absolute ethanol and 1/10 volume of 3 mol/L sodium acetate solution to the tube to precipitate the DNA samples at −20 °C for 60 min. Centrifuge the tube at 4 °C, 8,228 × *g* for 10 min and remove the supernatant from the tube.
6. Add 75% ethanol to the DNA sample, centrifuge at 4 °C, 12,857 × *g* for 10 min, and remove the supernatant. Repeat the rinsing procedure of DNA sample with 75% ethanol three times. Add 30 μL of sterilized dd-water to the precipitate to solubilize the DNA sample. Store the DNA samples at −20 °C before being used.
7. Prepare 50 μL of PCR reaction mixture consisting of 5.0 μL 10× PCR buffer, 1 μL of dNTPs (25 mM), 2 μL of the forward primer F27 (5′-AGAGTTTGATCATGGCTC AG-3′), 2 μL of the reverse primer R1492 (5′-TACGGTTACCTTGTTAC-GACTT-3′), 0.4 μL of *Taq* DNA polymerase (5 U/μL), 4 μL of $MgCl_2$ (25 mM), 1 μL of DNA template, and 34.6 μL of autoclaved dd-water. Perform PCR on an automated thermo-cycler device using the following parameters: denaturation at 94 °C for 4 min followed by 35 cycles of denaturation at 94 °C for 40 s, annealing at 50 °C for 50 s, elongation at 72 °C for 80 s, and a final extension at 72 °C for 8 min.
8. Load 5 μL of PCR product on a 1% agarose gel immersing in 1× TAE buffer with DL2000 as DNA marker. Perform electrophoresis at 100 V for approximately 40 min, and confirm the existence of a target PCR product of approximately 1.4–1.5 kb under a UV gel transmission apparatus (Fig. 3).
9. Cut out the agarose with target PCR product and purify the product with a PCR product purification kit based on the specifications of the maker. Send the purified PCR product to a commercial biotechnology company for sequencing (both strands sequenced) of the amplified 16S rDNA.
10. Compare the 16S rDNA sequence of the bacterial endophyte with those of bacterial strains in GenBank using NCBI BLAST searching tool. Construct a phylogenetic tree of the bacterial endophyte based on 16S rDNA sequences with the neighbor-joining algorithm in the MEGA software (such as version 4.0) to identify the bacterium (*see* **Note 6**).

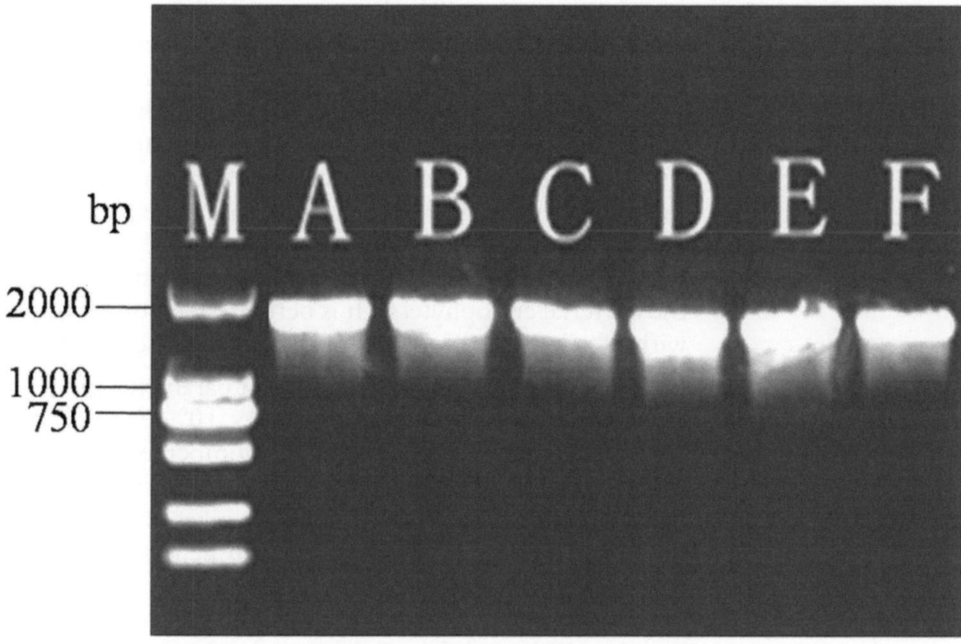

Fig. 3 Electropherogram for PCR products of 16S rDNA of bacterial endophytes isolated from wheat plants Lane M: marker DNA(DL-2000); Lanes A to B are the PCR products of 16S rDNA of strains RC79, RB132, RA135, SB127, LD161, and JD204, respectively

4 Notes

1. As the mixture of medium turns to be semitransparent, it can be considered as the indicator of complete melting of agar. Perform split charging of the melting medium into test tubes (5 mL per tube) with a 5-mL pipette, and into 300-mL conical flasks through a funnel (200 mL per flask) setting on the top of the flask.
2. To confirm a bacterial colony developed on a NA plate was an endophyte rather than an epiphyte, use three NA plates containing the last rinsing water of the surface-disinfected samples as controls. Simultaneously incubate both the NA plates with tissue-grinding fluid and control plates with the last rinsing water at 28 °C for 4–5 days. If no colonies developed on the control plates, the colonies that occurred on the NA plates with the tissue grinding fluid can be considered as endophytes [3, 11].
3. The primary goal of field screening is to test the efficacies of selected bacterial endophytes with high growth-promoting activities in controlling wheat rusts and enhancing yields on wheat cultivars. Based on the results of a previous study [3], the sensitivity of wheat cultivars to a bacterial endophyte greatly

varied from one to another. Some bacterial isolates have the ability to enhance disease tolerance and/or resistance to stripe rust on certain cultivars. In the case of enhanced disease tolerance on wheat plants by a bacterial endophyte, the disease indices of the endophyte-treated plants might be higher than the control plants, however, the yields of the former are significantly higher than the latter. For these reasons, it is also one of the purposes for the field screening to find those cultivars on which the disease tolerance could be significantly enhanced by the bacterial endophyte(s). It is better to include more cultivars with excellent agronomical traits in the field screening trials.

4. Basic seedling number is an important parameter for determining seeding rate, and is about $1.2–3 \times 10^6$ per ha for a winter wheat cultivar. The basic seedling number is determined by multifactors, especially the tillering capacity of a cultivar. Approximately $5–8 \times 10^6$ productive spikes per ha are required for normal wheat production [15]. In general, approximately 5×10^6 productive spikes per ha are suitable for the cultivars with a large spike (dry kernel weight per spike $\geq$2 g), and $6–6.5 \times 10^6$ productive spikes per ha for the cultivars with a moderate spike (1–2 g dry kernel weight per spike), and approximately 8×10^6 productive spikes per ha for the cultivars with a small spike (dry kernel weight per spike $\leq$1 g). A large-spike cultivar commonly has the lowest level of tillering capacity, and needs more basic seedlings. A moderate-spike cultivar has a medium level of tillering capacity, and needs moderate basic seedlings. A small-spike cultivar has the highest level of tillering capacity, and thus needs lesser basic seedlings. A winter wheat cultivar is normally sown at the middle to late autumn season. The seeding rate of the cultivar should be properly increased as the sowing date delay in the late autumn. Besides the agronomical characters of cultivars, the soil fertility also affects the numbers of productive spikes. Seeding rate should be properly reduced in the soil with higher fertility or vice versa.

5. Water contents of wheat seeds can be rapidly determined with a portable type of water tester with two parallel sensors. Transfer the wheat kernels of a plot into a polyethylene bag (20–23 cm length, 3–5 cm diameter). Set an appropriate parameter for testing the water content for wheat kernel based on the specifications of the maker. Insert the sensors into the wheat kernels to test the water content of the kernels.

6. Analysis of 16S rDNA sequence is frequently carried out for bacterial identification. Most of bacterial species can be identified based on their 16S rDNA sequences, however, some bacterial species cannot be clearly differentiated from one to another merely based on their 16S rDNA sequences. Other

DNA (gene) sequences specific for the identification of certain bacterial species should also be amplified and analyzed as well as the 16S rDNA sequences to identify the species more accurately. For example, *rpo*D gene is considered as one of core housekeeping genes and is used for the identification of *Psedomonas* spp. [3, 16]. Both *Bacillus subtilis* and *B. amyloliquefacient* have been frequently isolated as bacterial endophytes from winter wheat plants [13], however, the two *Bacillus* spp. cannot be clearly differentiated from each other merely based on their 16S rDNA sequences. Analysis of *gyr*A gene sequence is needed for discriminating the two species from one to another [17].

Acknowledgments

This work was supported by Nanyang Normal University [the projects for supporting the university science and technology innovation team of Henan Province (No.2010JRTSTHNO12) and for the construction of Henan Province key disciplines in biochemical and molecular biology].

References

1. Wang Y-F (2011) Biocontrol of wheat leaf rust caused by *Puccinia triticina* and induction of systemic acquired resistance by salicyclic acid. Agricultural University of Hebei (Master Dissertation)
2. Li H, Zhao J, Feng H, Huang L, Kang Z (2013) Biological control of wheat stripe rust by an endophytic *Bacillus subtilis* strain E1R-J in greenhouse and field trials. Crop Prot 43:201–206
3. Pang F, Wang T, Zhao C, Yu G, Tao A, Yu Z, Huang S (2016) Novel bacterial endophytes isolated from winter wheat plants as biocontrol agents against stripe rust of wheat. BioControl 61:207–219
4. Durán P, Acuña JJ, Jorquera MA, Azcón R, Paredes C, Rengel Z, de la Luz Mora M (2014) Endophytic bacteria from selenium-supplemented wheat plants could be useful for plant-growth promotion, biofortification and *Gaeumannomyces graminis* biocontrol in wheat production. Biol Fertil Soils 50:983–990
5. Pan D, Mionetto A, Tiscornia S, Bettucci L (2015) Endophytic bacteria from wheat grains as biocontrol agents of *Fusarium graminearum* and deoxynivalenol production in wheat. Mycotoxin Res 31:137–143
6. Liu B, Huang L, Kang Z, Buchenauer H (2011) Evaluation of endophytic bacterial strains as antagonists of take-all in wheat caused by *Gaeumannomyces graminis* var. *tritici* in greenhouse and field. J Pest Sci 84:257–264
7. Herrera SD, Grossi C, Zawoznik M, Groppa MD (2016) Wheat seeds harbour bacterial endophytes with potential as plant growth promoters and biocontrol agents of *Fusarium graminearum*. Microbiol Res 186:37–43
8. Cheng J, Sun X, Ding T, Ma Y (2015) Biocontrol effect of endophytic bacteria DZBG07 from *Eucommia ulmoides* Olivers on wheat take-all. J Biol 32:105–112
9. Peng L, Zhang S, Peng J, Zhang Y, Chai H, Wang G (2011) Biological control of take-all of wheat by endophytic bacteria. J Henan Agric Sci 40:74–77
10. Gong YF (2007) Biological control of powdery mildew of wheat by endophytic bacteria. Northwest A & F University (Doctoral Dissertation)
11. Tao A, Pang F, Huang S, Yu G, Li B, Wang T (2014) Characterization of endophytic *Bacillus thuringiensis* strains isolated from wheat plants as biocontrol agents against wheat flag smut. Biocontrol Sci Tech 24:901–924

12. Zhang Y, Xu Y, Wang S, Wang G (2013) Screening and identification of endophytic bacteria against wheat leaf blotch. J Henan Univ (Nat Sci) 43:423–426
13. Pang F (2016) Studies on the population diversity of bacterial endophytes in wheat plants and their application in the control of wheat stripe rust. Guangxi University (Doctoral Dissertation)
14. Pang F, Du R, Tang W, Huang S (2016) Screening of wheat growth-promoting bacterial strains and correlation analysis of factors influencing wheat growth. J China Agric Univ 21:8–21
15. Yu ZW (2015) Crop cultivation theory (North), 2nd edn. China Agriculture Press, Beijing
16. Mulet M, Lalucat J, García-Valdés E (2010) DNA sequence-based analysis of the *Pseudomonas* species. Environ Microbiol 12:1513–1530
17. Ma Y, Tan X, Huang S, Zhang X, Zang X, Niu X (2016) Identification of a biocontrol strain Z2 against pomegranate and optimization of its cultural conditions. Acta Phytopathol Sin 45:425–437

INDEX

A

Accelerated growth conditions (AGC)............... 184, 188
Agarose gel 51, 54, 55, 65, 69, 122, 156, 212, 220, 221, 223, 253, 254, 259, 260, 262, 278, 284
Aggressiveness 5, 41, 42, 44, 46, 137
Agroinfiltration.....................................85, 87, 90, 96, 97
Amphiploids 164–167
Analysis of variance (ANOVA) 144

B

Backcrossing 167–169, 171, 240
Bacteria 90, 96, 119, 245, 259, 277–287
Biocontrol...277–283, 285, 287
Bioinformatics 14, 111, 112, 239–241, 246, 248–251
Bioresources 277
Biotrophic..............................41, 50, 116, 125–131, 133
Breeding population (BP) 174, 175, 185, 189, 194
Bulked-segregant analysis (BSA)........................... 86, 88, 127, 156–158

C

Cartographer 139, 145, 147
Chromatogram.............................. 66, 68, 106, 108–110
Chromatography
 2D-liquid chromatography (2D-LC) 99–103, 105–108, 111, 112
Chromosome
 alien...164, 165, 167–171
 assembly.. 245–250, 252–254
 bacterial artificial chromosome (BAC) 245
 engineering.. 163–171
 flow-sorting 158, 232, 247
 gametocidal 171
 homoeologous 168, 171
 pairing.. 168, 169
 substitution..169
 suspension.. 233, 235, 236
 suspension.. 236, 237, 242
 walking...245
Coding sequence (CDS)...........79, 80, 89, 96, 231, 246, 248–250
Coefficient of infection (CI)....................................7, 194
Coiled-Coiled (CC) 86
Co-immunoprecipitation 86
Comparative genomics 216, 227
Confocal microscope 86, 87, 91
Controlled environmental conditions (CEC) 184

D

Database 16, 26, 27, 95, 112, 239, 246, 248, 249, 252–254
De novo assembly 224, 239, 246, 247, 250, 252
Differential................................... 29–31, 33–36, 39, 74, 108–110, 138–141, 145, 152, 153
Dikaryotic 49, 65
Diploid...165, 166, 216, 246
Disease
 efficacy...277
 evaluation.. 142
 incidence 282
 index 282
 management 277
 response 186, 190–192
 severity (DS)............3, 141, 142, 144, 145, 194, 282
DNA
 amplification 235, 237, 242
 isolation 63, 154, 247, 250
 quantification...51, 52, 54
 polymerase 253, 284
 purification 237
Dosage curve 203, 204
Double-Stranded RNAs (dsRNAs)..................... 116, 257

E

Effector 15, 73, 75, 76, 78–80, 85–97
Electrophoresis.................... 51, 54, 55, 88, 92, 94, 100, 117, 118, 123, 133, 208, 279, 284
Endophytes................................ 277–279, 282–285, 287
Epidemiology 8, 115
Ethyl methanesulfonate (EMS)..........199–204, 216, 227

F

Fertilizer 138, 140, 167, 186, 193, 282
Fluorescence in situ hybridization (FISH)....................................234, 236, 237, 242
Fluorescent resonance energy transfer (FRET).. 152, 157

Sambasivam Periyannan (ed.), *Wheat Rust Diseases: Methods and Protocols*, Methods in Molecular Biology, vol. 1659, DOI 10.1007/978-1-4939-7249-4, © Springer Science+Business Media LLC 2017

G

Gene pool
- primary ... 164
- secondary ... 164
- tertiary ... 164

Gene silencing
- host-induced (HIGS) ... 116, 121, 123
- virus-induced (VIGS) ... 115–120, 122, 123, 257

Generation sequencing ... 14
Genomic DNA (gDNA) ... 139, 207–210, 212, 217–219, 221
Genomic in situ hybridization (GISH) ... 169–171
Genomic selection (GS) ... 173, 188, 190
Genotyping-by-sequencing (GBS) ... 139, 143
Germplasm ... 152, 185, 189, 194
Green fluorescent protein (GFP) ... 85–97
Growth promotion ... 280, 281, 285

H

Haemocytometer ... 268, 273
Hairpin RNA (hpRNA) ... 116
Heterokaryotic ... 18, 19
Heterologous ... 265–267, 269, 271–273
Hexaploid ... 163, 166, 216, 218, 232, 245, 246, 265
Hexose transporter ... 265–267, 269, 271–273
High molecular weight (HMW) ... 49–52, 55, 207, 246, 247, 250
High performance chromatofocusing fractionation (HPCF) ... 103, 105
High performance reverse phase (HPRP) ... 102, 107

Host-pathogen interactions
- compatible ... 29
- incompatible ... 29

Hybridization ... 165, 169, 234, 236, 242

I

Illumina ... 18, 156, 217, 219, 221, 239, 247, 250, 252
Immunoprecipitation (IP) ... 87, 93
In planta ... 74, 76, 85, 265, 266
Inoculum ... 42, 120, 138, 139, 186, 191, 263
Interspecific ... 79, 167, 170
Introgression ... 163–171

K

Kompetitive allele specific PCR (KASP) ... 139, 152, 157–159, 161

L

Leucine-rich repeats (LRRs) ... 215
Linkage drag ... 163

M

Machine learning ... 76, 174, 180
Mapping ... 80, 139, 142–145, 151–153, 155–158, 160, 161, 207–210, 212, 224, 225, 228, 231, 232, 239–241, 245, 246, 248, 250, 252, 254
Marker assisted selection (MAS) ... 152
Mass spectrometry (MS) ... 7, 86, 88, 95, 109, 111, 112, 142
Materials safety data sheets (MSDS) ... 201
Meristem ... 204, 235
Metabolism ... 265
Microsatellite ... 8, 59, 234, 237
Molecular genotyping ... 3–5, 7, 8, 10
Molecular markers ... 59, 142, 145, 168–171, 245, 250
Monocot ... 116, 216, 257
Mutagenesis ... 199–203, 216, 227, 232, 242

N

Near isogenic lines (NILs) ... 31, 35, 140, 185, 189, 250
Next generation sequencing (NGS) ... 161, 200, 207, 232
Nuclear localization signals (NLSs) ... 76, 78
Nucleotide-binding (NB) ... 215, 280, 282

O

Optical density (OD) ... 268, 280

P

Pathogenesis ... 116
Pathogenomics ... 13
Photoperiod ... 167, 188, 193, 263
Photosynthetically active radiation (PAR) ... 186
Phylogenetic analysis ... 14–16, 18, 21–23
Phytopathogens ... 99, 277
Pipeline ... 15, 18, 19, 21–27, 75, 79, 86, 151, 215, 224, 239–241, 247
Polymerase chain reaction (PCR) ... 14, 26, 32, 60–62, 64–66, 68, 69, 89, 90, 96, 118, 122, 156–161, 169, 207, 212, 234, 237, 240, 253, 254, 258, 259, 262, 263, 279, 284
Polyploid ... 216, 232, 245
Population structure ... 8, 14–16, 18, 22, 24, 26, 27, 59
Promoters ... 86, 89, 273
Proton motive force (pmf) ... 266, 267

Q

Quantitative trait loci (QTL) ... 139, 143, 145, 147
Quarantine ... 8, 10, 165, 168

R

Read data 224, 239
Recombinant inbred line (RIL) 139, 141, 143, 153–155, 158, 216
Recombination
  homoeologous 169
  homologous 163
Resistance
  adult plant resistance (APR)138, 173, 183, 185, 186, 188, 190, 193, 194
  all stage resistance (ASR) 138
Reverse genetics 116, 200, 257
RNA interference (RNAi) 116, 121

S

Secretome 74–76, 78
Self-fertile 204
Self-pollination 166, 167
Simple sequence repeats (SSR)........ 61, 62, 64, 66, 139, 142, 143, 158, 160, 248
Single locus segregating population (SLSP) 155
Single nucleotide polymorphism (SNP)... 14–16, 18, 19, 21–25, 139, 142, 143, 148, 156–158, 161, 226, 250, 252–254
Single seed descent (SSD) 194
Sodium azide 199–204
Soil fertility 282, 286
Sonicator 93, 217
Spectrophotometer 118, 268, 280
Speed breeding 184–186, 188, 191, 193, 194
Sporulation 5, 30, 43, 45, 141, 142, 153, 190
Subcellular localization 74, 76, 80
Surveillance 3–5, 7, 8, 10, 13
Symptom .. 7, 30, 43, 45, 120, 121, 123, 140, 261–263, 282
Synteny 168

T

Targeting induced local lesions IN genomes (TILLING) 200
Tetraploid 164, 166
Thermocycler 60, 89, 90, 161, 218, 222, 284
Thousand kernal weight (TKW) 282
Tissue culture 168
Transgenics 200
Transient
  expression 85
  transformation 85
Transition 76, 78, 199
Translocation
  radiation-induced 168
  spontaneous 168

U

Ultrasonic 87, 102, 105, 107, 140, 186, 188
Urediniospore32, 33, 42, 43, 46, 49, 120, 138–142, 186, 188, 191

V

Varieties.4–7, 10, 32, 38, 137, 139–141, 164, 166, 167, 171, 200, 204
Vector...... 85–87, 89, 90, 116–119, 121–123, 176, 234, 257–260, 263, 267–269, 273

MIX
Papier aus verantwortungsvollen Quellen
Paper from responsible sources
FSC® C105338

If you have any concerns about our products,
you can contact us on
ProductSafety@springernature.com

In case Publisher is established outside the EU,
the EU authorized representative is:
Springer Nature Customer Service Center GmbH
Europaplatz 3, 69115 Heidelberg, Germany

Printed by Libri Plureos GmbH
in Hamburg, Germany